Mechanisms
of
Protease
Action

Author
László Polgár, Ph.D., D.Sc.
Supervisory Research Enzymologist
Biological Research Center
Hungarian Academy of Sciences
Budapest, Hungary

CRC Press, Inc.
Boca Raton, Florida

Library of Congress Cataloging-in-Publication Data

Polgár, László, 1930-
   Mechanisms of protease action / author, László Polgár.
      p. cm.
   Bibliography: p.
   Includes index.
   ISBN 0-8493-6901-0
   1. Proteolytic enzymes.   I. Title
QP609.P78P65 1989   88-3778
574.19'256—dc19   CIP

Direct all inquiries to CRC Press, Inc., 2000 Corporate Blvd., N.W., Boca Raton, Florida, 33431.

© 1989 by CRC Press, Inc.
Second Printing, 1990

International Standard Book Number 0-8493-6901-0

Library of Congress Card Number 88-3778
Printed in the United States

PREFACE

The contents of this book are focused on the four basic mechanisms associated with protease action. Particular emphasis has been placed on the chemistry and stereochemistry of the mechanisms, allowing a deeper insight into the catalysis than is obtainable from the more commonly discussed kinetic mechanisms. Therefore, only those proteases are examined in detail for which the steric structures are known at atomic resolution. As a result, little emphasis is given to large, multidomain proteases, in spite of their considerable regulatory importance.

The author wishes to express his gratitude to many of his colleagues for their comments on the manuscript. In particular, the contributions of Mr. B. Asbóth of the Institute of Enzymology, Biological Research Center, Budapest, Professor M. Kajtár of the Institute of Organic Chemistry, L. Eötvös University, Budapest, and Professor V. N. Schumaker of the Molecular Biology Institute, University of California, Los Angeles, are greatly acknowledged. The author is indebted to Professor J. S. Furton of Yale University, New Haven, Connecticut for sending his review article before publication.

László Polgár

THE AUTHOR

László Polgár, Ph.D., D.Sc. had a basic training in organic chemistry and got a diploma at the L. Eötvös University, Budapest. After graduation he was employed by the Institute of Medical Chemistry, University of Budapest, Hungary, where he took part in teaching and research work.

Since 1960 he has been working in the Institute of Enzymology of the Hungarian Academy of Sciences. In 1965 he obtained the degree of Candidate of Sciences. In the 1965/66 academic year he worked in the laboratory of Professor Myron L. Bender at Northwestern University, Evanston, Illinois. There he prepared thiolsubtilisin, an enzyme with a synthetically formed active site. This was actually the first site specific mutation at the active site of an enzyme. In 1970 he obtained the degree of D.Sc. For the last two decades he has been working on the mechanisms of action of serine and cysteine proteases. He wrote several reviews on this subject, and attended international meetings as an invited speaker and discussion leader. He is also working as an editorial advisor.

TABLE OF CONTENTS

Chapter 1

THE CHEMISTRY OF THE PEPTIDE BOND

I. NOTATION AND SYMBOLISM FOR AMINO ACIDS

Amino acids are the basic structural units of proteins. An amino acid contains an amino group, a carboxyl group, a hydrogen atom, and a distinctive side chain (often denoted by R), all linked to the same central carbon atom. The tetrahedral array of four different groups about the central carbon atom, which is called the α-carbon, confers optical activity on amino acids. The two mirror image isomers are designated by an L or D prefix. Only the L-amino acids are encountered in proteins. In the Fischer-Rosanoff convention the chiral center is projected onto the plane of the paper. The central carbon atom is then considered to lie in the plane, two groups are behind the plane, and the remaining two groups are in front of the plane, as illustrated below for an L-amino acid.

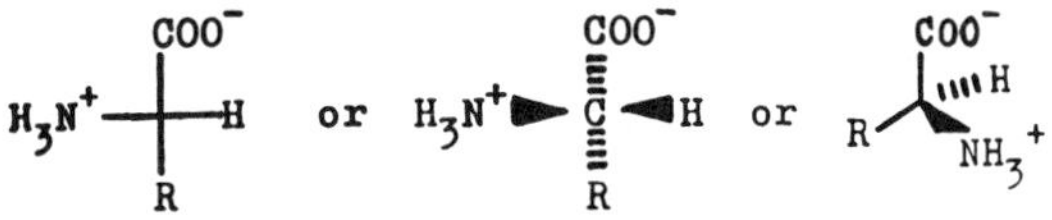

In the Cahn, Ingold, and Prelog convention,[1,2] the substituents to the central carbon atom are ordered according to their rank based largely on the atomic number. The central carbon is then viewed from the side opposite to the substituent of lowest priority. If the order of the remaining three substituents is clockwise when starting with the one of highest priority, the chiral center is R; if counterclockwise, it is S. In this R,S system, the configuration of L-amino acids corresponds to S. Cysteine, however, is an exception because of the high atomic number of its sulfur atom.

Besides configuration, conformation is also important when studying the stereochemical reactions of amino acid derivatives. Conformation is defined by the torsional angles in molecules of the same constitution and configuration. The relevant nomenclature suggested by Klyne and Prelog[3] is widely used in the literature and may be illustrated by the example of substituted ethane. If two vicinal groups of forms [1] to [4] have a torsional angle less than 30°, the conformation is *syn-periplanar* [1], whereas it is *anti-periplanar* [2] if the corresponding angle is 180° ± 30°. In *syn-clinal* [3] and *anti-clinal* [4] conformations the groups considered have torsional angles of 60° ± 30° and 120° ± 30°, respectively.

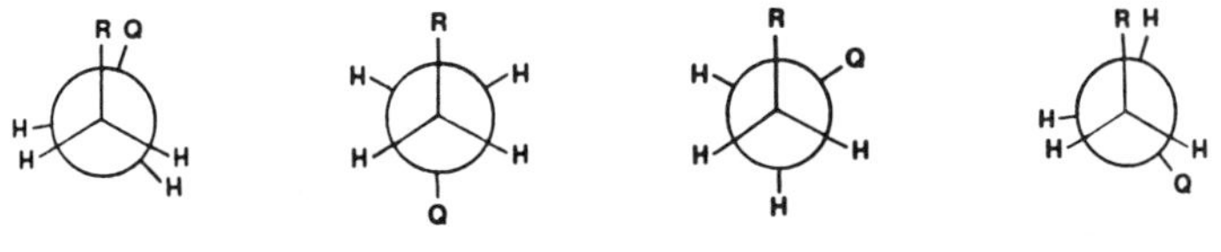

[1]	[2]	[3]	[4]

There are 20 kinds of amino acids that can be incorporated into proteins under mRNA direction. Their abbreviations, which are extensively used in delineation of amino acid sequences of polypeptides and proteins, are shown in Table 1.

Table 1
ABBREVIATION AND SYMBOLS FOR
AMINO ACIDS

Amino acid	Three-letter abbreviation	One-letter symbol
Alanine	Ala	A
Arginine	Arg	R
Asparagine	Asn	N
Aspartic acid	Asp	D
Asparagine or aspartic acid	Asx	B
Cysteine	Cys	C
Glutamine	Gln	Q
Glutamic acid	Glu	E
Glutamine or glutamic acid	Glx	Z
Glycine	Gly	G
Histidine	His	H
Isoleucine	Ile	I
Leucine	Leu	L
Lysine	Lys	K
Methionine	Met	M
Phenylalanine	Phe	F
Proline	Pro	P
Serine	Ser	S
Threonine	Thr	T
Tryptophan	Trp	W
Tyrosine	Tyr	Y
Valine	Val	V
Unspecified amino acid	Xaa	X

Notation of the individual atoms of amino acids is important in dealing with the three-dimensional structure of proteins. The carbon atom of the carboxyl group next to the α-carbon atom is numbered 1. The remaining carbon atoms in acyclic amino acids may be designated with numbers or Greek letters as illustrated by the example of lysine [5].

$$\begin{array}{ccccccc} 6 & 5 & 4 & 3 & 2 & & 1 \end{array}$$

$$H_3N^+-CH_2-CH_2-CH_2-CH_2-CH(NH_2)-COO^-$$

$$\begin{array}{ccccc} \epsilon & \delta & \gamma & \beta & \alpha \end{array}$$

$$[5]$$

A heteroatom has the same number as the carbon atom to which it is attached, e.g., N-2 is on C-2. The hydroxyl group of serine, which is known by organic chemists as the β-hydroxyl group, will be of particular interest in the discussion of catalysis by serine proteases (Chapter 3). According to the rule just mentioned this oxygen atom should be designated as O-3, but the reader finds O^γ and OG in the earlier and recent X-ray crystallographic literature, respectively.

Another important amino acid in protease catalysis is histidine. The nitrogen atoms of the imidazole ring are denoted by *pros* ("near", abbreviated π) and *tele* ("far", abbreviated τ) to show their position relative to the side chain of the ring. Again, in X-ray diffraction studies on protein N^π is denoted by $N^{\delta 1}$ or ND1, whereas N^τ is designated by $N^{\epsilon 2}$ or NE2 [6].

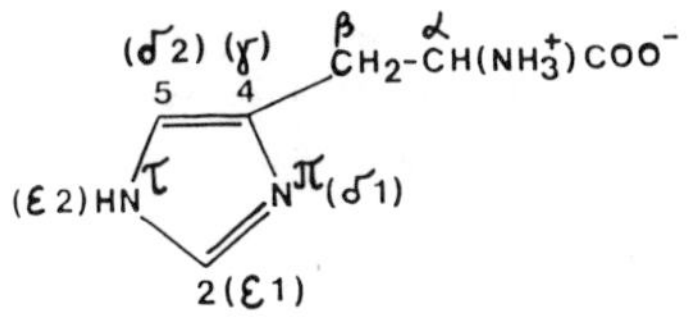

$$(\sigma 2)\ (\gamma) \quad \overset{\beta}{C}H_2-\overset{\alpha}{C}H(NH_3^+)COO^-$$

(structure [6])

[6]

II. SOME PROPERTIES OF THE PEPTIDE BOND

The central feature of the main chain (backbone) of a protein molecule is the succession of peptide linkages. The peptide link is rigid, i.e., there is no freedom of rotation about the C–N bond. This is a consequence of a high degree of double bond character of the peptide link which results from delocalization of the nitrogen lone pair into the carbonyl group (Equation 1).

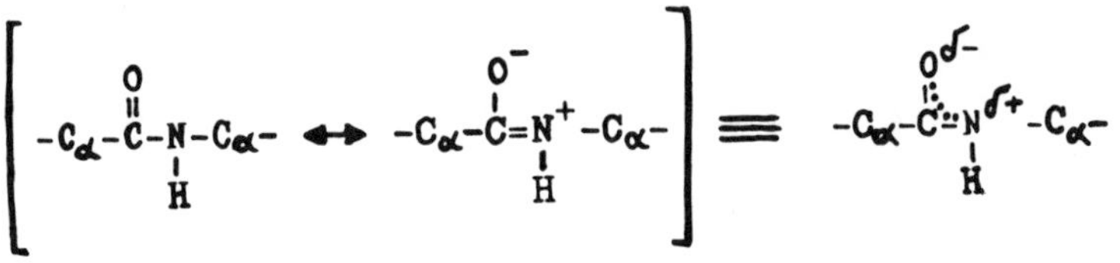

$$\left[\ -C_\alpha-\overset{\overset{O}{\|}}{C}-\underset{H}{N}-C_\alpha-\ \longleftrightarrow\ -C_\alpha-\overset{\overset{O^-}{|}}{C}=\overset{+}{\underset{H}{N}}-C_\alpha-\ \right] \equiv\ -C_\alpha-\overset{\overset{O^{\delta-}}{|}}{C}\overset{\cdots}{=}\overset{\delta+}{\underset{H}{N}}-C_\alpha- \tag{1}$$

The double bond character of the peptide bond is reflected in its bond length which is 132 pm — shorter than a C–N single bond (149 pm) and longer than a C=N double bond (127 pm).

The peptide bond is a planar structure, i.e., the four atoms of the peptide unit [7] are in the same plane. For steric reasons the hydrogen atom of the peptide unit is nearly always *trans* to the carbonyl oxygen.

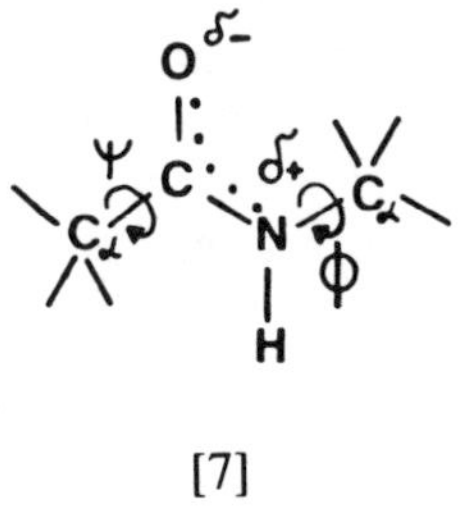

(structure [7])

[7]

In contrast to the rigid C–N bond, there is a large rotational freedom about the C_α–C and N–C_α bonds on either side of the peptide unit. Rotations about these bonds are denoted by the angles ψ and ϕ, respectively. The conformation of the polypeptide backbone is unambiguously defined if the ψ and ϕ values are known for each amino acid residue.

Besides the peptide bond, the ester bond deserves mention here because it is also a good substrate of proteases. Esters are stabilized by resonance structures analogous to those proposed for amides. However, the C=O$^+$ form in esters is relatively less than the C=N$^+$ form in amides, so that the ester bond is considerably weaker than the amide bond.

III. HYDROLYSIS OF AMIDES AND ESTERS

The hydrolysis of carboxylic amides takes place by C–N bond fission. On the other hand, in the case of carboxylic esters there are two alternatives: acyl-oxygen fission (RCO $\not\!\!\!+$ OR′)

and alkyl-oxygen fission (RCOO$\div$R'). Only the first process is of significance in enzyme-catalyzed hydrolysis and in acyl-transfer reactions. The alkyl-oxygen cleavage occurs in special cases, such as the acid-catalyzed hydrolysis of *t*-butyl acetate (Equation 2).

$$CH_3CO-\underset{\underset{CH_3}{|}}{\overset{\overset{O\quad CH_3}{\parallel\quad|}}{C}}CH_3 + H_2O \overset{H^+}{\rightleftharpoons} CH_3COOH + CH_3\underset{\underset{CH_3}{|}}{\overset{\overset{CH_3}{|}}{C}}-OH \qquad (2)$$

Labeling experiments have demonstrated the existence of a nucleophilic addition-elimination mechanism for ester hydrolysis (Equation 3). This mechanism involves a metastable tetrahedral intermediate that is apparent from ^{18}O exchange experiments.[4,5] The exchange technique requires only a partial hydrolysis of the ester in ^{18}O-enriched water, and the ^{18}O content of the carbonyl oxygen of the remaining ester is then determined. An exchange taking place concurrently with the hydrolysis has indicated the existence of an intermediate on the reaction path and thus, it has eliminated a direct S_N2 (bimolecular nucleophilic substitution) mechanism.

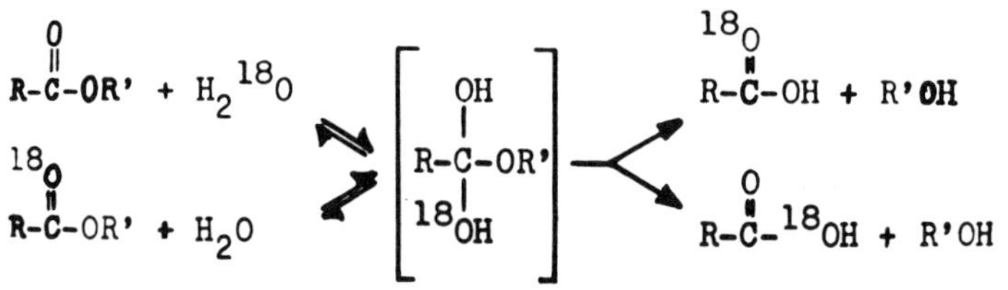

$$
\begin{array}{l}
R-\overset{\overset{O}{\parallel}}{C}-OR' + H_2{}^{18}O \\[6pt]
R-\overset{\overset{{}^{18}O}{\parallel}}{C}-OR' + H_2O
\end{array}
\rightleftharpoons
\left[
\begin{array}{c}
\overset{OH}{\underset{{}^{18}OH}{R-C-OR'}}
\end{array}
\right]
\rightarrow
\begin{array}{l}
R-\overset{\overset{{}^{18}O}{\parallel}}{C}-OH + R'OH \\[6pt]
R-\overset{\overset{O}{\parallel}}{C}-{}^{18}OH + R'OH
\end{array}
\qquad (3)
$$

The existence of a tetrahedral intermediate may also be inferred from breaks in catalyst concentration-rate curves or in structure-reactivity correlations.[6] For example, if a decrease in the rate constant occurs upon increasing the catalyst concentration, then there must be a consecutive two-step process involving an intermediate. Such an example, namely the hydrolysis of *o*-carboxyphthalimide, will be discussed in Section XII of this chapter.

Direct demonstration of the tetrahedral addition intermediate could only be achieved under special conditions. For instance, a stable addition compound was formed in the reaction of ethoxide ion with ethyl trifluoroacetate containing strongly electron-withdrawing substituents (Equation 4).[7] The reaction was performed in dibutyl ether under conditions in which breakdown to hydrolytic products was not possible.

$$CF_3-\overset{\overset{O}{\parallel}}{C}-OC_2H_5 + {}^-OC_2H_5 \rightleftharpoons CF_3-\underset{\underset{OC_2H_5}{|}}{\overset{\overset{O^-}{|}}{C}}-OC_2H_5 \qquad (4)$$

It may be noted that the reverse of the addition-elimination mechanism, i.e., elimination-addition, also takes place in certain nonenzymic acyl transfers including hydrolytic reactions (Equation 5).[8]

$$R-NH-\overset{\overset{O}{\parallel}}{C}-X \underset{\rightleftharpoons}{\overset{-HX}{}} R-N=C=O \underset{\rightleftharpoons}{\overset{+HY}{}} R-NH-\overset{\overset{O}{\parallel}}{C}-Y \qquad (5)$$

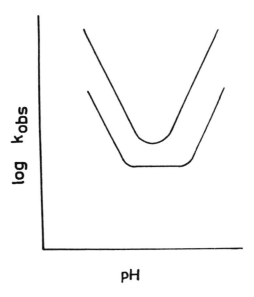

FIGURE 1. Typical pH-log k profiles of specific acid-base catalysis of hydrolyses of carboxylic derivatives such as esters and amides.

In this example, the departure of the leaving group (X) of a carbamic acid derivative precedes the addition of the acyl acceptor compound (Y).

IV. SPECIFIC ACID-BASE CATALYSIS

The hydrolysis of acyl derivatives usually occurs through oxonium ion-catalyzed, pH-independent, and hydroxide ion-catalyzed processes,[9] as shown in Figure 1 and Equation 6. The catalyses by oxonium and hydroxide ions are called specific catalyses to distinguish them from general catalyses effected by Brönsted or Lewis acids and bases.

$$\frac{v}{[\text{ester}]} = k_{obs} = k_w + k_{H^+}[H^+] + k_{OH^-}[OH^-] \tag{6}$$

where k_{obs} is the observed pH-dependent rate constant, k_w is the pseudo first-order rate constant for the pH-independent "water" reaction and k_{H^+} and k_{OH^-} are the first-order rate constants for the specific acid and base catalysis, respectively. At low and high pH the plot is linear with slope -1.0 and $+1.0$, respectively. The minimum of the curve is usually encountered in the acidic pH range because $k_{H^+} \ll k_{OH^-}$.

A. Oxonium Ion Catalysis

The carbonyl oxygen of carboxylic acid derivatives has basic properties although the pK_a values of the conjugate acids are quite low, e.g., the pK_a is 0.0 for acetamide and -6.5 for ethyl acetate. The conjugate acid plays an important role as an intermediate in the acid-catalyzed reactions of these compounds.

The acid-catalyzed hydrolysis of esters of primary and secondary alcohols is an equilibrium process, but can be driven practically to completion by using a large excess of water. The reaction path is relatively complex involving probably five intermediates (Equation 7).[9]

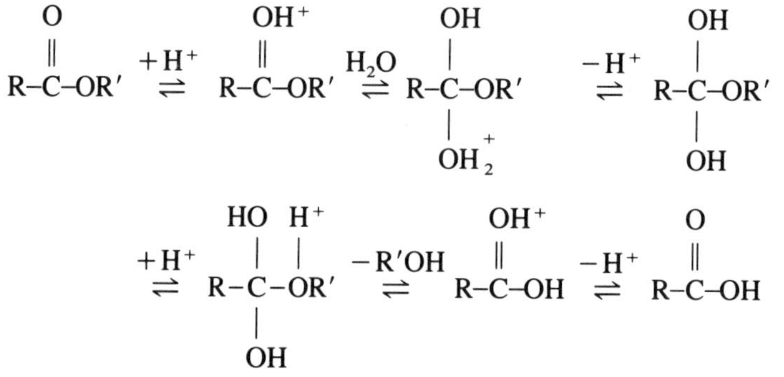

$$
\underset{\substack{\| \\ \text{R--C--OR}'}}{\text{O}} \; \overset{+\text{H}^+}{\rightleftharpoons} \; \underset{\substack{\| \\ \text{R--C--OR}'}}{\overset{\text{OH}^+}{\;}} \; \overset{\text{H}_2\text{O}}{\rightleftharpoons} \; \underset{\substack{| \\ \text{R--C--OR}' \\ | \\ \overset{+}{\text{OH}}_2}}{\overset{\text{OH}}{\;}} \; \overset{-\text{H}^+}{\rightleftharpoons} \; \underset{\substack{| \\ \text{R--C--OR}' \\ | \\ \text{OH}}}{\overset{\text{OH}}{\;}}
$$

$$
\overset{+\text{H}^+}{\rightleftharpoons} \; \underset{\substack{| \\ \text{R--C--OR}' \\ | \\ \text{OH}}}{\overset{\text{HO H}^+}{\;}} \; \overset{-\text{R}'\text{OH}}{\rightleftharpoons} \; \underset{\substack{\| \\ \text{R--C--OH}}}{\overset{\text{OH}^+}{\;}} \; \overset{-\text{H}^+}{\rightleftharpoons} \; \underset{\substack{\| \\ \text{R--C--OH}}}{\overset{\text{O}}{\;}} \qquad (7)
$$

Due to their substantial resonance stabilization, amides undergo hydrolysis much more slowly than esters. The reaction is essentially irreversible because under the acidic conditions ammonium ion is produced which is a much worse nucleophile than ammonia.

B. Hydroxide Ion Catalysis

In contrast to oxonium ion catalysis, acceleration of ester or amide hydrolysis by hydroxide ion is not a true catalysis since the hydroxide ion is usually not regenerated. The hydroxide ion is actually neutralized by the carboxylic acid product so that it should be employed in stoichiometric amount. Thus, the hydroxide ion may be regarded as a reagent rather than a catalyst.[9] The mechanism for base-catalyzed hydrolysis is shown by Equation 8.

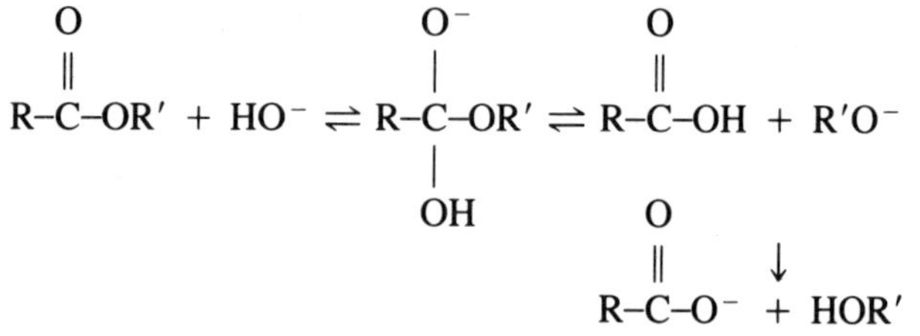

$$
\underset{\substack{\| \\ \text{R--C--OR}'}}{\text{O}} + \text{HO}^- \rightleftharpoons \underset{\substack{| \\ \text{R--C--OR}' \\ | \\ \text{OH}}}{\overset{\text{O}^-}{\;}} \rightleftharpoons \underset{\substack{\| \\ \text{R--C--OH}}}{\overset{\text{O}}{\;}} + \text{R}'\text{O}^- \qquad (8)
$$

$$
\underset{\substack{\| \\ \text{R--C--O}^-}}{\overset{\text{O}}{\;}} + \text{HOR}' \; \downarrow
$$

It is worthy of note that in this reaction (Equation 8) a strong nucleophile (OH$^-$) adds to the weakly electrophilic C=O bond, whereas in the acid-catalyzed reaction (Equation 7) a weak nucleophile (H$_2$O) adds to the strongly electrophilic C=OH$^+$ bond.

V. GENERAL ACID-BASE CATALYSIS

Specific catalysis usually occurs at extreme pH values where the concentration of oxonium or hydroxide ion is appreciable. Therefore, enzymes working around neutrality cannot readily utilize specific catalysis. Instead, they operate with use of general acid-base catalysis.[6,10-12] A general acid catalyst can be any proton donor, e.g., the neutral carboxyl group or the positively charged imidazolium ion, which is indeed involved in protease catalysis. A general base is a proton acceptor such as the carboxylate ion and the imidazole group, i.e., the conjugates of general acids.

Weak acids and bases used as components of buffers often fulfill the role of general catalysts. Specific acid- or base-catalyzed reactions are independent of the concentration of buffer, i.e., the buffer concentration does not significantly affect the concentrations of the oxonium and hydroxide ions. On the other hand, the rates of hydrolyses of a number of compounds such as acetylimidazole[13] or various haloacetates[14] have been shown to depend on the buffer concentrations. Figure 2 shows the dependence of the rate constant on the buffer concentration. The slope represents the apparent second-order rate constant of the

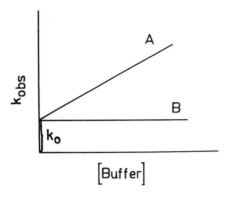

FIGURE 2. Dependence of the reaction rate on buffer concentration at constant pH. A, general catalysis, the slope represents k_{HA} or k_{A^-}; B, specific catalysis including the water reaction, k_0 is independent of the buffer concentration.

buffer-catalyzed reaction and the intercept at zero buffer concentration gives the contribution of the specific ion and the water catalyses to the reaction. At a different pH where the ratio of the acidic and basic components of the buffer is different, it can be inferred from the concomitant change in the slope which component of the buffer is the more active species for catalysis. For example, if the slope increases at a higher pH where the concentration of the basic component becomes higher, then the buffer operates as a general base catalyst.

When general catalysis is also considered, Equation 6 extends as shown by Equations 9 and 10 for acidic and basic buffers, respectively.

$$k_{obs} = k_w + k_{H^+}[H^+] + k_{OH^-}[OH^-] + k_{HA}[HA] + k_{A^-}[A^-] \tag{9}$$

$$k_{obs} = k_w + k_{H^+}[H^+] + k_{OH^-}[OH^-] + k_{BH^+}[BH^+] + k_B[B] \tag{10}$$

The complicated systems of Equations 9 and 10 can be analyzed experimentally by varying the conditions so that some of the terms can be negligible such as at highly acidic and alkaline pH or with and without buffer.

An example of general base catalysis is provided by the hydrolysis of acetylimidazole in imidazole buffer.[13] The reaction depicted in Equation 11 shows that the imidazole removes a proton from the attacking water molecule. This increases the nucleophilicity of the water oxygen atom.

Transition state

$$\tag{11}$$

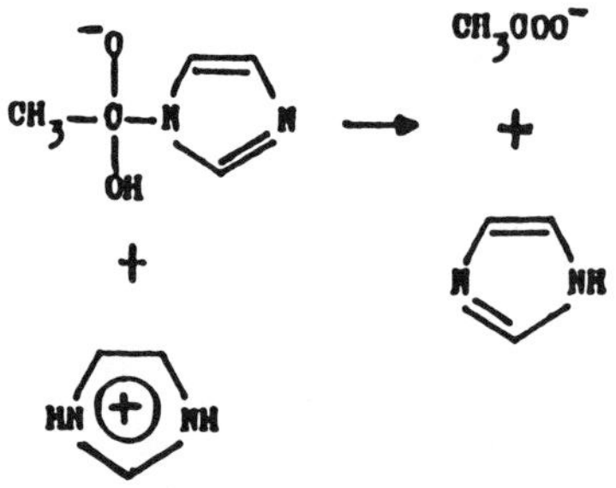

(11 cont.)

It is important that in a true general acid- or base-catalyzed reaction the proton transfer occurs at the rate-determining transition state and not at a pre-equilibrium. As to the position of the proton in the transition state, general and specific catalyses are two limiting cases as we shall see later in the discussion of the Brönsted catalysis law (Section VIII.A).

VI. NUCLEOPHILIC CATALYSIS

In general base catalysis, a Brönsted or Lewis base abstracts a proton from the reactant. The same base may alternatively react as a nucleophile with an electrophilic center such as the carbonyl carbon of an ester or amide. This is an ordinary chemical reaction, not catalysis. The nucleophilic attack becomes a catalytic step only if the initial reaction product is an unstable intermediate that rapidly decomposes to give the final product and to regenerate the catalyst.[6,10,15,16] In general, Equation 12 illustrates a simple nucleophilic displacement. If RCO–B$^+$ of Equation 12 is more prone to hydrolysis than RCO–X, then nucleophilic catalysis is accomplished according to Equation 13.

$$RCO-X + B \rightarrow RCO-B^+ + X^- \qquad (12)$$

$$RCO-B^+ + H_2O \rightarrow RCO-OH + B + H^+ \qquad (13)$$

Nucleophilic catalysis may be illustrated by the imidazole catalysis of the hydrolysis of *p*-nitrophenyl acetate (Equation 14).[17]

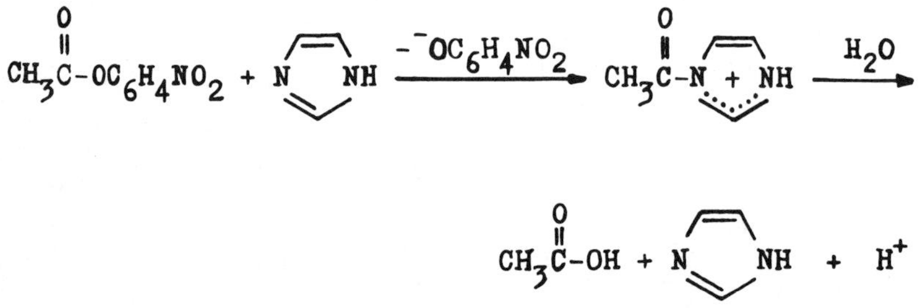

(14)

The acetylimidazole intermediate can be detected by its absorption at 245 nm. The slow step of the overall reaction is the hydrolysis of the acetylimidazole as indicated by the accumulation of the intermediate. Some compounds, such as thiols or amines, are much better nucleophiles than water. Thus, the transfer of the acetyl group from acetylimidazole to a thiol compound takes place readily in aqueous solution and results in the formation of a thiolester.

Imidazole is an effective nucleophilic catalyst — more nucleophilic than an oxygen nucleophile of comparable basicity. The high reactivity is conceivable since the positive charge developing during the reaction is distributed between the two nitrogens of the imidazole ring. Its pK_a value of 7 is ideal for reactions taking place at neutral pH. Specifically, a weaker base would have a smaller nucleophilicity, whereas a stronger base would not have its reactive free base form at an appreciably high concentration at neutrality. Furthermore, an acylimidazole intermediate displays high susceptibility to nucleophilic attack. This can be attributed in a large part to its relatively low resonance stabilization compared with ordinary amides. Notably, the nonbonding electron pair of the nitrogen atom becomes part of the π-electron system and cannot, therefore, contribute effectively to the double bond character of the C–N bond of the amide group. As a consequence, acylimidazoles are thermodynamically much less stable than most other amides and esters. Despite these excellent properties of the imidazole as a nucleophilic catalyst, in serine and cysteine protease catalyses it operates as a general acid-base catalyst rather than a nucleophilic catalyst. An explanation of this finding may be that imidazole is an effective nucleophile only in the reactions with activated acyl compounds such as the nitrophenyl esters. In these reactions, the good leaving group of the reactant is easily displaced by the imidazole. In contrast, in the reactions with physiological substrates which possess poor leaving groups, the imidazole becomes the better leaving group. Thus, even if the tetrahedral intermediate formed, it would decompose to the reactants rather than to the products.

VII. STRUCTURE-REACTIVITY RELATIONSHIPS

The overall rate constant for the reaction of a carboxylic ester or amide with a nucleophile (N) can be expressed as a function of two parameters, namely, the rate constant for the formation of the tetrahedral intermediate (k_1) and the ratio of the rate constants characteristic of the breakdown of the intermediate (k_{-1}/k_2). The rate constants are defined by Equation 15.

$$
\underset{\text{R–C–X}}{\overset{\text{O}}{\overset{\|}{}}} + N^- \underset{k_{-1}}{\overset{k_1}{\rightleftharpoons}} \underset{\underset{N}{|}}{\overset{O^-}{\overset{|}{\text{R–C–X}}}} \overset{k_2}{\rightarrow} \underset{\text{R–C–N}}{\overset{O}{\overset{\|}{}}} + X^- \tag{15}
$$

It is obvious that structural changes in the attacking nucleophile and in the acyl and leaving groups of the carboxylic acid derivative will affect both k_1 and k_{-1}/k_2. This will be discussed in Sections VII.A and B.

A. The Nucleophile

Many nucleophiles can react with carboxylic acid derivatives. Such compounds may contain, for example, an oxygen atom with negative charge (hydroxide, alkoxide, phenoxide, carboxylate, oximate, hypochlorite anions), uncharged oxygen atom (water), nitrogen atom (various amines), sulfur atom (thiolate ions), and others. A nucleophilic reaction, as well as nucleophilic catalysis, depends to a great extent on the relative efficacy of the nucleophile. Edwards[18] has proposed two factors that determine the nucleophilic character of a reactant: (1) basicity, which is generally expressed quantitatively in terms of the pK_a of the conjugate acid and (2) polarizability. The polarizability of a nucleophile reflects the ability of its electrons to respond to changing electric field during reaction. The smaller and more compact an atom, the lower the polarizability of its electrons. Consequently, third- and fourth-period elements are invariably more nucleophilic than second-period elements of comparable basicity. Furthermore, nonbonding electron pairs are usually held more loosely than bonding electrons, and thus, can be more polarizable.

Table 2
NUCLEOPHILIC
**REACTIVITY TOWARDS *P-*
NITROPHENYL ACETATE**[19]

Nucleophile	pK$_a$	k(M^{-1} min^{-1})
HOO$^-$	11.5	2×10^5
Acetoximate	12.4	3×10^3
ClO$^-$	7.2	1.6×10^3
OH$^-$	15.7	9×10^2
C$_6$H$_5$O$^-$	10.0	1×10^2
NH$_2$OH	6.0	1×10^2
NH$_3$	9.2	16
CN$^-$	10.4	11
C$_5$H$_5$N	5.4	0.1
CH$_3$CO$_2^-$	4.8	5×10^{-4}
H$_2$O	-1.7	6×10^{-7}

Another factor determining nucleophilicity is seen in nucleophiles possessing a pair of unshared electrons on the atom α to the nucleophilic atom ($\ddot{X}$–$\ddot{Y}$). Such compounds including hydroxylamine [8], hydrazine [9], isonitrosoacetone, the anions of peroxides, and hypochlorite ion [10], exhibit an enhanced reactivity which is referred to as the α effect.[19] The exceptional reactivity of these nucleophiles has been attributed to the stabilization of the transition state of the reaction by electron donation from the α atom.

$$H\ddot{O}\text{–}\ddot{N}H_2 \qquad\qquad H_2\ddot{N}\text{–}\ddot{N}H_2 \qquad\qquad :\ddot{C}l\text{–}O^-$$

[8]	[9]	[10]

In addition to basicity, polarizability, and α effects, other factors may also be of significance in determining nucleophilicity. Solvation is such an additional factor. This accounts, for example, for the poor reactivity of the hydroxide ion toward the ester bond in water when compared to other oxyanions such as the methoxide or ethoxide ion.[20,21]

The relative nucleophilicity of various nucleophiles, as manifested in the second-order rate constants, is illustrated by the data of Table 2. It is seen that basicity is an important component of the nucleophile in some reactions with *p*-nitrophenyl acetate (C$_6$H$_5$O$^-$, CH$_3$CO$_2^-$, H$_2$O). In other reactions, the α effect is predominant, which renders an exceptionally good reactant of the nucleophile (ClO$^-$, NH$_2$OH). Table 2 also shows that all correlation between basicity and nucleophilicity vanishes when the nucleophilic atom is altered. Thus, the nitrogen nucleophiles are considerably more reactive than the oxygen nucleophiles of the same pK.

B. The Carboxylic Acid Derivative

The effects of structural changes in the acyl moiety (R) of a carboxylic acid derivative (RCOX) is clearly illustrated by the relative reactivities of ethyl chloroacetates toward the hydroxide ion: acetate (1) < chloroacetate (761) < dichloroacetate (16,000) < trichloroacetate (100,000).[22] A comparison of the relative rates shown in parentheses indicates that an increase in the electron-attracting character of R, which makes the carbonyl carbon atom more positive, results in an increased hydrolysis rate. This is mainly a consequence of the increase of the k$_1$ in Equation 15.

To assess the effects of structural changes in X of RCOX is not as straightforward as in

R. In most cases, an increase in the electron-attracting power of X increases k_1 and k_2 of Equation 15, which in turn tends to increase the overall rate constant. The effect of an inductive change in X is usually less pronounced than a similar change in R because the substituent in X is farther from the reaction center. Resonance effects in group X are of particular importance with the amide substrates of proteases. Increased resonance interaction between X and the carbonyl group tends to stabilize the ground state relative to the transition state of the reaction which leads to a decreased k_1.

In some other examples, the overall rate constant is practically independent of structural changes. Thus, it has been found that in the alkaline hydrolysis of *p*-substituted acetanilides, the effect of structural changes on k_1 is canceled by an opposite effect on k_2.[23] Specifically, the formation of the tetrahedral intermediate is promoted by electron-attracting groups; its decomposition, however, is facilitated by electron-donating groups. Electron-donating groups promote the protonation of the nitrogen atom which is required for expulsion of anilines (Equation 16).

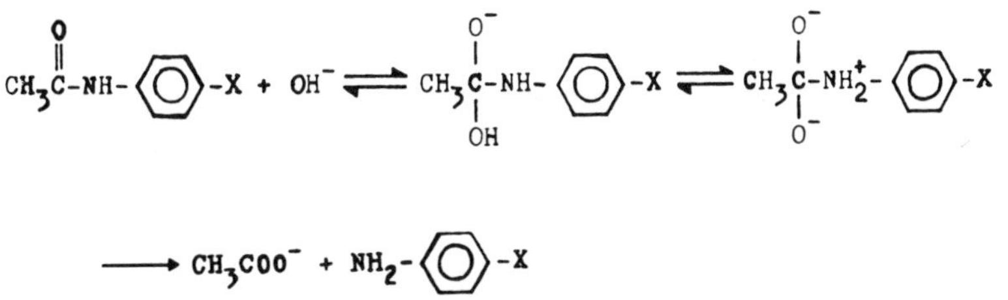

$$(16)$$

VIII. LINEAR FREE ENERGY CORRELATIONS

Structural changes in the reactants can often be correlated with one or more parameters at least within certain limits. This is a helpful method for estimation of the transition state structures of organic and to a more limited extent, enzymic reactions. Free energy correlations involve comparisons of the free energies between the studied and the standard equilibrium or rate processes which are proportional to the logarithms of the equilibrium or rate constants. Several parameters are used in such comparisons, all are related to a particular model, which should be kept in mind in the interpretation of correlations. A detailed account on free energy correlations has recently been published.[24]

A. The Brönsted Catalysis Law

An important factor that governs both general acid-base catalyses and nucleophilic reactions is the pK_a of the catalyst. Brönsted and Pedersen[25] proposed and verified a relationship between the pK_a of the catalyst and the catalytic rate constant. This relationship is defined by Equations 17 and 18 for general base and general acid catalysis, respectively.

$$\log k_B = \beta pK_a + \text{constant} \tag{17}$$

$$\log k_A = -\alpha pK_a + \text{constant} \tag{18}$$

The Brönsted equations indicate that a plot of $\log k_B$ against the pK_a of the catalyst should be linear with a slope $+\beta$, while a plot of $\log k_A$ against pK_a should be linear with a slope $-\alpha$. As an example, a plot for the catalysis of the hydrolysis of ethyl dichloroacetate by general base catalysts including amines, carboxylate ions, and phosphate dianion[14] is shown in Figure 3. In this example, bases of different types fit a single plot with $\beta = 0.47$, indicating a similar catalytic mechanism for all species. This value of β (~ 0.5) is charac-

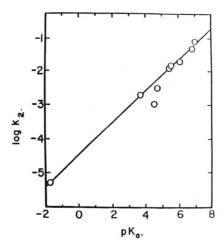

FIGURE 3.   Brönsted plot of the second-order rate constants for general base-catalyzed hydrolysis of ethyl dichloroacetate. The bases shown are water, formate, aniline, acetate, pyridine, succinate, 4-picoline, phosphate, and imidazole in the order of increasing pK_a. (From Jencks, W. P. and Carriulo, J., *J. Am. Chem. Soc.,* 83, 1743, 1961. With permission.)

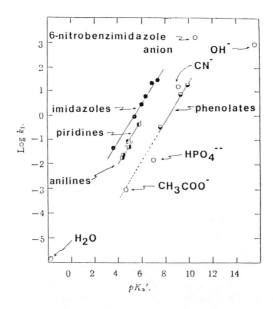

FIGURE 4.   The effect of basicity on the reactivity toward *p*-nitrophenyl acetate of substituted imidazoles, pyridines, anilines, and oxygen anions in 28.5% ethanol at 30°C. (From Bruice, T. C. and Lapinski, R., *J. Am. Chem. Soc.,* 80, 2265, 1958. With permission.)

teristic of general base-acid catalysis although some separation of the grossly different base families can be observed.[26] By contrast, in nucleophilic catalysis the lines representing particular base types are separated considerably (see Section VII.A). For example, imidazole and phosphate dianion, both possessing similar pK_a values and similar rate constants in the general base-catalyzed hydrolysis of ethyl dichloroacetate (Figure 3), exhibit vast difference in the nucleophilic reactivities toward *p*-nitrophenyl acetate (Figure 4), imidazole being about 4000 times more reactive than the phosphate dianion.[27] This difference in reactivity may be employed as a diagnostic tool to distinguish between nucleophilic and general catalyses.

The value of Brönsted α or β is a useful indication of the position of the proton in the transition state, in as much that α or β approximates the fractional displacement of the transition state along the reaction coordinate from reactants to products. For example, in a general base-catalyzed reaction the position of the proton is dependent on the basicity of both the catalyst (B) and the reactant (HA). Typical general base catalysis occurs when the pK_a values of the donor and the acceptor are similar (Figure 5A). The proton is equally bonded to the base and the reactant, yielding $\beta \sim 0.5$. If the pK_a of the base is greater than that of the reactant, the H–A bond will be stretched to a lesser extent in the transition state (Figure 5B), i.e., the transition state will resemble the reactant state and β will be between 0.0 and 0.5. On the other hand, if the pK_a of the base is lower than that of the reactant, the transition state will resemble the product (Figure 5C), the proton being closer to the catalyst, and β will be between 0.5 and 1.0. When α or β approaches zero, the reactant will not show discrimination toward acids or bases. Thus, the acid or base being at the largest concentration, e.g., water in an aqueous system, will catalyze the reaction. When α or β approaches 1, the reactant will exhibit a great discrimination and only the most potent catalyst will be operative, such as the oxonium ion or the hydroxide ion in aqueous systems. This implies that general catalysis will change to specific catalysis at high values of the Brönsted coefficients.

The above consideration refers to a simple proton transfer reaction. However, in organic reactions one seldom solely measures proton transfer reactions, but the proton transfer is accompanied by other chemical processes such as a nucleophilic attack by water in the case of peptide hydrolysis. This may significantly affect the position of the proton in the transition state.

From the reaction coordinate diagrams for proton transfer (Figures 5B and 5C), it is apparent that in the transition state the proton is always closer to the weaker base. In other words, the transition state bears a greater resemblance to the less stable of the species in the reaction. This sort of relationship has been noted in several cases,[28-30] but is most commonly referred to as the Hammond postulate. The Brönsted linear free energy relationship is essentially a quantitative statement of the Hammond postulate. This principle is most obvious in the limiting case of a strongly exothermic reaction with almost no activation energy (Figure 6) where the starting materials and the transition state are energetically very similar. It is apparent from the foregoing that in an exothermic reaction the Brönsted slope will be >0.5, whereas in an endothermic reaction it will be <0.5.

The Brönsted linear free energy relationship is not valid over a very large range of basicity. The nonlinearity was predicted by Brönsted and Pedersen.[25] In principle, it varies from 1.0 to 0.0, but measurements extending over the entire range may not be achieved because of practical time limitations. In fact, the linear range may be appreciably large as seen from Figure 3.

Finally, it should be noted that Brönsted's original correlation for proton transfer reactions was extended to nucleophilic reactions. In this case, the symbol β_N or β_{nuc} is used to designate substituent variation in the nucleophile, and β_L or $\beta_{l.g.}$ is employed to denote variation in the leaving group.[24]

B. The Hammett Equation

The correlation postulated by Hammett[31] (Equations 19 and 20) considers the effects of polarity of a substituent (X) on reaction rates and equilibria, in particular on ester hydrolyses including enzyme-catalyzed reactions.

$$X\text{-}\langle\bigcirc\rangle\text{-COOR} + H_2O \longrightarrow X\text{-}\langle\bigcirc\rangle\text{-COOH} + ROH \qquad (19)$$

$$\log\frac{k}{k_0} = \rho\sigma \qquad (20)$$

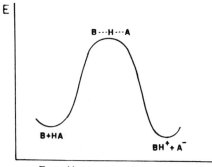

A

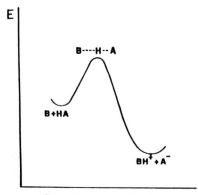

B

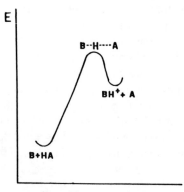

C

FIGURE 5.   Hypothetical energy diagrams for the movement of proton in the course of general base catalysis. (A) $pK_{BH+} \sim pK_{HA}$ and $\beta \sim 0.5$, typical general base catalysis; (B) $pK_{BH+} > pK_{HA}$ and $0 < \beta < 0.5$. If $\beta \to 0$, then general base catalysis $\to$ solvent catalysis (C) $pK_{BH+} < pK_{HA}$ and $0.5 < \beta < 1$, If $\beta \to 1$, then general base catalysis $\to$ specific catalysis.

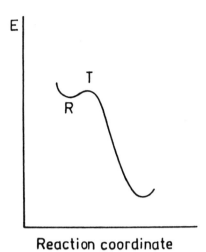

Reaction coordinate

FIGURE 6. Energy profile of a highly exothermic reaction. Note that the energies of
the reactant state (R) and the transition state
(T) are nearly identical.

where k and k_0 are the rate constants of the substituted and unsubstituted compound, respectively. The substitution occurs at *meta* or *para* position of a benzenoid compound. *Ortho*
substituents are generally disregarded because of their possible steric effects on reactions.
The value of σ represents the polar effects relative to the benzoic acid standard (Equation
21).

$$\sigma = \log\frac{K}{K_0} \qquad (21)$$

where K and K_0 are the dissociation constants of the substituted and unsubstituted benzoic
acid, respectively. The substituent constant σ is positive for electron-attracting substituents,
whereas it is negative for electron-releasing substituents. The value of the reaction constant,
ρ, shows the overall sensitivity of rate constants or equilibria to polar effects. For instance,
the alkaline hydrolysis of substituted ethyl benzoates is more sensitive to electronic effects
than the dissociation of the corresponding benzoic acid derivatives (ρ = 2.5). By contrast,
the acid hydrolysis of ethyl benzoates is virtually insensitive to polar effects (ρ = 0.03).
In the case of the alkaline hydrolysis of benzamides, the sensitivity is similar to that of the
dissociation of benzoic acids (ρ = 1.06).

The original values of σ constants do not always give a satisfactory correlation when a
large series of substituents are used. Therefore, a number of special σ values have been
introduced to correlate better the data. For example, even Hammett[31] noted that p-NO_2 and
p-CN substituents exhibit exalted effects in the reactions of phenol and aniline derivatives.
This may be a consequence of an extra resonance interaction between the strongly electron-
attracting substituents and the aromatic hydroxyl or amino groups.

Taft extended the Hammett-type linear free energy relationship to the hydrolysis of aliphatic esters and amides.[32] Equation 22 provided by Taft includes steric effects in addition
to polar effects.[32]

$$\log(k/k_0) = \delta E_s + \rho^*\sigma^* \qquad (22)$$

The parameter, σ^*, depends only on the net polar effects whereas the steric constant, E_s, accounts for the steric effects. These values for different reactions can be obtained from appropriate tables[32] and the sensitivity of a given reaction to steric (δ) or polar (ρ^*) effects may then be evaluated.

It is important to note that the ρ^* is sensitive to the total charge around the acyl carbon atom in the transition state relative to the ground state, but it is expected to be relatively insensitive to the redistribution of this charge among the nucleophile, the carbonyl oxygen, and the leaving group. By contrast, the Brönsted coefficient is sensitive to the changes in charge localized on the nucleophile or on the leaving group.[33] Thus, ρ^* will be less sensitive than β to the charge redistribution that occurs upon going from the ground state to the transition state.

The $\rho\sigma$ treatment has often been applied to enzymatic reactions. However, in this case the additional interactions between the enzyme and substrate may adversely affect the results, compared to the case of simple model reactions.

For the ligand binding to enzymes, a more extended linear free energy correlation has been introduced by Hansch.[34] According to his treatment, the effects of structural changes in a set of congeners is a multivariable problem. Besides the electronic effects, the hydrophobicity, the molar refractivity, and the steric properties of the substituent are also taken into account. (The molar refractivity is related to polarizability.) Although it provides a better correlation, the consideration of several parameters, say more than two, obscures the physical meaning of the models to be tested.

IX. KINETIC ISOTOPE EFFECTS

Like linear free energy relationships, the study of isotope effects can be rewarding in the diagnosis of reaction mechanism. There are several valuable reviews[35-37] on kinetic solvent isotope effects, usually expressed as ratios of the rate constants in water and deuterium oxide. Deuterium and in some cases tritium isotope effects have most often been employed for demonstrating the occurrence of proton transfer in the rate-limiting step of a reaction.

In protease catalysis, the proton transfer is associated with general acid-base catalysis involving oxygen or nitrogen atoms which readily exchange their proton(s) with those of the solvent (H_2O or ^2H_2O) The magnitude of the deuterium isotope effect of such reactions is considerable, usually between 2 and 4. In proton transfers to or from a heavy atom such as carbon, oxygen, or nitrogen, the kinetic isotope effect is ascribed mainly to the difference in zero point energies of the heavy atom-hydrogen bond (X–H) vs. the heavy atom- deuterium bond (X–^2H). The hydrogen atom bound to a heavy atom has different energy levels and the lowest energy state of an X–H bond, the zero point energy, is above the minimum of the potential curve (Figure 7). As the frequency of vibration is lower with an atom of greater mass, the zero point energy of the X–^2H bond is lower than that of the X–H bond. This energy difference is of primary importance in determining the kinetic isotope effects because the zero point stretching vibrations are virtually lost in the transition state as the hydrogen or deuterium atom moves toward the acceptor (Figure 7). In general, whenever an isotopic nucleus encounters less bonding and thus, less restriction in the transition state than in the reactant state, the lighter isotopic molecule will react faster than the heavier one. Conversely, if the isotopic nucleus becomes more tightly bound in the transition state, the molecule containing the heavier nucleus will react faster. The direction of an isotope effect is normal when $k_{light}/k_{heavy} > 1$ and inverse when $k_{light}/k_{heavy} < 1$.

General acid-base catalyses commonly produce normal isotope effects. In these catalytic processes, the isotopic atom is translated during the reaction (bond to the isotope breaks in the transition state) resulting in *primary* isotope effects. By contrast, *secondary* isotope effects, which are substantially smaller, refer to reactions in which bond cleavages occur at

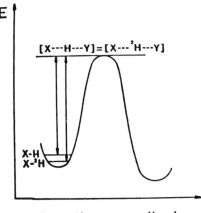

Reaction coordinate

FIGURE 7. Illustration of kinetic deuterium isotope effect. The difference in the zero point energies of X–H and X–^{2}H bonds is lost in the transition state of proton transfer to an acceptor, Y.

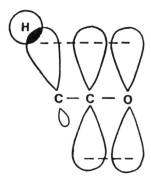

FIGURE 8. Illustration of hyperconjugation between a C–H bond of a methyl group and the carbonyl double bond.

positions not containing isotopic atom. This is clearly illustrated by the β-hydrogen isotope effects in acyl-transfer reactions.[38] As seen from the acetyl derivative of Figure 8, there is an overlap of σ bonds in the alkyl substituent (the CH_3 or C^2H_3 group) with the π system (the carbonyl group). This interaction is called hyperconjugation. Nucleophilic bonding to the carbonyl function reduces hyperconjugation from the adjacent C–H (^{2}H) bonds. This strengthens the bonds in the transition state, leading to an inverse isotope effect. The magnitude of the β-hydrogen secondary isotope effect seems to be maximally 0.955 per hydrogen atom (0.87 for the methyl group),[38] as judged from the equilibrium effect for ketone hydration. Because ketone hydration gives rise to a tetrahedral compound, a kinetic isotope effect of this magnitude (0.955) should signal a completely tetrahedral transition state while a completely trigonal transition state should produce no isotope effect. The β-hydrogen secondary isotope effect thus provides a means of studying the "tetrahedrality" of the transition states of various protease reactions (Chapter 3, Section VII).

The magnitude of a solvent isotope effect is also affected by the fact that the isotope enrichment of the solute (SH) may not have the same value as that of the solvent (ROH), even if the isotope exchange is completely free. The preference in the hydrogen site of the solute for deuterium over protium relative to that for the hydrogen site in the solvent is measured by the fractionation factor ϕ as shown by Equation 23.[37]

$$\phi = \frac{[S^2H]/[SH]}{[RO^2H]/[ROH]} \tag{23}$$

The ϕ values for various compounds relative to water as solvent have been found between 0.4 and 1.3. The ratio of the fractionation factors for reactant (ϕ^R) and product (ϕ^P) gives the isotope equilibrium effect in the case of a simple proton-transfer reaction (Equation 24). For rate constants the isotope effect is dependent on the ratio of the fractionation factors of the reactant state (ϕ^R) and the transition state (ϕ^T) (Equation 25).

$$K_H/K_{2_H} = \phi^R/\phi^P \tag{24}$$

$$k_H/k_{2_H} = \phi^R/\phi^T \tag{25}$$

Finally, we should mention proton inventory studies which are directed to determine the number of protons translating in the transition state, as well as the contributions by the individual protonic sites to the overall isotope effect. Such a separation has an obvious value in generating and testing mechanistic hypotheses. Indeed, proton inventory has been employed in a number of cases to study the mechanism of protease catalysis. In such experiments the reaction rate is measured in various mixtures of light and heavy water. A linear relationship between the observed rate constant and the mole fraction of deuterium oxide in water indicates the participation of a single proton. If the rate changes with the second, third, or higher power of the fraction of the isotope in the solvent, the effect will arise from two, three, or more protons, in principle. In practice, it is almost impossible to distinguish between two or more protons because of the small deviation from linearity relative to the large errors in the determination of enzymic rates.

Although kinetic isotope effects are very helpful in testing reaction mechanisms, it should be kept in mind that interpretation of the results is often difficult, especially in the case of enzyme reactions.[35-37] Even the readily measurable, large primary isotope effects may be reduced substantially by inverse secondary isotope effects leading to erroneous conclusions about the existence or nonexistence of general acid-base catalysis. In some cases, there is no sufficient data for constructing proper models for the enzymic reactions. Such a model, the hydrated ketone used for studying the tetrahedrality of transition states in protease catalysis, has already been mentioned. It is obvious that this neutral compound is only a crude approximation of the negative tetrahedral intermediate that interacts with the oxyanion binding site of serine proteases. It may be also difficult to find an appropriate model of ϕ^T for the peculiar transition state of an enzyme reaction. It appears then that it is the qualitative rather than the quantitative character of the isotopic data that is significant. Because of the unpredictable effects arising from the interaction between the reactant and the enzyme, the isotope effects should be examined with a number of reactants to minimize the possibility of drawing erroneous conclusions.

Considering the values of quantitative data, it is interesting to note that the magnitude of the kinetic isotope effect is dependent on the difference between the pK values of the donor and acceptor atoms.[39-41] Maximum isotope effect is obtained when the pK difference is zero. This implies a nearly symmetrical transition state when the proton is half-transferred from the donor to the acceptor. It was shown in Section VIII that in the same situation the Brönsted α or β is equal to 0.5.

Table 3
BOND DISTANCES AND BOND ENERGIES

Bond	Distance (pm)	Energy (kJ/mol)
C–S	182	272
C–O	143	360
S–H	133	347
O–H	96	464

X. THE REACTIVITY OF THIOLS, THIOLESTERS, AND THIONESTERS

In the catalysis by cysteine proteases, the nucleophile is a thiol group and the intermediate acyl-enzyme is a thiolester. It is, therefore, of considerable interest to discuss the chemistry of these groups and to make a comparison between the sulfur and the corresponding oxygen compounds. In addition to thiols and thiolesters, thionesters (thionoesters) are also important in the study of the mechanism of protease action (see the Chapters 3 and 4).

Thiols (R-SH) bear an obvious relationship to alcohols (R-OH). The principal structural differences between methanethiol and methanol are that the C–S bond is about 40 pm longer than the C–O bond (Table 3) and that the C–S–H bond angle is smaller than the C–O–H bond angle. The energy of the sulfur bond is considerably less than that of the corresponding oxygen bond (Table 3). Thiols are stronger acids than alcohols. Thus, the pK_a value of ethanethiol is only 10.6 requiring its conversion into the anion when reacting with the hydroxide ion (Equation 26).

$$C_2H_5SH + OH^- \rightleftharpoons C_2H_5S^- + H_2O$$

$$pK_a \ 10.6 \qquad\qquad pK_a \ 15.7 \qquad\qquad (26)$$

Although the thiols are more acidic than alcohols, sulfur is less electronegative than oxygen. Hence, hydrogen bonding between thiols is much less important than between alcohols. However, hydrogen bonding from the acidic thiol proton to water oxygen is significant. Indeed, thiols are fairly soluble in water.

Detailed studies on the reactions of a number of thiols with p-nitrophenyl acetate have been reported.[42,43] The results have established that the thiolate anion is the nucleophilic species. A Brönsted plot for these reactions is presented in Figure 9. In the case of thiol amines, the microscopic pK_a values are plotted. Equation 27 describes the Brönsted plot of Figure 9.

$$\log k = 0.38 \ pK_a - 0.75 \qquad\qquad (27)$$

An interesting characteristic of Equation 27 is the value (0.38) of the Brönsted coefficient. This is about half of the corresponding value found for the oxygen[27] and nitrogen[44] bases (see the corresponding plots for imidazoles and phenols in Figure 9). The low slope for the thiol compounds may be of significance since a thiol group of low pK_a would still be an efficient nucleophile, at least for the substrate p-nitrophenyl acetate. It can be estimated from Figure 9 that a thiol of pK_a 7 is about four times more reactive toward p-nitrophenyl acetate than the imidazole of pK_a 7.

The high nucleophilicity of the thiolate anion is best interpreted in terms of its high polarizability which offsets its lower basicity as compared with that of the corresponding oxygen nucleophile. Pearson distinguishes two classes of bases: those which have a strong

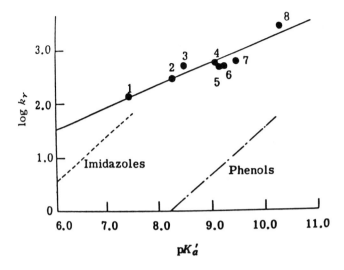

FIGURE 9.   Brönsted plot for the reaction of thiols with *p*-nitrophenyl acetate. 1, Cysteine ethyl ester [R(NH$_3$$^+$)S$^-$]; 2, 2-mercaptoethylamine [R(NH$_3$$^+$)S$^-$]; 3, cysteine [R(NH$_3$$^+$)S$^-$]; 4, cysteine ethyl ester [R(NH$_2$)S$^-$]; 5, homocysteine; 6, glutathione; 7, 2-mercaptoethanol; 8, 2-mercaptoacetate. Plots for imidazoles (- - -) and phenols (-.-.-). (From Ogilvie, J. W., Tildon, J. T., and Strauch, B. S., *Biochemistry*, 3, 754, 1964. With permission.)

tendency to bind protons are called hard and those which preferentially combine with heavy metal ions are called soft.[45] Typical examples of hard bases are water or the fluoride anion, whereas thiolate or iodide ion is soft. A hard base has a strong tendency to bind with hard acids and vice versa. Thus, the relative nucleophilicity of oxygen and sulfur compounds depends on the type of substrate. For example, for substitutions at a saturated carbon atom, reactivity of the nucleophile is mainly governed by polarizability, whereas for reactions at a carbonyl group the basicity is more important.

Thiolesters are less stable compounds than the corresponding oxygen esters. This is apparent from the equilibrium studies which indicate that an acyl group of a thiolester favors transfer to an alcohol (Equation 28) by a factor of about 50.[46]

$$
\begin{array}{ccc}
\overset{\displaystyle O}{\overset{\displaystyle \|}{R\text{--}C\text{--}SR'}} + HOR'' \rightleftharpoons \overset{\displaystyle O}{\overset{\displaystyle \|}{R\text{--}C\text{--}OR''}} + HSR'
\end{array}
\qquad (28)
$$

This can be accounted for by the relatively small resonance stabilization of thiolesters (forms [11] and [12]) compared with the resonance stabilization of oxygen esters (form [13]). However, a significant contribution by form [12] was inferred from the low carbonyl stretching frequency and the small basicity of the carbonyl oxygen of thiolesters.[47] Yet, the importance of form [12] may be questioned because of the lack of an electron octet of the oxygen atom.

$$
\underset{[11]}{\overset{\displaystyle O}{\overset{\displaystyle \|}{R\text{--}C\text{--}SR'}}} \leftrightarrow \underset{\ }{\overset{\displaystyle O^-}{\overset{\displaystyle |}{R\text{--}C=}}} \overset{+}{S}R' \leftrightarrow \underset{[12]}{\overset{\displaystyle O^+}{\overset{\displaystyle |}{R\text{--}C=SR'}}} \overset{-}{\ }
$$

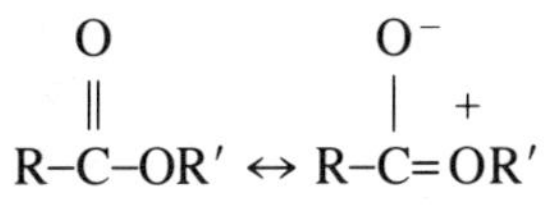

$$
\begin{array}{ccc}
\text{O} & & \text{O}^- \\
\parallel & & \mid \quad + \\
\text{R-C-OR}' & \leftrightarrow & \text{R-C=OR}'
\end{array}
$$

[13]

An additional factor concerning the reactivity of thiolesters is the low electronegativity of sulfur showing about the same value as that of carbon. Hence, a thiolester is expected to be more polarizable than oxygen esters and form [14] should be more important in thiolesters than in oxygen esters.

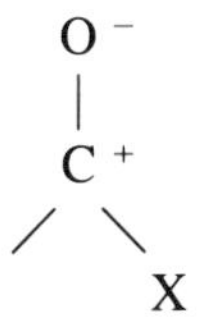

$$
\begin{array}{c}
\text{O}^- \\
\mid \\
\text{C}^+ \\
\diagup \quad \diagdown \\
\text{X}
\end{array}
\qquad\qquad \text{X = O or S}
$$

[14]

In contrast to the presumed higher reactivity of the thiolesters as compared to that of the oxygen esters, the hydrolysis rates are similar or even lower for the thiolester reactions.[48,49] Inspection of the data of Table 4 reveals that the rate constants of alkaline hydrolysis of the corresponding members of the two series are of the same order of magnitude. Larger activation energies are found with the thiolester reactions which become offset, however, by the greater entropies of activation. The more positive entropies of activation for the thiol compounds than those for the corresponding oxygen compounds can be interpreted in terms of more rigid, exactly oriented transition state structures for the oxygen compared to the sulfur series. This is reasonable in the light of the larger size of the sulfur atom compared to the oxygen and of the probability that the oxygen esters are solvated to a larger extent than the thiolesters.

Under acidic conditions, the thiolesters are hydrolyzed much more slowly than the oxygen esters. This may be expected if, as mentioned above, the oxygen esters have a greater contribution from the resonance form [13] which makes them more susceptible to an attack by the proton. From the study of the hydrolysis of methyl thiolformate in the pH range 0 to 4.6, it was concluded that the mechanism involves the participation of cationic and neutral tetrahedral addition intermediates (Equation 29) with a change in the rate-determining step occurring at pH 1.2.[50] The cationic intermediate expels H_2O and CH_3SH about equally well while the neutral intermediate decomposes with predominant expulsion of CH_3SH.

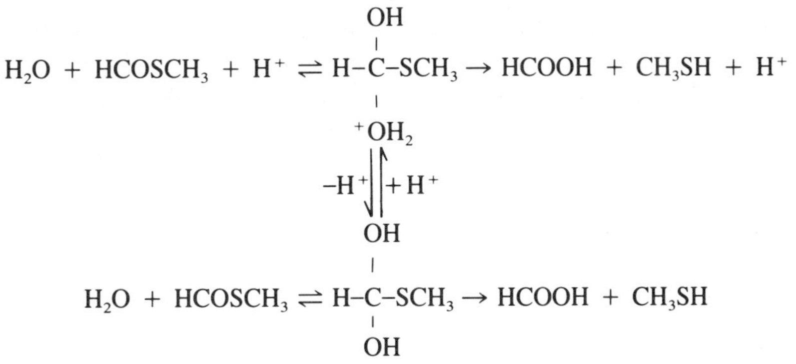

$$
\begin{array}{c}
\text{OH} \\
\mid \\
\text{H}_2\text{O} + \text{HCOSCH}_3 + \text{H}^+ \rightleftharpoons \text{H-C-SCH}_3 \rightarrow \text{HCOOH} + \text{CH}_3\text{SH} + \text{H}^+ \\
\mid \\
{}^+\text{OH}_2
\end{array}
$$

$$
-\text{H}^+ \big\updownarrow +\text{H}^+
$$

$$
\begin{array}{c}
\text{OH} \\
\mid \\
\text{H}_2\text{O} + \text{HCOSCH}_3 \rightleftharpoons \text{H-C-SCH}_3 \rightarrow \text{HCOOH} + \text{CH}_3\text{SH} \\
\mid \\
\text{OH}
\end{array}
\qquad (29)
$$

Table 4
HYDROLYSIS RATE CONSTANTS
FOR THIOLESTERS AND OXYGEN
ESTERS[a] [48]

R	$k_{alkaline}$[b] $(M^{-1}\ min^{-1})$	$k_{acid} \times 10^3$[c] $(M^{-1}\ min^{-1})$
Thiolesters ($CH_3\ COSR$)		
CH_3-	2.41	1.0
$(CH_3)_2CH-$	0.82	0.57
$(CH_3)_2CHCH_2-$	0.72	0.51
Oxygen esters ($CH_3\ COOR$)		
CH_3-	3.96	49.2
$(CH_3)_2CH-$	0.29	19.8
$(CH_3)_2\ CHCH_2-$	0.68	29.2

[a]   In 62% (weight) acetone.
[b]   20°C.
[c]   30°C.

In the reactions with nitrogen nucleophiles, thiolesters are considerably more reactive than oxygen esters. It has been known for a long time that acetyl-CoA reacts readily with hydroxylamine at neutrality,[51] whereas oxygen esters react with hydroxylamine only at strongly alkaline conditions.[52] In both cases, hydroxamic acids are formed (Equation 30) which can be determined as colored complexes with $FeCl_3$. Butylaminolysis also supports the greater reactivity of thiolesters, in as much as *n*-butylamine reacts with ethyl *p*-nitro-thiolbenzoate but not with ethyl *p*-nitrobenzoate.[53]

$$R-COXR + NH_2OH \rightarrow R-CONHOH + RXH \qquad (30)$$

$$(X = O, S)$$

Of the nitrogen nucleophiles, the imidazole derivatives have been studied most extensively.[54,55] Several different mechanisms have been observed in the various reactions of imidazole with thiolesters. Most of these reactions represent nucleophilic catalyses, which in some cases are facilitated by general base and/or specific base catalysis. With oxygen esters having poor leaving groups, the nucleophilic catalysis by the imidazole is much less probable.

Acyl substituent effects on rates of acyl transfer to thiolate and hydroxide ions have also been compared.[56] It was found that the ρ^* value for the thiolysis is only slightly larger than the corresponding value for the alkaline hydrolysis. These kinetic ρ^* values are essentially identical with the equilibrium ρ^* for hydroxide or thiolate addition to aldehydes forming an anionic gem-diolate (Equation 31) and thiohemiacetalate (Equation 32), respectively.[56] Hence, the magnitude of the kinetic ρ^* values suggests that the transition states for the alkaline hydrolysis and for the thiolysis reactions are similar in geometry and charge distribution to the tetrahedral intermediates.

$$
R-C\!\!\begin{array}{c} O \\ \diagdown \\ H \end{array}\ +\ OH^- \rightleftharpoons R-\!\!\begin{array}{c} O^- \\ | \\ C-OH \\ | \\ H \end{array} \qquad (31)
$$

Table 5
PRODUCT DISTRIBUTION AND RATE CONSTANTS (M^{-1} sec^{-1}) OF THE HYDROLYSIS OF ETHYL THIONBENZOATE IN 40% AQUEOUS ACETONE[a] [59]

	NaOH (25°C)	HCl (125°C)
% Thiobenzoic acid	78	15
% Ethyl benzoate	22	85
$10^4 k_1$	57.2	0.96
$10^4 k_2$	15.7	5.3
$10^4 k_3$	54.9	5.9

[a] The rate constants are defined by Equation 34.

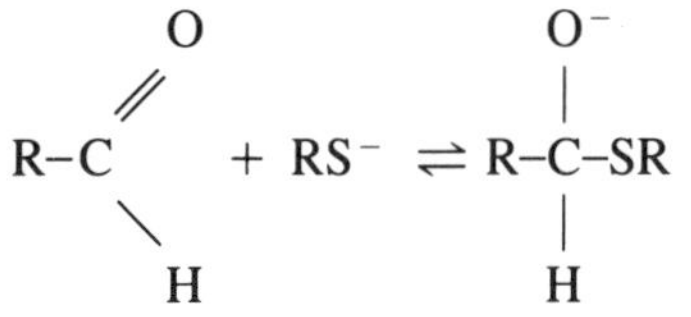

$$\begin{array}{cc}
\text{O} & \text{O}^- \\
\diagup\!\!\!\diagup & | \\
\text{R--C} \quad + \text{RS}^- \rightleftharpoons \text{R--C--SR} \\
\diagdown & | \\
\text{H} & \text{H}
\end{array} \qquad (32)$$

Thionesters [15] also proved to be useful in the studies of the mechanism of action of proteases. They are isomers of thiolesters [16]. Indeed, the free acids exist in a fast tautomeric equilibrium which is shifted towards the thiol acid. The salt derived from either tautomeric form contains the mesomeric anion [17] which exhibits an infrared (IR) band in the carbonyl region.[58] This implies that most of the negative charge resides on the sulfur atom. A more negative sulfur, however, is not expected on the basis of the higher electronegativity of oxygen (3.5) than that of sulfur (2.6). Hence, the tendency of the carbon-sulfur bond to remain a single bond appears to prevail.

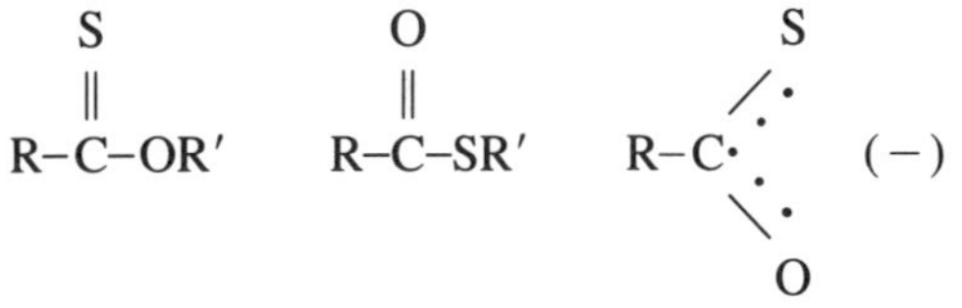

$$\begin{array}{ccc}
\text{S} & \text{O} & \text{S} \\
\| & \| & \diagup\cdot \\
\text{R--C--OR}' & \text{R--C--SR}' & \text{R--C}\cdot \quad (-) \\
& & \diagdown\cdot \\
& & \text{O}
\end{array}$$

$$\quad [15] \qquad\qquad [16] \qquad\qquad [17]$$

A detailed study on the hydrolysis of ethyl thionbenzoate has shown the formation of an addition intermediate which decomposes in two ways according to Equation 33.[59] It is seen that the intermediate gives rise to either ethyl benzoate and hydrogen sulfide or to thiobenzoic acid and ethanol. The quantitative data are given in Table 5 where the rate constants correspond to those of Equation 34.

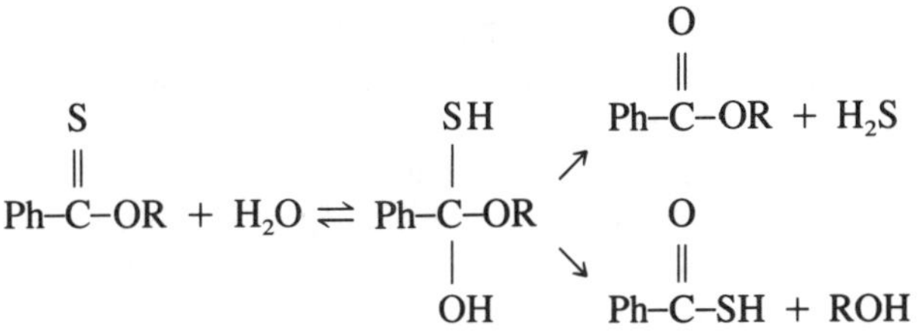

$$\begin{array}{l}
\qquad\qquad\qquad\qquad\qquad\qquad \text{O} \\
\qquad\qquad\qquad\qquad\qquad\qquad \| \\
\text{S} \qquad\qquad\qquad \text{SH} \quad \text{Ph--C--OR} + \text{H}_2\text{S} \\
\| \qquad\qquad\qquad\quad | \quad \nearrow \\
\text{Ph--C--OR} + \text{H}_2\text{O} \rightleftharpoons \text{Ph--C--OR} \qquad \text{O} \qquad\qquad (33)\\
\qquad\qquad\qquad\qquad | \quad \searrow \quad \| \\
\qquad\qquad\qquad\qquad \text{OH} \qquad \text{Ph--C--SH} + \text{ROH}
\end{array}$$

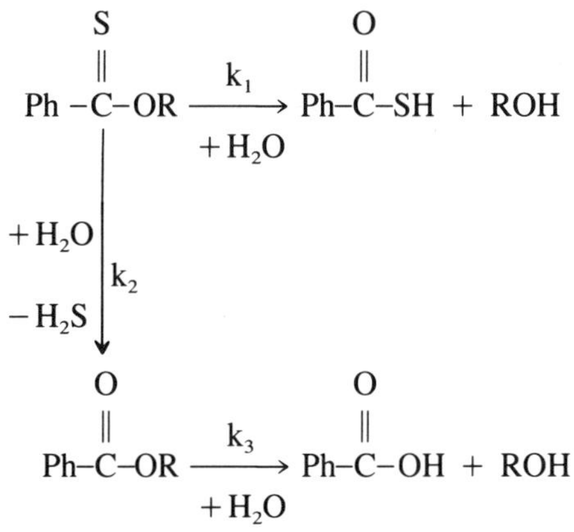

$$
\begin{array}{c}
\overset{\displaystyle S}{\overset{\displaystyle \|}{Ph-C-OR}} \xrightarrow{\ k_1\ } \overset{\displaystyle O}{\overset{\displaystyle \|}{Ph-C-SH}} + ROH \\[2mm]
+H_2O
\end{array}
$$

$$
\begin{array}{c}
+H_2O \\
\Big\downarrow k_2 \\
-H_2S
\end{array}
$$

$$
\overset{\displaystyle O}{\overset{\displaystyle \|}{Ph-C-OR}} \xrightarrow{\ k_3\ } \overset{\displaystyle O}{\overset{\displaystyle \|}{Ph-C-OH}} + ROH \qquad (34)
$$
$$
+H_2O
$$

It is seen from Table 5 that under alkaline conditions the rates of loss of ethoxide from ethyl thionbenzoate and from ethyl benzoate are the same within experimental error. Under acidic conditions, however, ethyl benzoate is hydrolyzed six times faster than ethyl thion-benzoate is converted to thiobenzoic acid.

The hydrolysis of ethyl thionbenzoate was also studied at high concentrations of sulfuric acid.[60] It was found that the breakdown of the tetrahedral intermediate depended on the acid concentration. In 32% sulfuric acid, ethyl benzoate and hydrogen sulfide were formed exclusively, whereas in 66% acid, thiobenzoic acid and ethanol became the preferred products and the thiobenzoic acid subsequently was hydrolyzed to benzoic acid. The partitioning of the tetrahedral intermediate was also affected by the concentration of sulfuric acid in the hydrolysis of thionacetanilide.[61] In diluted sulfuric acid (1%), the hydrolysis of thionacet-anilide occurred chiefly by C–S cleavage of the tetrahedral intermediate (Equation 35). As the concentration of sulfuric acid was raised, C–N cleavage became more important and was the exclusive pathway in 48% acid. The results were explained by postulating that C–S cleavage involved the neutral and C–N cleavage the amino-protonated tetrahedral interme-diate (Equation 35).

$$
\begin{array}{c}
\overset{\displaystyle SH}{\underset{\displaystyle OH}{\overset{\displaystyle |}{\underset{\displaystyle |}{CH_3-C-NHPh}}}} \xrightarrow{\ k_{C-S}\ } SH^- + \overset{\displaystyle NHPh}{\underset{\displaystyle OH}{CH_3-\overset{..}{C}\!\cdot\!\overset{+}{}}} \xrightarrow{\ -H^+\ } \overset{\displaystyle NHPh}{\underset{\displaystyle O}{CH_3-C\!\!\diagdown\!\!\|}}
\end{array}
$$

$$
-H^+ \Big\updownarrow +H^+
$$

$$
\overset{\displaystyle SH}{\underset{\displaystyle OH}{\overset{\displaystyle |}{\underset{\displaystyle |}{CH_3-C-{}^+NH_2Ph}}}} \xrightarrow{\ k_{C-N}\ } PhNH_2 + \overset{\displaystyle SH}{\underset{\displaystyle OH}{CH_3-\overset{..}{C}\!\cdot\!\overset{+}{}}} \xrightarrow{\ -H^+\ } \overset{\displaystyle SH}{\underset{\displaystyle O}{CH_3-C\!\!\diagdown\!\!\|}} \qquad (35)
$$

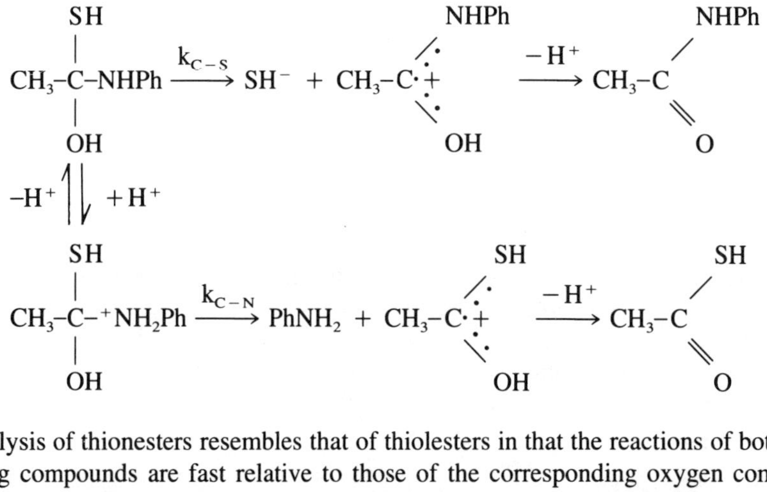

Aminolysis of thionesters resembles that of thiolesters in that the reactions of both sulfur-containing compounds are fast relative to those of the corresponding oxygen compounds. Structure-reactivity correlations have indicated a change in the rate determining step of the aminolysis of *p*-nitrophenyl thionbenzoate.[62] The results have been interpreted in terms of rate-limiting breakdown of a zwitterionic tetrahedral intermediate [18] with moderately basic amines and in terms of rate-limiting attack with highly basic amines. As compared to the oxygen esters, the higher reactivity of thionesters appears to stem from the enhanced ability of sulfur to expel leaving groups.

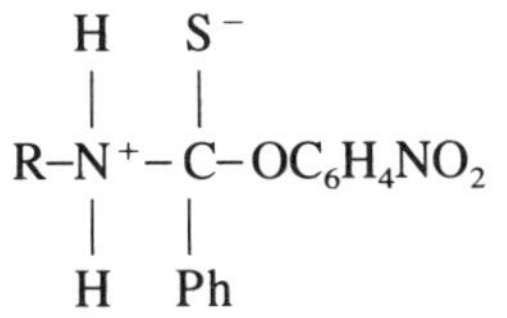

$$
\begin{array}{cc}
\text{H} & \text{S}^- \\
| & | \\
\text{R--N}^+\text{--C--OC}_6\text{H}_4\text{NO}_2 \\
| & | \\
\text{H} & \text{Ph}
\end{array}
$$

[18]

In summary, we have seen that there are some important distinguishing features between the sulfur and oxygen compounds which should be considered in the discussion of the serine and cysteine protease catalysis. These features are (1) around neutrality the stability and reactivity of the thiolate ion are greater than those of the hydroxide ion; (2) the thiolester (acyl-enzyme in cysteine proteases) is less stable than the corresponding oxygen ester (acyl-enzyme of serine proteases); (3) the carbonyl oxygen of an ester exhibits more propensity to associate with the proton than the carbonyl oxygen of a thiolester does; and (4) the transition state of thiolester reaction is looser than that of an oxygen ester.

XI. METAL IONS IN THE HYDROLYSIS OF ESTERS AND AMIDES

A number of proteases including carboxypeptidase A, angiotensin-converting enzyme, collagenases, and leucine aminopeptidase have a catalytically competent metal ion at their active site. The metal ion is Zn^{2+} in the well-known instances. The Zn^{2+} can be replaced by other metal ions such as Co^{2+} and Cu^{2+} with more or less alteration in the catalytic activity. For example, substitution of Co^{2+} for Zn^{2+} in carboxypeptidase A does not affect the enzymic activity to an appreciable extent. Moreover, the considerable absorbency of the cobalt enzyme in the visible spectrum can be exploited in ligand binding and mechanistic studies.

As the same catalytic features of metal ions may prevail in enzymic and simple organic reactions, the metal ion-promoted organic reactions have been extensively studied. In many reactions the role of metal ion is akin to that of the proton. It has been proposed that metal ions possessing more than one positive charge can operate as super acid catalysts in a more efficient way.[63] However, this assumption is not supported by compelling experimental evidence. Although metal ions can greatly increase reaction rates, they are usually no more effective than the oxonium ion. This is conceivable in the light of the tighter binding of the proton which is due to its more concentrated charge as compared with the more diffuse charges of the large metal ions. For example, coordination of the proton to a water molecule decreases the basicity of water from a pK_a of 15.7 to -1.7, whereas coordination with Cu^{2+} ion decreases the basicity of water only to about 8.[64,65] The metal ions, however, are superior to the proton in that they are capable of acting in neutral medium where the concentration of proton is very low. This feature of metal ions is of particular importance in enzymic reactions.

It has been known for a long time that metal ions effectively catalyze the hydrolysis of amino acid esters.[66] For example, Co^{2+} and Cu^{2+} ions can promote the hydrolysis of glycine ethyl ester at pH 7 to 8 at 25°C where the esters are ordinarily stable. The presence of at least one donor atom in the substrate such as the amino nitrogen in the glycine ethyl ester is an essential requirement for specific complexation or chelation to occur with the metal ion. The reactive complex contains one substrate and one metal component.

In glycine buffer, which also coordinates with the metal ion, the hydrolysis of glycine methyl ester was proposed to take place according to Equation 36.[67]

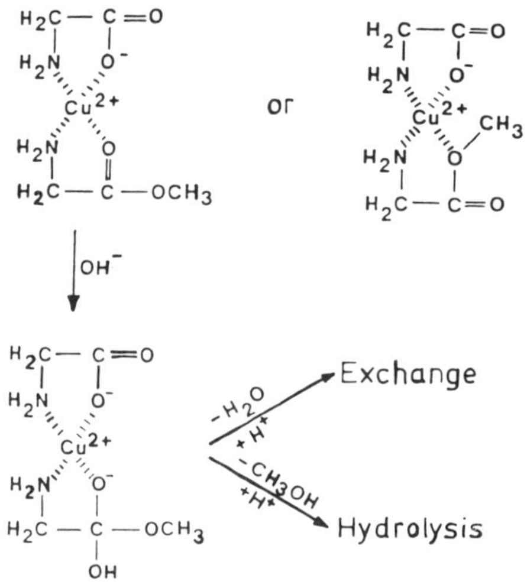

(36)

It is seen that Cu^{2+} is complexed with both the buffer and the ester. The ester group may interact with the metal ion through the carbonyl oxygen or alternatively through the alkoxy oxygen. Of course, stabilization of the tetrahedral intermediate would be more effective in the former case. Formation of a tetrahedral addition intermediate on the reaction path is supported by the observation that carbonyl oxygen exchange accompanies the hydrolytic process.[67] An alternative mechanism may involve an intramolecular attack by the hydroxide ion bound to the metal ion (form [19]).

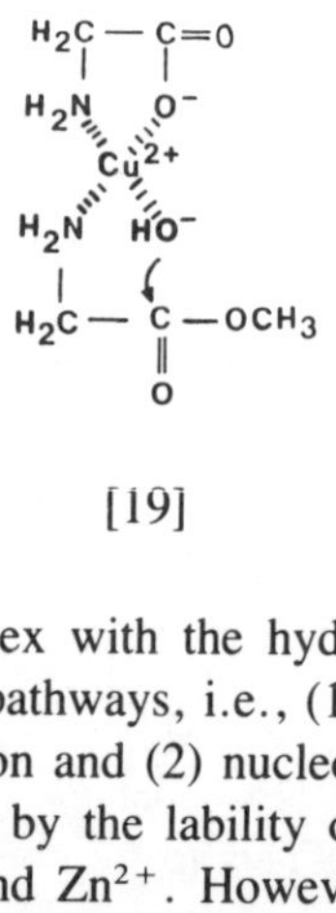

[19]

Indeed, cupric ion can form a complex with the hydroxide ion around neutrality.[68] To distinguish between the two principal pathways, i.e., (1) direct coordination of the ester of amide carbonyl oxygen to the metal ion and (2) nucleophilic attack by metal coordinated hydroxide ion, is usually complicated by the lability of complexes formed with divalent cations such as Cu^{2+}, Co^{2+}, Ni^{2+}, and Zn^{2+}. However, in the case of the stable Co^{3+} complexes, distinction between the two mechanisms can be made on the basis of ^{18}O experiments.[69] Interestingly, both mechanisms occur in the various Co^{3+} systems.[70-73] For example, glycine reacting with the metal coordinated hydroxide ion is hydrolyzed 10^7 to 10^{11} times faster than the uncomplexed species. In the case of the carbonyl activation mechanism, the rate enhancement is 10^4-fold.[71]

An interesting complex ion catalyst is *cis*-hydroxoaquatriethylenetetraminecobalt[3+] (Equation 37), abbreviated $[Co(trien)(H_2O)OH]^{2+}$, which selectively hydrolyzes N-terminal amino acid residues and thus, it may be regarded as an exopeptidase model.[73] However, this metal complex is not a true catalyst because the product, the N-terminal amino acid, remains bound to the metal ion (Equation 37) so that the product complex must be reduced to remove the N-terminal amino acid.

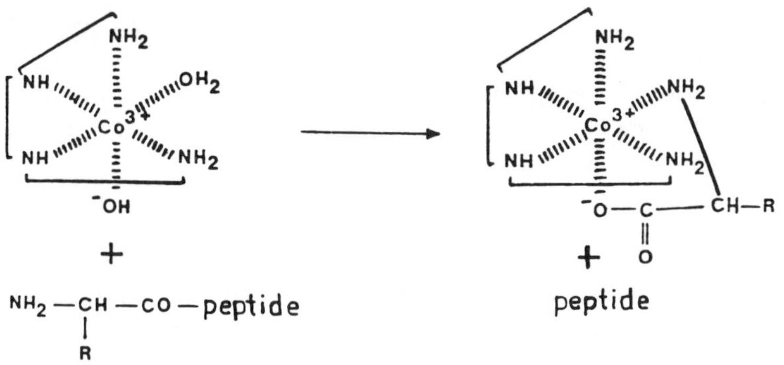

(37)

Nucleophilic addition of metal bound hydroxide ion was also concluded from ester hydrolysis by divalent ions, Co^{2+} and Ni^{2+}.[74] Various esters of 2-(2'-hydroxyphenol)-4(5)-methyl-5(4)-(2'',2''-dimethylacetate)imidazole [20] complexed with metal ion exhibited a large rate enhancement (10^3 to 10^5) over hydroxide ion-mediated hydrolysis. The metal was bound to the unprotonated imidazolyl moiety and the dissociated phenolic hydroxyl group. The reaction involved (1) pre-equilibrium formation of a tetrahedral intermediate accompanying nucleophilic addition of metal bound hydroxide ion [21], (2) the formation of an intermediate metal carboxylate capable of acting as an acylating agent ("carboxyl-metal anhydride") [22], and (3) acid catalysis of the "carboxyl-metal anhydride" intermediate leading to the free acid of ester [20].

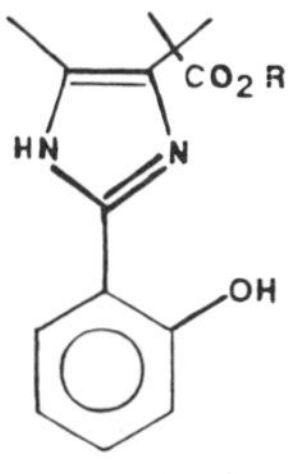

[20] 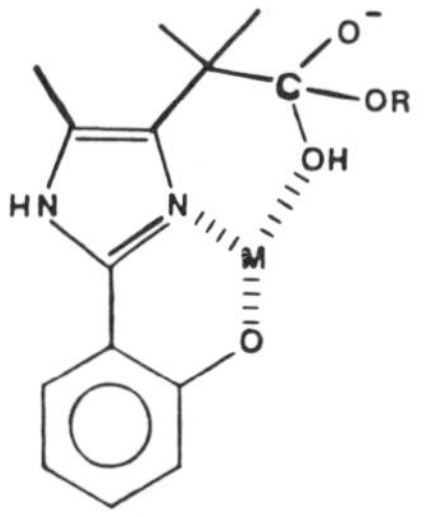   [21] 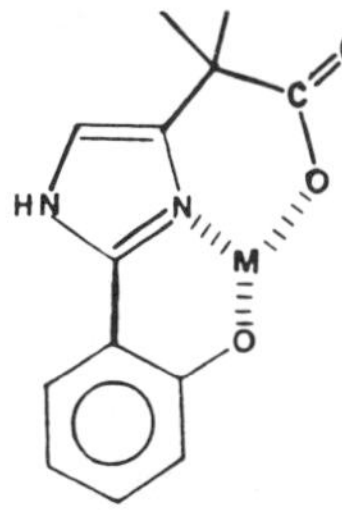   [22]

The zinc ion is essential for the action of metalloproteases. Therefore, its role in simple model reactions, namely in the ester and amide hydrolysis, is of particular interest. From the kinetic studies[75] on the hydrolysis of 8 quinolyl hydrogen glutarate [23], it has been concluded that the metal ion-promoted catalysis prevails above pH 6. This contribution, however, becomes unimportant relative to the intramolecular attack by the carboxylate ion below pH 6.

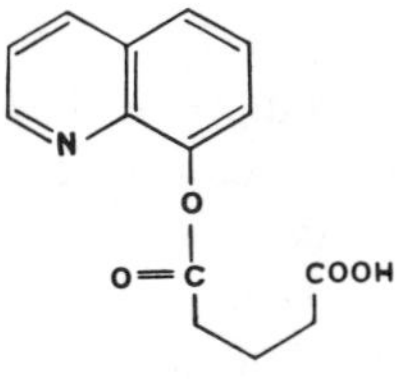

[23]

As with the hydrolysis of [23], the substantial rate acceleration observed in some metal ion-promoted reactions may not be attributed simply to the electronic distortion brought about by the positive charge(s). The metal ion complexes bear many of the features of compounds reacting intramolecularly. The advantage of intramolecularity of a reaction will be discussed in the next chapter.

XII. INTRAMOLECULAR REACTIONS

Intramolecular and enzymic reactions are similar in that the reacting groups are held together in both processes: in the former by a covalent bond and in the latter by the formation of a noncovalent enzyme-substrate complex. Therefore, the study of the much simpler intramolecular reactions should provide useful data for understanding the more complex enzymic catalysis. Intramolecular reactions have been reviewed in considerable detail.[10,76-79] By showing a few examples, this section is focused on the rate enhancement attainable by bringing together the reacting groups in intramolecular reactions.

Before comparing the efficiency of intramolecular reactions relative to their intermolecular counterparts, one should be certain about the identity of the mechanisms of the two processes. Clearly, the bringing together of the functional groups in an intramolecular reaction may alter the mechanism or even render a mechanism possible, which does not prevail in the corresponding intermolecular reaction.

A further problem emerges when the values of rate constants are to be compared. Direct comparison between the intra- and intermolecular reactions is not permissible because the units of the rate constants are different: sec^{-1} for the former monomolecular reaction (Equation 38) and $M^{-1} sec^{-1}$ for the latter bimolecular process (Equation 39).

$$V = k_1[\overarc{AB}] \tag{38}$$

$$v = k_2[A][B] \tag{39}$$

The rate enhancement in the intramolecular reaction relative to the intermolecular process may be expressed by the ratio of the first-order rate constant to the second-order rate constant. The ratio has the dimension of molarity and represents formally the concentration of the reactant in excess in the intermolecular reaction. For example, at a particular concentration of B in excess the pseudo first-order rate constant for the bimolecular reaction can be equal to the first-order rate constant for the intramolecular reaction of $\overarc{AB}$.[76,79] Therefore, the ratio k_1/k_2 is considered as the effective concentration or effective molarity of B in the vicinity of A in the molecule $\overarc{AB}$. The values measured for effective molarity often represent unattainably high concentrations. Therefore, in addition to the local concentration effect, other factors, e.g., orientation, strain, and changes in solvation, must be invoked to account for the rapid rate of an intramolecular reaction.

Studies of the nucleophilic reactions of esters with amines or carboxylate ions offer the possibility of comparing intramolecular and intermolecular reactions. Thus, the second-order rate constant ($8 \times 10^{-3} M^{-1} min^{-1}$ at 20°C) for the reaction of phenyl acetate with trimethylamine (Equation 40) may be compared with the first-order rate constant ($10 min^{-1}$) for the reaction of phenyl 4-(*N, N*-dimethylamino) butyrate (Equation 41) under similar conditions.[80] The effective molarity of the trialkyl amino group in the intramolecular reaction of the amino ester is $k_1/k_2 = 1250 M$. If one considers that the concentration of pure water is 55 M, it is immediately apparent that the 1250 M is a physically unattainable high concentration. Consequently, the 1250 M effective concentration reflects a substantial rate enhancement which can be attributed to the existence of the two reacting groups in a single molecule. In other words, this large rate enhancement is accounted for by the favorable entropy term of the intramolecular system.

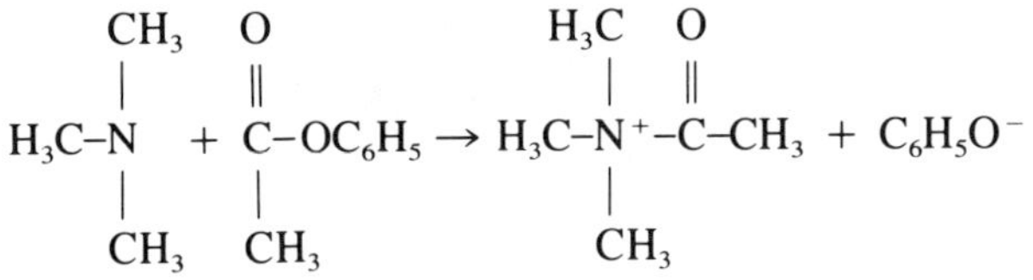

$$\begin{array}{ccccc}
\text{CH}_3 & \text{O} & & \text{H}_3\text{C} & \text{O} \\
| & \| & & | & \| \\
\text{H}_3\text{C-N} & + \text{C-OC}_6\text{H}_5 & \rightarrow & \text{H}_3\text{C-N}^+\text{-C-CH}_3 & + \text{C}_6\text{H}_5\text{O}^- \\
| & | & & | & \\
\text{CH}_3 & \text{CH}_3 & & \text{CH}_3 &
\end{array} \tag{40}$$

$$\begin{array}{ccccc}
\text{CH}_3 & \text{O} & & \text{CH}_3 & \text{O} \\
| & \| & & | & \| \\
\text{H}_3\text{C-N} & \text{C-OC}_6\text{H}_5 & \rightarrow & \text{H}_3\text{C-N}^+\text{—C} & + \text{C}_6\text{H}_5\text{O}^- \\
\diagup \quad \diagdown & & & \diagup \quad \diagdown & \\
\text{H}_2\text{C} \qquad \text{CH}_2 & & & \text{H}_2\text{C} \qquad \text{CH}_2 & \\
\diagdown \quad \diagup & & & \diagdown \quad \diagup & \\
\text{CH}_2 & & & \text{CH}_2 &
\end{array} \tag{41}$$

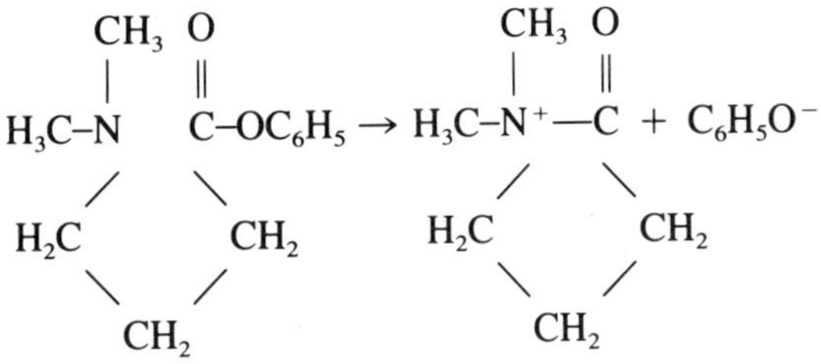

The importance of the relative positions of the reacting groups in acceleration of intramolecular processes has been shown by the hydrolysis of a series of monoaryl succinates and glutarates.[81-83] These reactions proceed by intramolecular nucleophilic catalysis according to Equation 42.

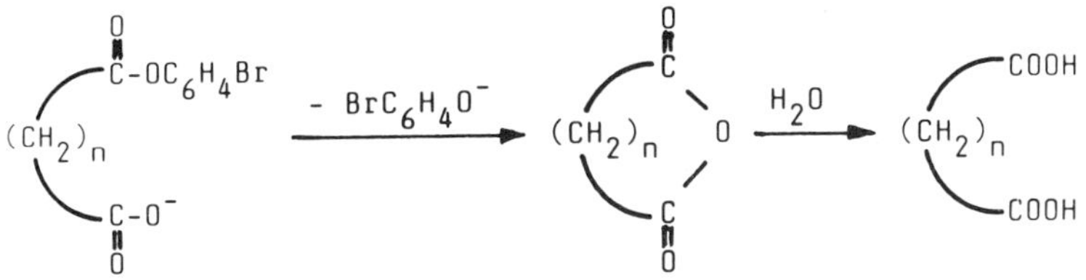

$$(\text{CH}_2)_n \begin{array}{c} \text{O} \\ \| \\ \text{C-OC}_6\text{H}_4\text{Br} \\ \\ \text{C-O}^- \\ \| \\ \text{O} \end{array} \xrightarrow{- \text{BrC}_6\text{H}_4\text{O}^-} (\text{CH}_2)_n \begin{array}{c} \text{O} \\ \| \\ \text{C} \\ \diagdown \\ \qquad \text{O} \\ \diagup \\ \text{C} \\ \| \\ \text{O} \end{array} \xrightarrow{\text{H}_2\text{O}} (\text{CH}_2)_n \begin{array}{c} \text{COOH} \\ \\ \\ \text{COOH} \end{array} \tag{42}$$

The hydrolysis of some mono-*p*-bromophenyl esters of dicarboxylic acids is shown in Table 6. It is seen that there is a progressive increase in rate as the structure of the molecule becomes more rigid. In proceeding from the glutarate monoester (compound 1) to the succinate monoester (compound 3), which involves a loss of one degree of rotational freedom, the rate constant increases by 230-fold. Proceeding from the succinate monoester (compound 3) to the bicyclic compound (compound 5) results in a further increase of 230-fold. This factor is equivalent to a change of 12.5 kJ/mol in the activation entropy.

Addition of gem-dimethyl substituents (Table 6, compare compounds 1 and 2) also increases considerably the reaction rate presumably by tending to keep the reacting groups in proximity. This lessens the probability for the molecule to take up a multitude of extended conformations which would not lead to reaction. Other alkyl substitutions in this system,[84] as well as in many other systems,[79] also result in considerable rate enhancements. For example, tetramethylsuccinanilic acid undergoes hydrolysis (Equation 43) 1200 times faster than unsubstituted succinanilic acid.[85]

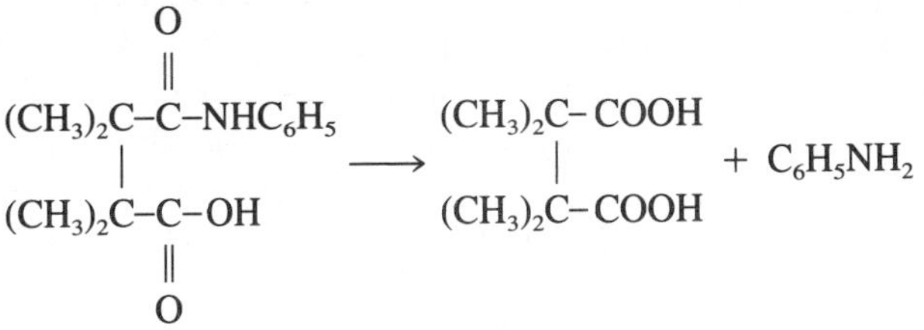

$$\begin{array}{ccc}
\text{O} & & \\
\| & & \\
(\text{CH}_3)_2\text{C-C-NHC}_6\text{H}_5 & & (\text{CH}_3)_2\text{C- COOH} \\
| & \longrightarrow & | \qquad\qquad + \text{C}_6\text{H}_5\text{NH}_2 \\
(\text{CH}_3)_2\text{C-C-OH} & & (\text{CH}_3)_2\text{C- COOH} \\
\| & & \\
\text{O} & &
\end{array} \tag{43}$$

Table 6
RELATED RATE CONSTANTS FOR THE INTRAMOLECULAR REACTIONS OF MONO-*p*-BROMOPHENYL ESTERS OF DICARBOXYLIC ACIDS

No.	Ester[a]	$\dfrac{k_{hydrolysis}}{k_{hydrolysis}\ \text{(glutarate)}}$
1	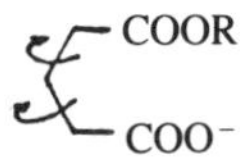	1
2	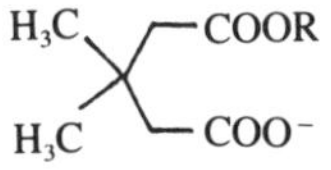	20
3	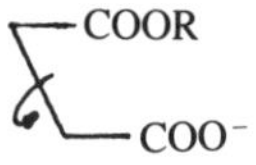	230
4	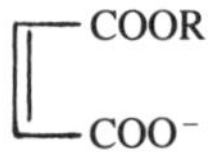	10,000
5	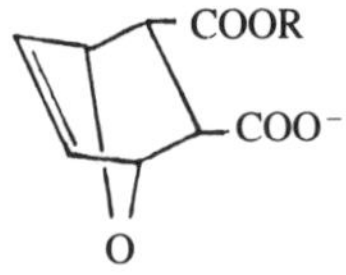	53,000

[a] $R = p\text{–}Br\text{–}C_6H_4\text{–}$.

From Bruice, T. C. and Pandit, U. K., *Proc. Natl. Acad. Sci. U.S.A.*, 46, 402, 1960. With permission.

It is of particular interest that the difference in rates between the substituted and unsubstituted succinanilic acids is associated with a change in the enthalpy of activation rather than in the entropy of activation. There is not enough mechanistic information available to account for the energetics of this reaction. Steric compression brought about by the alkyl substituents may be a factor that causes a strained ground state which is relieved in the transition state.

When the hydrolysis of tetramethylsuccinanilic acid is compared with the intermolecular hydrolysis of acetanilide, an effective molarity of 1.6×10^8 is obtained. Such a great value is fairly common in intramolecular nucleophilic reactions.[78,79] Even higher values were observed in the lactonization of a series of alkyl substituted *o*-hydroxy-hydrocinnamic acid.[86] In compound [24], the reacting groups are free to move away from each other by rotation of the side chain. However, this rotation is prevented by introduction of methyl groups into both the aromatic ring and the side chain and this leads to a rate enhancement factor in compound [25] as high as 5×10^{10}. In comparison with the bimolecular esterification of phenol with acetic acid, the rate enhancement is greater than 10^{15} *M*. The achievement of such a great rate enhancement requires, in addition to an almost complete entropy loss of the reacting groups, a considerable destabilization of the starting material, i.e., by steric compression, relative to the transition state. [87,88]

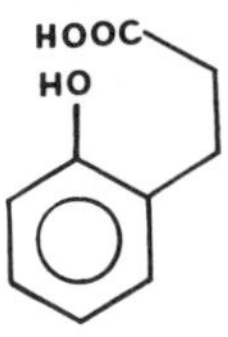

[24] 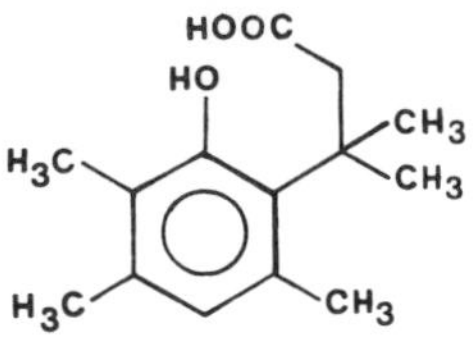   [25]

From a survey of a large number of intramolecular reactions, Kirby has concluded that the effective molarity is greatly dependent upon as to whether the reaction is nucleophilic or general acid-base catalyzed.[79] He has suggested that an effective molarity >80 is associated with a nucleophilic mechanism provided that the system contains unstrained (five- or six-membered) rings. If the effective molarity is <80, the mechanism is almost certainly general acid or general base catalysis. There are, however, exceptions to this generalization.[79]

It is an important question as to why the effective molarities for general acid-base-catalyzed intramolecular reactions are so much lower relative to the nucleophilic intramolecular reactions. This phenomenon may be rationalized in terms of the different complexities of the transition states. The nucleophilic reactions accompanied with ring closures lead to constrained transition state structures possessing low residual entropy of activation. On the other hand, the general acid-base-catalyzed intramolecular reactions involve a hydrogen-bonded water molecule in their transition state where several covalent and hydrogen bonds are being made and broken simultaneously. This results in a much looser transition state. The advantage of the intramolecular reaction, which is primarily entropic, is thus reduced for general acid-base-catalyzed processes.[79] Nevertheless, intramolecular general base-catalyzed reactions may occur as in the hydrolysis of acetylsalicylic acid (aspirin [26]).[89,90]

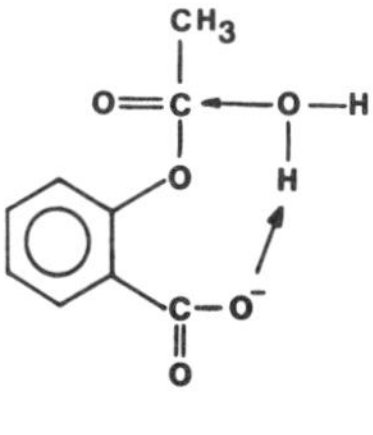

[26]

The intramolecular reactions discussed so far involved only a single catalytic group to facilitate hydrolysis. It is possible, however, to build two or more catalytic groups into a single molecule. Such a system resembles more of the sophisticated enzymic reactions and may show an even higher rate enhancement than that obtained with simple intramolecular reactions. An early example of such catalysis has been the hydrolysis of succinylsalicylic acid [27].[91,92] The hydrolysis of this compound exhibits a bell-shaped pH-rate profile, indicating the participation of both a carboxyl group and a carboxylate ion in the reaction (Figure 10). On the other hand, acetylsalicylic acid [28] and the monomethyl ester [29] are hydrolyzed at rates proportional to the concentration of a single carboxylate ion (Figure 10).

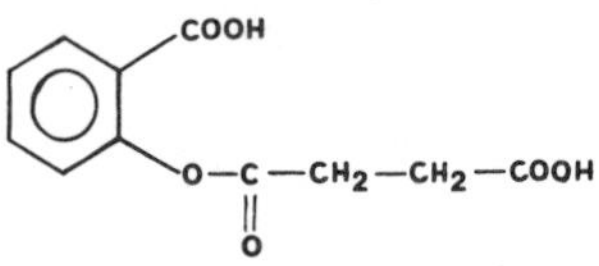

[27]

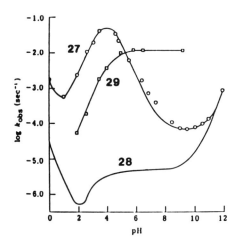

FIGURE 10.   Hydrolysis rate constants of salicylic acid derivatives: compounds [27], [28], and [29] at 25°C. (From Morawetz, H. and Oreskes, I., *J. Am. Chem. Soc.*, 80, 2591, 1958. With permission.)

[28]

[29]

Two kinetically indistinguishable pathways are possible: (1) the salicylic carboxylate ion is a nucleophile and the succinyl carboxyl group is a general acid catalyst (Equation 44); (2) the un-ionized salicylic carboxyl group is a general acid catalyst and the succinyl carboxylate ion is the nucleophile (Equation 45).

(44)

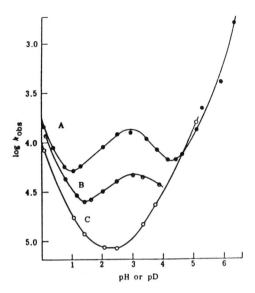

FIGURE 11. The pH (p²H) dependence of the hydrolyses of *o*-carboxyphthalimide in (A) water, (B) *o*-car-
boxyphthalimide in deuterium oxide, and (C) phthalimide
in water at 100°C. (From Zerner, B. and Bender,
M. L., *J. Am. Chem. Soc.*, 83, 2267, 1961. With
permission.)

$$ \text{(45)} $$

We have seen above that ring closure with succinyl derivatives is very easy, whereas the carboxylate group in acetylsalicylic acid is not a good nucleophile. Consequently, the second mechanism (Equation 45) involving nucleophilic catalysis by the succinyl carboxylate and general acid catalysis by the salicylic carboxyl group is more probable. Indeed, the contribution to rate enhancement by the succinyl carboxylate group is characteristic of an intramolecular nucleophilic reaction. Specifically, as compared to acetylsalicylic acid [28], the extra carboxyl group of succinylsalicylic acid [27] leads to a rate enhancement of 24,000-fold, a value too high for general acid catalysis, but reasonable for nucleophilic catalysis. Also, the contribution by the salicylic carboxyl group is only 66-fold, as calculated from the hydrolysis rates of the monomethyl ester [29] and succinylsalicylic acid [27]. This value is acceptable for a general acid-catalyzed process such as illustrated by Equation 45.

As with the hydrolysis of succinylsalicylic acid, a bell-shaped pH-rate profile has also been found in the hydrolysis of *o*-carboxyphthalimide (Figure 11).[93] However, this compound has only one ionizable group which cannot participate simultaneously as a general acid and a general base. To account for the bell-shaped pH dependence, one should assume at least two distinct consecutive steps in the reaction, namely an acid-catalyzed step and a base-catalyzed step. Figure 11 shows that kinetic isotope effects in heavy water are consistent with general acid-base catalysis in the pH region 1 to 4. Here, the hydrolysis of *o*-carboxy-phthalimide is more rapid than that of phthalimide, not containing the catalytically competent carboxyl group. However, at high and low pH values, where hydroxide and oxonium ion catalysis, respectively, becomes important, the two compounds are hydrolyzed at similar rates. The two-step reaction sequence implies that a tetrahedral intermediate must lie on the reaction path. Again, two kinetically indistinguishable mechanisms are possible. One involves general base-catalyzed addition of water to form an anionic intermediate that breaks down by some form of general acid catalysis (Equation 46). With the other possible mechanism, the formation of the tetrahedral intermediate is catalyzed by the neighboring carboxyl group as a general acid and the intermediate decomposes by general base catalysis. The maximum of the pH-dependence curve (Figure 11) arises from a change in the rate-limiting step with pH. Specifically, at higher pH than the rate maximum where the concentration of the un-ionized carboxyl group is low, the acid-catalyzed process becomes rate limiting; at lower pH values where the carboxylate ion is limited in supply, the base-catalyzed process becomes the rate-determining step. This phenomenon is an extremely valuable indication of an intermediate on the reaction path.

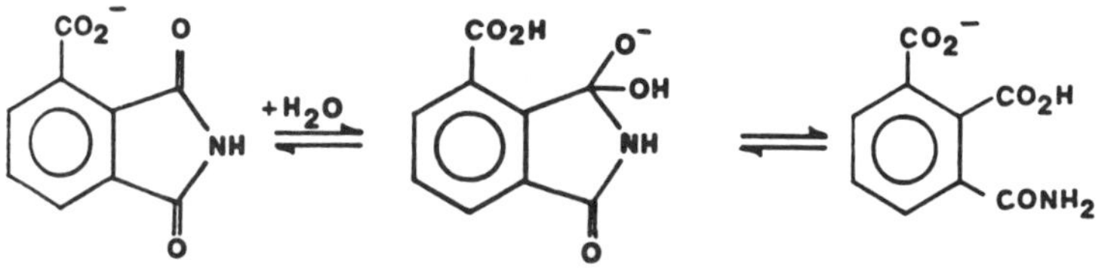

$$(46)$$

XIII. CONCERTED GENERAL ACID AND GENERAL BASE CATALYSES

In the preceding section, we have seen that general acid or general base catalysis is usually much less effective than nucleophilic catalysis. Nevertheless, some enzymes with high catalytic power, like aspartic proteases, do not form a covalent intermediate with the substrate, but instead utilize general acid-base catalysis. Examples, long known from organic chemistry, have shown that concerted action of two catalysts may indeed be very efficient, particularly if the catalysts are built into the same molecule, providing a bifunctional catalyst.

The classical example of bifunctional catalysis is the catalysis of the mutarotation of tetramethylglucose by 2-pyridone or by its tautomer, 2-hydroxypyridine.[94,95] Mutarotation is the interconversion of the two anomers of a sugar and proceeds by a mechanism of ring opening and ring closure according to Equation 47.

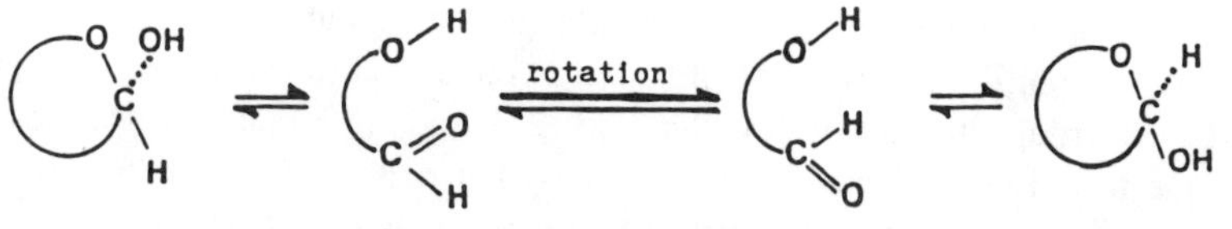

$$(47)$$

In aprotic solvents, mutarotation of tetramethylglucose does not proceed at an appreciable rate. The addition of either pyridine or phenol results in virtually no rate increase. However, if pyridine and phenol are added together, the reaction proceeds rapidly. The reaction was found to be third-order, indicating that both pyridine and phenol as general base and general acid, respectively, are required in the transition state.[94]

A substantially more effective catalyst is obtained if the acidic and basic groups are combined as in 2-pyridone (Equation 48) or carboxylic acids such as benzoic acid or trichloroacetic acid. For example, with a reaction mixture of 1 mM in 2-pyridone and 100 mM in tetramethylglucose, the rate is 7000 times that obtained with the combination of 1 mM pyridine and 1 mM phenol even though the basicity and acidity of 2-pyridone are 10^4

and 10^2 less than the basicity and acidity of pyridine and phenol, respectively. Such a comparison, of course, depends very much on the concentrations of the reactants because second-order and the third-order reactions are compared. Therefore, the rate differential would decrease at higher catalyst concentrations and would increase at lower concentrations. A "push-pull" catalysis by 2-pyridone is illustrated by Equation 48. It is seen that the basic carbonyl group and the acidic N–H group promotes cleavage of the hemiacetal to give free aldehyde. The ring closure takes place through the reverse reaction, leading to a mixture of the two anomers.[95] A "push-pull" mechanism appears to be associated with the catalysis of aspartic proteases (Chapter 5, Section VI.C).

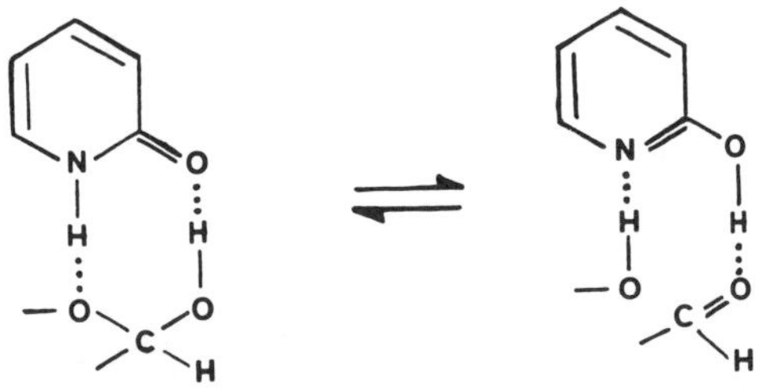

(48)

The precise mechanism of the catalysis by 2-pyridone is not known. It has been suggested that it is the neutral characteristics of 2-pyridone that facilitates the exchange of the two protons without forming a dipolar ion.[96] In this respect, the mechanism may be different from that associated with the reaction in the presence of phenol-pyridine mixture.

XIV. SYMMETRICAL AND UNSYMMETRICAL MECHANISMS

It is clear from the preceding sections that the hydrolysis of carboxylic acid derivatives proceeds, in general, according to a two-step mechanism involving the formation and decomposition of a tetrahedral intermediate. Thus, either hydration of the carbonyl group or expulsion of the leaving group can be rate limiting, and general catalysis can occur in either of the two steps. Accordingly, there may be two main types of mechanism: symmetrical and unsymmetrical. Symmetrical mechanisms involve a similar type of catalysis in partitioning the tetrahedral intermediate in the k_{-1} and k_2 steps of Equation 49. Unsymmetrical mechanisms involve dissimilar partitioning of the tetrahedral intermediate.

$$\mathrm{R{-}\overset{\overset{\displaystyle O}{\|}}{C}{-}X} + H_2O \underset{k_{-1}}{\overset{k_1}{\rightleftharpoons}} \mathrm{R{-}\underset{\underset{\displaystyle OH}{|}}{\overset{\overset{\displaystyle OH}{|}}{C}}{-}X} \underset{k_{-2}}{\overset{k_2}{\rightleftharpoons}} \mathrm{RCOOH + XH} \qquad (49)$$

In a few cases, the existence of these two types of catalysis has unequivocally been demonstrated. A symmetrical catalytic partitioning of a tetrahedral intermediate is exemplified most lucidly by the general base-catalyzed ethanolysis of ethyl trifluoroacetate (Equation 50).[97]

$$\underset{+\mathrm{H}\bullet\mathrm{C_2H_5}}{\mathrm{CF_3{-}\overset{\overset{\displaystyle O}{\|}}{C}{-}OC_2H_5}} \underset{BH}{\overset{B^-}{\rightleftharpoons}} \mathrm{CF_3{-}\underset{\underset{\displaystyle \bullet C_2H_5}{|}}{\overset{\overset{\displaystyle O^-}{|}}{C}}{-}OC_2H_5} \underset{B^-}{\overset{BH}{\rightleftharpoons}} \underset{+\mathrm{HOC_2H_5}}{\mathrm{CF_3{-}\overset{\overset{\displaystyle O}{\|}}{C}{-}\bullet C_2H_5}} \qquad (50)$$

Equation 50 represents a completely symmetrical isotopic (●) exchange reaction. The decomposition of the tetrahedral anionic intermediate is catalyzed by general acid in both directions. Similarly, the formation of the tetrahedral intermediate is catalyzed by general base in both directions. This mechanism is prevalent in the hydrolytic reactions, in which the entering and leaving groups are of similar nature. When they are grossly different, the mechanism may be unsymmetrical as in the case of the hydrolysis of ethyl trifluorothiol-acetate.[98,99] Equation 51 shows that the reaction pathway is unsymmetrical because the removal of ethyl thiolate from the intermediate is not assisted by general acid catalysis.

$$\underset{\|}{\overset{O}{C}}F_3\text{-}C\text{-}SC_2H_5 + H_2O \underset{BH}{\overset{B^-}{\rightleftharpoons}} \underset{|}{\overset{O^-}{C}}F_3\text{-}C\text{-}SC_2H_5 \rightarrow CF_3\text{-}COOH + {}^-SC_2H_5 \qquad (51)$$
$$\underset{OH}{}$$

It is conceivable that the mechanism is dependent upon the basicity of the leaving group. In the case of less basic leaving groups such as thiolates and aryloxides, general acid assistance of departure from the tetrahedral intermediate would not be expected to be necessary. In the reverse reaction, the attack of thiolate rather than a general base-catalyzed thiol reaction should occur. In fact, the reaction between thiols and *p*-nitrophenyl acetate is dependent only upon the thiolate ion concentration.[43]

In connection with symmetrical and unsymmetrical reactions, consideration should be given to the principle of microscopic reversibility or detailed balance. According to this law, reversible reactions must pass through the same transition state in both directions. Thus, if a reaction leading to a tetrahedral intermediate occurs by general base catalysis, the reverse reaction from the tetrahedral intermediate to the reactants must proceed by general acid catalysis. It is not uncommon, however, to reach incorrect mechanistic conclusions by disregarding the principle of microscopic reversibility. This mistake amounts to the violation of the second law of thermodynamics. Consideration of principle of microscopic reversibility is very rewarding in mechanistic investigations. Specifically, if a reaction mechanism has been explored in one direction, the mechanism of the reverse reaction can be postulated. The important contribution by microscopic reversibility to the elucidation of the mechanism of action of serine proteases will be discussed in Chapter 3.

XV. STEREOELECTRONIC CONTROL IN ESTER AND AMIDE REACTIONS

The tetrahedral adduct is a key intermediate in the hydrolysis of esters and amides. An important postulation concerning its reactivity states that its decomposition, either to reactants or to products, is controlled by the orientation of the nonbonded electron pairs of the heteroatoms.[100,101]

The tetrahedral intermediate can assume several conformations depending on the orientation of the lone pairs of the heteroatoms. For example, form [30] represents such a conformation which is a relatively low energy *trans* conformer, the two R groups being away from each other. The shaded electron pairs are antiperiplanar to the C–X bond.

[30]

According to the theory of stereoelectronic control,[100,101] preferential cleavage of a tetrahedral intermediate occurs when there are two lone pairs oriented antiperiplanar (shaded electron pairs of form[30]) to the bond to be cleaved (C–X bond of form [30]). The theory seems to be plausible because (1) the elimination that involves X and a shaded electron pair of form [30] is analogous to an antielimination which is generally preferred[102] and (2) resonance stabilization of the ester or amide product requires the involvement of both non-bonded electron pairs antiperiplanar to X of form [30]. Indeed, molecular orbital calculations support the validity of the theory.[103]

The classical example of stereoelectronic control is the hydrolysis of cyclic hemi-orthoesters and related species. It was found by Deslongchamps and his co-workers[100,101] that the hemi-orthoester of Equation 52 decomposes in a completely specific manner, yielding the hydroxy ester as the only product. No lactone was produced in the reaction. To explain this result, it was proposed that the hemi-orthoester assumes a conformation similar to that depicted in Equation 52. This is a low energy conformer with an equatorial OR group. The C–OR bond, however, is antiperiplanar to only one lone pair so its cleavage to lactone is disfavored. In contrast, the ring C–O bond is antiperiplanar to two lone pairs, and this permits the breakdown to the hydroxy ester.

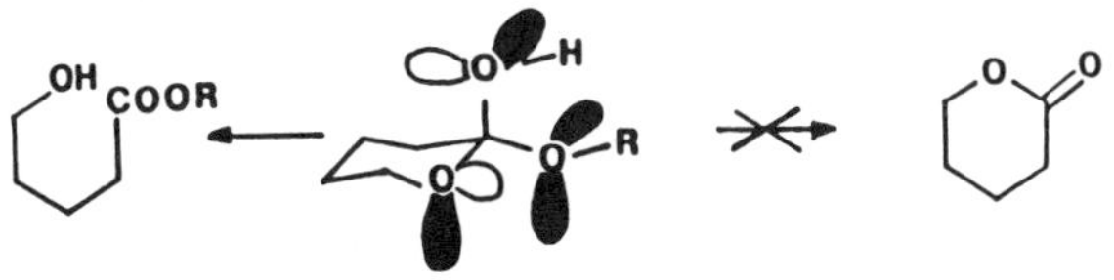

$$(52)$$

The theory of stereoelectronic control has recently been confirmed by a study on the hydrolysis of cyclic amidines where the lactams in contrast to lactones are not significantly destabilized.[104]

It can be concluded that the conformation of the reactants may determine the outcome of the reaction: (1) formation of a new C–O or C–N bond takes place in such a way that the bond must be antiperiplanar to the lone pair on each the neighboring O or N atoms, (2) a C–O or C–N bond breaks only if it lies antiperiplanar to the lone pair on each of the neighboring O or N atoms, and (3) when the conformation of the tetrahedral intermediate is not favored energetically, then the reaction proceeds at a decreased rate as it does in *syn*-elimination. Problems of applying the stereoelectronic theory to the catalysis by serine proteases will be discussed in Chapter 3.

REFERENCES

1. **Cahn, R. S., Ingold, C., and Prelog, V.**, Spezifikation der molekularen Chiralität, *Angew. Chem.*, 78, 413, 1966.
2. International Union of Biochemistry, *Biochemical Nomenclature and Related Documents*, The Biochemical Society, London, 1978, 1.
3. **Klyne, W. and Prelog, V.**, Description of steric relationships, *Experientia*, 16, 521, 1960.
4. **Bender, M. L.**, Oxygen exchange as evidence for the existence of an intermediate in ester hydrolysis, *J. Am. Chem. Soc.*, 73, 1626, 1951.
5. **Bender, M. L.**, Mechanisms of catalysis of nucleophilic reactions of carboxylic acid derivatives, *Chem. Rev.*, 60, 53, 1960.
6. **Johnson, S. L.**, General base nucleophilic catalysis of ester hydrolysis and related reactions, *Adv. Phys. Org. Chem.*, 5, 237, 1967.

7. **Bender, M. L.,** Intermediates in the reactions of carboxylic acid derivatives. II. Infrared absorption spectra as evidence for the formation of addition compounds of carboxylic acid derivatives, *J. Am. Chem. Soc.,* 75, 5986, 1953.

8. **Williams, A. and Douglas, K. T.,** Elimination-addition mechanisms of acyl transfer reactions, *Chem. Rev.,* 75, 627, 1975.

9. **Bender, M. L.,** *Mechanisms of Homogeneous Catalysis from Protons to Proteins,* Wiley-Interscience, New York, 1971, chap. 3.

10. **Bruice, T. C. and Benkovic, S. J.,** *Bioorganic Mechanisms,* Vol. 1, W. A. Benjamin, Inc., New York, 1966, chap. 1.

11. **Jencks, W. P.,** *Catalysis in Chemistry and Enzymology,* McGraw-Hill, New York, 1969, chap. 3.

12. **Bender, M. L.,** *Mechanisms of Homogeneous Catalysis from Protons to Proteins,* Wiley-Interscience, New York, 1971, chaps. 4, 5.

13. **Jencks, W. P. and Carriulo, J.,** Imidazole catalysis. II. Acyl transfer and the reaction of acetyl imidazole with water and oxygen anions, *J. Biol. Chem.,* 234, 1272, 1959.

14. **Jencks, W. P. and Carriulo, J.,** General base catalysis for ester hydrolysis, *J. Am. Chem. Soc.,* 83, 1743, 1961.

15. **Jencks, W. P.,** *Catalysis in Chemistry and Enzymology,* McGraw-Hill, New York, 1969, chap. 2.

16. **Bender, M. L.,** *Mechanisms of Homogeneous Catalysis from Protons to Proteins,* Wiley-Interscience, New York, 1971, chap. 6.

17. **Bruice, T. C. and Schmir, G. L.,** Imidazole catalysis. I. The catalysis of the hydrolysis of phenyl acetates by imidazole, *J. Am. Chem. Soc.,* 79, 1663, 1957.

18. **Edwards, J. O.,** Polarizability, basicity and nucleophilic character, *J. Am. Chem. Soc.,* 78, 1819, 1956.

19. **Edwards, J. O. and Pearson, R. G.,** The factors determining nucleophilic reactivities, *J. Am. Chem. Soc.,* 84, 16, 1962.

20. **Tommila, E. and Murto, M. L.,** The influence of solvent on reaction velocity. XXIII. The alkaline hydrolysis of ethyl acetate in dimethyl sulphoxide-water mixtures, *Acta Chem. Scand.,* 17, p.1947, 1963.

21. **Tommila, E. and Palenius, I.,** The influence of solvent on reaction velocity. XXV. Dependence of the substituent effect on solvent composition in the alkaline hydrolysis of benzoic esters in dimethyl sulphoxide-water mixtures, *Acta Chem. Scand.,* 17, p.1980, 1963.

22. **Bender, M. L.,** Mechanisms of catalysis of nucleophilic reactions of carboxylic acid derivatives, *Chem. Rev.,* 60, 53, 1960.

23. **Bender, M. L. and Thomas, R. J.,** The concurrent alkaline hydrolysis and isotopic oxygen exchange of a series of p-substituted acetanilides, *J. Am. Chem. Soc.,* 83, 4183, 1961.

24. **Williams, A.,** Free-energy correlations and reaction mechanisms, in *The Chemistry of Enzyme Action,* Vol. 6, Page, M. I., Ed., in New Comprehensive Biochemistry, Elsevier, Amsterdam, 1984, chap. 5.

25. **Brönsted, J. N. and Pedersen, K.,** Die katalytische Zersetzung des Nitramids und ihre physikalisch-chemische Bedeutung, *Z. Phys. Chem. (Leipzig),* 108, 185, 1924.

26. **Hine, J.,** *Physical Organic Chemistry,* 2nd ed., McGraw-Hill New York, 1962, 115.

27. **Bruice, T. C. and Lapinski, R.,** Imidazole catalysis. IV. The reaction of general bases with p-nitrophenyl acetate in aqueous solution, *J. Am. Chem. Soc.,* 80, 2265, 1958.

28. **Evans, M. G. and Polányi, M.,** Inertia and driving force of chemical reactions, *Trans. Faraday Soc.,* 34, 11, 1938.

29. **Leffler, J. E.,** Parameters for the description of transition states, *Science,* 117, 340, 1953.

30. **Hammond, G. S.,** A correlation of reaction rates, *J. Am. Chem. Soc.,* 77, 334, 1955.

31. **Hammett, L. P.,** *Physical Organic Chemistry,* McGraw-Hill, New York, 1940, chap. 7.

32. **Taft, R. W., Jr.,** in *Steric Effects in Organic Chemistry,* Newman, M. S., Ed., John Wiley & Sons, New York, 1956, chap. 13.

33. **Hupe, D. J. and Jencks, W. P.,** Nonlinear structure-reactivity correlations. Acyl transfer between sulfur and oxygen nucleophiles, *J. Am. Chem. Soc.,* 99, 451, 1977.

34. **Hansch, C. and Leo, A.,** *Substituent Constants for Correlation Analysis in Chemistry and Biology,* Wiley-Interscience, New York, 1979, chaps. 1, 2, 3, and 5.

35. **Jencks, W. P.,** *Catalysis in Chemistry and Enzymology,* McGraw-Hill, New York, 1969, chap. 4.

36. **Klinman, J. P.,** Kinetic isotope effects in enzymology, *Adv. Enzymol. Relat. Areas Mol. Biol.,* 46, 415, 1977.

37. **Schowen, K. B. and Schowen, R. L.,** Solvent isotope effects on enzyme systems, *Methods Enzymol.,* 87, 551, 1982.

38. **Kovach, I. M., Hogg, J. L., Raben, T., Halbert, K., Rodgers, J., and Schowen, R. L.,** The β-hydrogen secondary isotope effect in acyl transfer reactions. Origins, temperature dependence, and utility as a probe of transition-state structure, *J. Am. Chem. Soc.,* 102, p.1991, 1980.

39. **Westheimer, F. H.,** The magnitude of the primary kinetic isotope effect for compounds of hydrogen and deuterium, *Chem. Rev.,* 61, 265, 1961.

40. **More O'Ferrall, R. A.,** in *Proton Transfer Reactions,* Caldin, E. F. and Gold, V., Eds., Chapman and Hall, London, 1975, 201.
41. **Bordwell, F. G. and Boyle, W. J., Jr.,** Kinetic isotope effects for nitroalkanes and their relationship to transition-state structure in proton-transfer reactions, *J. Am. Chem. Soc.,* 97, 3447, 1975.
42. **Whitaker, J. R.,** The reaction of p-nitrophenyl acetate with thiols, *J. Am. Chem. Soc.,* 84, 1900, 1962.
43. **Ogilvie, J. W., Tildon, J. T., and Strauch, B. S.,** A kinetic study of the reaction of thiols with p-nitrophenyl acetate, *Biochemistry,* 3, 754, 1964.
44. **Bruice, T. C. and Schmir, G. L.,** Imidazole catalysis. II. The reaction of substituted imidazoles with phenyl acetates in aqueous solution, *J. Am. Chem. Soc.,* 80, 148, 1958.
45. **Pearson, R. G.,** Hard and soft acids and bases, *J. Am. Chem. Soc.,* 85, 3533, 1963.
46. **Jencks, W. P., Cordes, S., and Carriuolo, J.,** The free energy of thiol ester hydrolysis, *J. Biol. Chem.,* 235, 3608, 1960.
47. **Baker, A. W. and Harris, G. H.,** Physical and chemical effects of substituent groups on multiple bonds. II. Thiolesters, *J. Am. Chem. Soc.,* 82, 1923, 1960.
48. **Rylander, P. N. and Tarbell, D. S.,** Cleavage of the carbon-sulfur bond. Rates of hydrolysis of some alkyl acetates and the corresponding thiolacetates in aqueous acetone, *J. Am. Chem. Soc.,* 72, 3021, 1950.
49. **Morse, B. K. and Tarbell, D. S.,** Cleavage of the carbon-sulfur bond. Rates of the basic and acid-catalyzed hydrolysis of allyl, benzyl and trityl thioacetates, and the corresponding acetates in aqueous acetone solution, *J. Am. Chem. Soc.,* 74, 416, 1952.
50. **Hershfield, R. and Schmir, G. L.,** The mechanism of the acid-catalyzed hydrolysis of methyl thiolformate, *J. Am. Chem. Soc.,* 94, 1263, 1972.
51. **Chou, T. C. and Lipmann, F.,** Separation of acetyl transfer enzymes in pigeon liver extract, *J. Biol. Chem.,* 196, 89, 1952.
52. **Hestrin, S.,** The reaction of acetylcholine and other carboxylic acid derivatives with hydroxylamine and its analytical application, *J. Biol. Chem.,* 180, 249, 1949.
53. **Connors, K. A. and Bender, M. L.,** The kinetics of alkaline hydrolysis and n-butylaminolysis of ethyl p-nitrobenzoate and ethyl p-nitrothiolbenzoate, *J. Org. Chem.,* 26, 2498, 1961.
54. **Bruice, T. C. and Benkovic, S. J.,** *Bioorganic Mechanisms,* Vol. 1, W. A. Benjamin Inc., New York, 1966, chap. 3.
55. **Fife, T. H. and DeMark, B. R.,** General-base-catalyzed intramolecular aminolysis of thiol esters. Cyclization of S-n-propyl o-(2-imidazolyl) thiolbenzoate. Relationship of the uncatalyzed and base-catalyzed nucleophilic reactions, *J. Am. Chem. Soc.,* 101, 7379, 1979.
56. **Shames, S. L. and Byers, L. D.,** Acyl substituent effects on rates of acyl transfer to thiolate, hydroxide, and oxy dianions, *J. Am. Chem. Soc.,* 103, 6170, 1981.
57. **Kanchunger, M. S. and Byers, L. D.,** Acyl substituent effects on thiohemiacetal equilibria, *J. Am. Chem. Soc.,* 101, 3005, 1979.
58. **Nyquist, R. A. and Potts, W. J.,** Characteristic infrared absorption frequencies of thiol esters and related compounds, *Spectrochim. Acta,* 15, 514, 1959.
59. **Smith, S. G. and O'Leary, M.,** The kinetics of the acidic and alkaline hydrolysis of ethyl thionbenzoate, *J. Org. Chem.,* 28, 2825, 1963.
60. **Edward, J. T. and Wong, S. C.,** Effect of sulfuric acid concentration on the rates of hydrolysis of ethyl benzoate, ethyl thiolbenzoate, and ethyl thionbenzoate, *J. Am. Chem. Soc.,* 99, 7224, 1977.
61. **Edward, J. T. and Wong, S. C.,** Effect of acid concentration on the partitioning of the tetrahedral intermediate in the hydrolysis of thioacetanilide, *J. Am. Chem. Soc.,* 101, 1807, 1979.
62. **Campbell, P. and Lapinskas, B. A.,** Aminolysis of thioesters, *J. Am. Chem. Soc.,* 99, 5378, 1977.
63. **Westheimer, F. H.,** The mechanisms of some metal-ion-promoted reactions, *Trans. N.Y. Acad. Sci.,* 18, 15, 1955.
64. **Pedersen, K. J.,** The cupric ion catalysis in the bromination of ethyl acetate, *Acta Chem. Scand.,* 2, 252, 1948.
65. **Chin, J. and Zou, X.,** Relationship between effective nucleophilic catalysis in the hydrolysis of esters with poor leaving groups and the life time of the tetrahedral intermediate, *J. Am. Chem. Soc.,* 106, 3687, 1984.
66. **Kroll, H.,** The participation of heavy metal ions in the hydrolysis of amino acid esters, *J. Am. Chem. Soc.,* 74, 2036, 1952.
67. **Bender, M. L. and Turnquest, B. W.,** The kinetics and oxygen exchange of the cupric ion-catalyzed hydrolysis of α-amino esters, *J. Am. Chem. Soc.,* 79, 1889, 1957.
68. **Koltun, W. L., Fried, M., and Gurd, F. R. N.,** Coordination complexes and catalytic properties of proteins and related substances. IV. Reactions of glycine-containing dipeptides with cupric ions and with p-nitrophenyl acetate, *J. Am. Chem. Soc.,* 82, 233, 1960.
69. **Buckingham, D. A., Foster, D. M., and Sargeson, A. M.,** Cobalt (III)-promoted hydrolysis of glycine esters. Kinetics, product analysis, and oxygen-18 exchange studies of the base hydrolysis of $[Co(en)_2X(GlyOR)]^{2+}$ ions, *J. Am. Chem. Soc.,* 91, 4102, 1969.

70. **Buckingham, D. A., Davis, C. E., Foster, D. M., and Sargeson, A. M.,** Cobalt(III)-promoted hydrolysis of chelated glycine amides, glycylglycine, and glycylglycine esters, kinetics and mechanism, *J. Am. Chem. Soc.,* 92, 5571, 1970.

71. **Buckingham, D. A., Foster, D. M., and Sargeson, A. M.,** Cobalt(III)-promoted hydrolysis of glycine amides. Intramolecular and intermolecular hydrolysis following the base hydrolysis of the cis-$[Co(en)_2Br(GlyNR_1R_2)]^{2+}$ ions, *J. Am. Chem. Soc.,* 92, 6151, 1970.

72. **Buckingham, D. A., Harrowfield, J. MacB., and Sargeson, A. M.,** Metal ion activation in the base hydrolysis of amides. Hydrolysis of the dimethylformamidepentaaminecobalt(III) ion, *J. Am. Chem. Soc.,* 96, 1726, 1974.

73. **Buckingham, D. A., Keene, F. R., and Sargeson, A. M.,** Facile intramolecular hydrolysis of dipeptides and glycinamide, *J. Am. Chem. Soc.,* 96, 4981, 1974.

74. **Wells, M. A. and Bruice, T. C.,** Intramolecular catalysis of ester hydrolysis by metal complexed hydroxide ion. Acyl oxygen bond scission in Co^{2+} and Ni^{2+} carboxylic acid complexes, *J. Am. Chem. Soc.,* 99, 5341, 1977.

75. **Fife, T. H. and Squillacote, V. L.,** Metal ion effects on intramolecular nucleophilic carboxyl group participation in amide and ester hydrolysis. Hydrolysis of N-(8-quinolyl) phthalamic acid and 8-quinolyl hydrogen glutarate, *J. Am. Chem. Soc.,* 100, 4787, 1978.

76. **Jencks, W. P.,** *Catalysis in Chemistry and Enzymology,* McGraw-Hill, New York, 1969, chap. 1.

77. **Bender, M. L.,** *Mechanisms of Homogeneous Catalysis from Protons to Proteins,* Wiley-Interscience, New York, 1971, chap. 9.

78. **Jencks, W. P.,** Binding energy, specificity, and enzymic catalysis: the Circe effect, *Adv. Enzymol. Relat. Areas Mol. Biol.,* 43, 219, 1975.

79. **Kirby, A. J.,** Effective molarities for intramolecular reactions, *Adv. Phys. Org. Chem.,* 17, 183, 1980.

80. **Bruice, T. C. and Benkovic, S. J.,** A comparison of the bimolecular and intramolecular nucleophilic catalysis of the hydrolysis of substituted phenyl acylates by the dimethylamino group, *J. Am. Chem. Soc.,* 85, 1, 1963.

81. **Gaetjens, E. and Morawetz, H.,** Intramolecular carboxylate attack on ester groups. The hydrolysis of substituted phenyl acid succinates and phenyl acid glutarates, *J. Am. Chem. Soc.,* 82, 5328, 1960.

82. **Bruice, T. C. and Pandit, U. K.,** The effect of geminal substitution, ring size and rotamer distribution on the intramolecular nucleophilic catalysis of the hydrolysis of monophenyl esters of dibasic acids and the solvolysis of the intermediate anhydrides, *J. Am. Chem. Soc.,* 82, 5858, 1960.

83. **Bruice, T. C. and Pandit, U. K.,** Intramolecular models depicting the kinetic importance of "fit" in enzymatic catalysis, *Proc. Natl. Acad. Sci. U.S.A.,* 46, 402, 1960.

84. **Bruice, T. C. and Bradbury, W. C.,** The *gem* effect. II. The influence of 3-mono- and 3,3-disubstitution on the rates of solvolysis of mono-p-bromophenyl glutarate, *J. Am. Chem. Soc.,* 87, 4846, 1965.

85. **Higuchi, T., Eberson, L., and Herd, A. K.,** The intramolecular facilitated hydrolytic rates of methyl-substituted succinanilic acids, *J. Am. Chem. Soc.,* 88, 3805, 1966.

86. **Milstien, S. and Cohen, L. A.,** Stereopopulation control. I. Rate enhancement in the lactonizations of o-hydroxyhydrocinnamic acids, *J. Am. Chem. Soc.,* 94, 9158, 1972.

87. **Danforth, C., Nicholson, A. W., James, J. C., and Loudon, G. M.,** Steric acceleration of lactonization reactions: an analysis of "stereopopulation control", *J. Am. Chem. Soc.,* 98, 4275, 1976.

88. **Winans, R. E. and Wilcox, C. F., Jr.,** Comparison of stereopopulation control with conventional steric effects in lactonization of hydrocoumarinic acids, *J. Am. Chem. Soc.,* 98, 4281, 1976.

89. **Fersht, A. R. and Kirby, A. J.,** Structure and mechanism in intramolecular catalysis. The hydrolysis of substituted aspirins, *J. Am. Chem. Soc.,* 89, 4853, 1967.

90. **Fersht, A. R. and Kirby, A. J.,** The hydrolysis of aspirin. Intramolecular general base catalysis of ester hydrolysis, *J. Am. Chem. Soc.,* 89, 4857, 1967.

91. **Morawetz, H. and Oreskes, I.,** Intramolecular bifunctional catalysis of ester hydrolysis, *J. Am. Chem. Soc.,* 80, 2591, 1958.

92. **Morawetz, H. and Shafer, J.,** Intramolecular bifunctional catalysis of amide hydrolysis, *J. Am. Chem. Soc.,* 84, 3783, 1962.

93. **Zerner, B. and Bender, M. L.,** The kinetics and mechanism of the hydrolysis of o-carboxyphthalimide, *J. Am. Chem. Soc.,* 83, 2267, 1961.

94. **Swain, C. G. and Brown, J. F.,** Concerted displacement reactions. VII. The mechanism of acid-base catalysis in non-aqueous solvents, *J. Am. Chem. Soc.,* 74, 2534, 1952.

95. **Swain, C. G. and Brown, J. F.,** Concerted displacement reactions. VIII. Polyfunctional catalysis, *J. Am. Chem. Soc.,* 74, 2538, 1952.

96. **Rony, P. R.,** Polyfunctional catalysis. I. Activation parameters for the mutarotation of tetramethyl-D-glucose in benzene, *J. Am. Chem. Soc.,* 90, 2824, 1968.

97. **Johnson, S. L.,** General base catalysed ethanolysis of ethyl trifluoroacetate, *J. Am. Chem. Soc.,* 86, 3819, 1964.

98. **Fedor, L. R. and Bruice, T. C.,** Kinetic evidence for the formation of a tetrahedral intermediate in the aqueous solvolysis of ethyl trifluorothiolacetate, *J. Am. Chem. Soc.,* 86, 5697, 1964.

99. **Bender, M. L. and Heck, H. d'A.,** Carbonyl oxygen exchange in general base catalyzed ester hydrolysis, *J. Am. Chem. Soc.,* 89, 1211, 1967.

100. **Deslongchamps, P., Atlani, P., Frehel, D., and Malaval, A.,** The importance of conformation of the tetrahedral intermediate in the hydrolysis of esters. Selective cleavage of the tetrahedral intermediate controlled by orbital orientation, *Can. J. Chem.,* 50, 3405, 1972.

101. **Deslongchamps, P.,** Stereoelectronic control in the cleavage of tetrahedral intermediates in the hydrolysis of esters and amides, *Tetrahedron,* 31, 2463, 1975.

102. **Lowry, T. H. and Richardson, K. S.,** *Mechanism and Theory in Organic Chemistry,* 2nd ed., Harper & Row, New York, 1981, 548.

103. **Lehn, J.-M. and Wipff, G.,** Stereoelectronic control in acid and base catalysis of amide hydrolysis. A theoretical study, *J. Am. Chem. Soc.,* 102, 1347, 1980.

104. **Perrin, C. L. and Arrhenius, G. M. L.,** A critical test of the theory of stereoelectronic control, *J. Am. Chem. Soc.,* 104, 2839, 1982.

Chapter 2

GENERAL ASPECTS OF PROTEASES

I. NOMENCLATURE

Proteases form a large group of enzymes, the nomenclature of which is rather difficult. This is mainly due to the fact that proteases exhibit a complicated substrate specificity which serves as a basis of the present classification. Although the chemical reaction they catalyze is essentially the same, the different proteases have different preferences for the various peptide bonds as modified by the other groups in their neighborhood. For example, some proteases have a preference to the C-terminus or the N-terminus of the peptide chain. These exopeptidases are specific for the free α-carboxyl group or the free α-amino group. Other proteases act inside the polypeptide chain. The specificity of these endopeptidases is mainly dependent on the amino acid side chains encountered in the vicinity of the scissile peptide bond.

Another criterion of classification beside specificity is the mechanism whereby a protease is acting. On this basis, serine proteases, cysteine proteases, aspartic proteases, and metalloproteases are distinguished. The specificity criterion which serves as a basis of dividing the enzymes into exo- and endopeptidases has a priority over the mechanism criterion in the classification proposed by the International Union of Biochemists.[1]

According to the *Enzyme Nomenclature,*[1] classes, subclasses, and sub-subclasses of enzymes are distinguished. These groups are designated by code numbers so that the first, second, and third code number refers to class, subclass, and sub-subclass, respectively. Proteases (peptide hydrolases: 3.4), as a subclass, belong to hydrolases (class 3) and are divided into two sets of sub-subclasses: peptidases (exopeptidases: 3.4.11—19 and proteinases (proteolytic enzymes, endopeptidases, peptidylpeptide hydrolyses: 3.4.21—24). Exopeptidases include sub-subclasses of enzymes hydrolyzing single amino acids from the N-terminus of the peptide chain (3.4.11), enzymes hydrolyzing single residues from the C-terminus (3.4.16—18), enzymes specific for dipeptide substrates (3.4.13), and enzymes cleaving off dipeptide units either from the N-terminus (3.4.14) or the C-terminus (3.4.15). A separate category (omega peptidases: 3.4.19) includes enzymes removing substituted N- and C-terminal amino acid residues.

Endopeptidases are divided into serine proteinases (3.4.21) having catalytically competent serine and histidine residues, cysteine proteinases (3.4.22) possessing cysteine and histidine residues at the active site, aspartic proteinases (3.4.23) involving two acidic residues in the catalytic process, and metalloproteinases (3.4.24) using a metal ion and a glutamic acid residue in the mechanism of action. As an interim measure, the code number 3.4.99 is given to the proteinases of unknown catalytic mechanism.

Examples of proteases grouped according to the latest recommendation of the Nomenclature Committee of the International Union of Biochemistry are given in Tables 1 and 2. First of all, the enzymes discussed in the following chapters are mentioned in the tables.

The problem of the two related terms, protease and proteinase, should be mentioned here. Although these words are synonyms in the sense that both imply an enzyme capable of hydrolyzing proteins, actually they have different meanings. Specifically, as in the present book, proteases refer to all enzymes acting on the peptide bond, i.e., they stand for both exo- and endopeptidases, whereas proteinases refer only to endopeptidases according to the *Enzyme Nomenclature.*[1] This classification[1] does not consider proteases which comprise both exo- and endopeptidases and thus, represent peptidases acting on the peptide bond irrespective of the bond position. As the recommendations in the *Enzyme Nomenclature*[1] imply that an

Table 1
SOME EXOPEPTIDASES GROUPED ACCORDING TO THE ENZYME NOMENCLATURE[1]

Number	Name	Reaction
3.4.11	α-Aminoacylpeptide hydrolases	
3.4.11.1	Cytosol aminopeptidase (leucine aminopeptidase)	Aminoacyl-peptide + H_2O = amino acid + peptide
3.4.11.2	Microsomal aminopeptidase (aminopeptidase M, aminopeptidase N)	Aminoacyl-peptide + H_2O = amino acid + peptide
3.4.11.9	Aminopeptidase P	Aminoacyl-peptide + H_2O = amino acid + peptide
3.4.13	Dipeptide hydrolases	
3.4.13.3	Aminoacyl-histidine dipeptidase (carnosinase)	Aminoacyl-histidine + H_2O = amino acid + histidine
3.4.14	Dipeptidylpeptide hydrolases	
3.4.14.1	Dipeptidyl peptidase I (cathepsin C)	Dipeptidyl-peptide + H_2O = dipeptide + peptide
3.4.14.5	Dipeptidyl peptidase IV (Xaa-pro-dipeptidyl-aminopeptidase)	Aminoacylprolyl-peptide + H_2O = aminoacylproline + peptide
3.4.15	Peptidyldipeptide hydrolases	
3.4.15.1	Dipeptidyl carboxypeptidase I (angiotensin converting enzyme, peptidase P, kininase II, carboxycathepsin)	Peptidyl-dipeptide + H_2O = peptide + dipeptide
3.4.16	Serine carboxypeptidases	
3.4.16.1	Serine carboxypeptidase (carboxypeptidase Y, carboxypeptidase C, cathepsin A, phaseoline)	Peptidyl-amino acid + H_2O = peptide + amino acid
3.4.17	Metallocarboxypeptidases	
3.4.17.1	Carboxypeptidase A	Peptidyl-amino acid + H_2O = peptide + amino acid
3.4.17.2	Carboxypeptidase B	Peptidyl-lysine/arginine + H_2O = peptide + lysine/arginine
3.4.18	Cysteine carboxypeptidases	
3.4.18.1	Lysosomal carboxypeptidase B	Peptidyl-amino acid + H_2O = peptide + amino acid
3.4.19	Omega peptidases	
3.4.19.3	5-Oxoprolyl-peptidase (pyroglutamyl amino-peptidase)	5-Oxoprolyl-peptide + H_2O = 5-oxoproline + peptide

endopeptidase is not a peptidase, it has been most recently suggested that peptidase should be synonymous with protease and not with exopeptidase.[2]

The greatest problem associated with the present concept of *Enzyme Nomenclature*[1] is that it gives priority to specificity over mechanism. The difficulty of this classification is immediately apparent if we consider that some proteases can act as both endo- and exo-peptidases. For example, cathepsin H is not only an endopeptidase but also an aminopeptidase; cathepsin B is an endopeptidase, as well as a peptidyldipeptide hydrolase. Because of the above difficulties, specificity may not serve as a good primary distinctive feature for proteases. The use of specificity in the classification stems from the early times when the four basic mechanisms of action, i.e., those of serine, cysteine, aspartic, and metallopro-teases, were not available. To date the mechanism can be established for most proteases (see the *Enzyme Nomenclature*[1]). Therefore, *a change from specificity to mechanism is warranted in classification of proteases.* Each mechanistic group may be divided into families which include evolutionarily related enzymes, and the enzymes within a family may be distinguished by their specificity characteristics.

Some confusion appears to exist in the literature about the different protease groups. The "class" and "family" terms are frequently used in referring to a group of proteases irrespective of the kind of relationship among the members of the group. For instance, for trypsins, chymotrypsins, elastase, and thrombin, each was considered as a separate family.[3]

Table 2
SOME ENDOPEPTIDASES GROUPED ACCORDING THE
ENZYME NOMENCLATURE[1]

Number	Name	Preferential cleavage at the carboxyl end of Xaa- or Xaa–Xaa bond
3.4.21	Serine Proteinases	
3.4.21.1	Chymotrypsin	Tyr⊥, Trp⊥, Phe⊥, Leu⊥, Met⊥
3.4.21.4	Trypsin	Arg⊥, Lys⊥
3.4.21.5	Thrombin	Arg⊥, converts fibrinogen to fibrin
3.4.21.6	Coagulation factor Xa (thrombokinase)	Arg⊥ Ile, Arg⊥ Gly, converts prothrombin to thrombin
3.4.21.7	Plasmin (fibrinolysin)	Lys⊥, > Arg⊥, converts fibrin into soluble products
3.4.21.9	Enteropeptidase (enterokinase)	Asp–Lys⊥ Ile in trypsinogen
3.4.21.12	*Myxobacter* α-lytic proteinase	Peptide portion as Xaa–Xaa–Xaa–Ala⊥
3.4.21.14	Microbial serine proteinases: subtilisin, *Aspergillus* alkaline proteinase, *Tritirachium* alkaline proteinase (proteinase K)	
3.4.21.19	*Staphylococcal* serine proteinase	Glu⊥, Asp⊥
3.4.21.20	Cathepsin G	Similar to chymotrypsin
3.4.21.26	Prolyl endopeptidase (post-proline cleaving enzyme, postproline endopeptidase)	Pro⊥ L-Xaa and Pro⊥ D-Xaa but neither Pro⊥ L-Pro nor Pro⊥ D-Pro bonds
3.4.21.34	Plasma kallikrein	Lys⊥ Arg and Arg⊥ Ser bonds in kininogen to produce bradykinin
3.4.21.35	Tissue kallikrein	Met⊥ Lys and Arg⊥ Ser bonds in kininogen to produce lysylbradykinin (kallidin)
3.4.21.36	Pancreatic elastase	A short segment as Xaa–Xaa–Xaa–Ala⊥
3.4.21.37	Leukocyte elastase (neutrophil elastase)	Xaa–Xaa–Xaa–(Ala, Val)⊥
3.4.21.39	Chymase (mast cell protease I)	Similar to chymotrypsin
3.4.22	Cysteine proteinases	
3.4.22.1	Cathepsin B	Phe–Xaa⊥ Phe–Xaa⊥
3.4.22.2	Papain	Phe–Xaa⊥
3.4.22.3	Ficin	
3.4.22.4	Bromelain	
3.4.22.6	Chymopapain	Phe–Xaa⊥
3.4.22.7	Asclepain	Phe–Xaa⊥, Phe–Xaa⊥
3.4.22.8	Clostripain	Arg⊥, especially Arg⊥ Pro bond
3.422.14	Actindin	Phe–Xaa⊥
3.4.22.15	Cathepsin L	
3.4.22.16	Cathepsin H	
3.4.22.17	Calpain (Ca²⁺-activiated neutral protease)	
3.4.23	Aspartic proteinases	
3.4.23.1	Pepsin A (pepsin)	Phe(Tyr, Leu)⊥ Trp(Phe, Tyr) bond
3.4.23.4	Chymosin (rennin)	A single bond in casein K
3.4.23.5	Cathepsin D	
3.4.23.6	Microbial aspartic proteinases: *Aspergillus saitoi* aspartic proteinase (aspergillopeptidase A), *Penicillium janthinellum* aspartic proteinase (penicillopepsin), *Rhisopus* aspartic proteinase, *Endothia* aspartic proteinase	
3.4.23.15	Renin (angiotensin-forming enzyme)	Leu⊥ Leu bond in angiotensinogen to generate angiotensin I
3.4.24	Metalloproteinases	
3.4.24.3	*Clostridium histolyticum collagenase*	Xaa⊥ Gly bond in the sequence –Pro–Xaa–Gly–Pro–

Table 2 (continued)
SOME ENDOPEPTIDASES GROUPED ACCORDING THE
ENZYME NOMENCLATURE[1]

Number	Name	Preferential cleavage at the carboxyl end of Xaa– or Xaa–Xaa bond
3.4.24.4	Microbial metalloproteinases: *Bacillus thermoproteolyticus* neutral proteinase (thermolysin), *Bacillus subtilis* neutral proteinase, *Myxobacter* β-lytic proteinase	Xaa$\perp$ Leu (Phe)
3.4.24.7	Vertebrate collagenase	
3.4.24.11	Membrane metalloendopeptidase (enkephalinase, neutral endopeptidase 24.11, kidney-brush-border neutral proteinase)	
3.4.24.14	procollagen *N*-proteinase	Xaa$\perp$ Gln in pro α1 and pro α2 chains of procollagen

In contrast, all these enzymes as well as those thought to originate from a common ancestor are proposed to be the members of the same family, i.e., the chymotrypsin family.[4]

Another approach using statistical analysis of the amino acid sequences defines family as a group of proteins that differ at fewer than half of their amino acid positions.[5] (According to this definition, one protein, very unfortunately, may belong to more than one family.) The family is further divided into subfamilies so that the sequences within a subfamily differ from each other at fewer than 20% of their amino acid positions. Furthermore, several families are organized into a superfamily which comprises those sequences that are demonstrably homologous, but differ at more than half of their amino acid positions.[5]

Because the sharp percent identity cutoffs used in the definitions of family, subfamily, and superfamily are too arbitrary and may not provide a significant information, a different usage for these terms was proposed.[6] Specifically, those sequences that are demonstrably related to each other belong to the same family regardless of the percent identity they exhibit. Superfamilies, on the other hand, are composed of two or more families — not all proteins in the superfamily being demonstrably homologous with all the members of the other families in the set.[6] The practical usefulness of defining superfamily in this way is not clear. However, the definition of the family implies a genuine relationship of the family members. This proposal has already been adapted to proteases[4] and is used in this book as well. The families of proteolytic enzymes that can clearly be distinguished by the amino acid sequences and in most cases by the tertiary structures of the family members are shown in Table 3, which is essentially an extension of the categories proposed by Neurath.[4] Each family is designated by a representative protease.

An interesting feature of Table 3 is that within a mechanistic group several families may exist. The best known example is the serine protease group comprising the chymotrypsin and the subtilisin families. Chymotrypsin and subtilisin have entirely unrelated amino acid sequences and steric structures, so that they are thought to have evolved independently, but they have acquired virtually identical catalytic sites. The papain and the virus protease groups also constitute different families as their polypeptide chains are basically unrelated inasmuch as the active site cysteine is located near the N-terminus and the C-terminus, respectively, in papain and the viral protease. Although it has recently been suggested that these two groups are evolutionarily related because of some similarities around the active site cysteine,[7] the homology is too low to demonstrate a meaningful relationship. Moreover, the catalytic cysteine and histidine residues are sequentially close to one another in the virus protease, whereas they are in different domains in papain (see Chapter 4). Of course, some apparently

Table 3
PROTEASE FAMILIES

Serine protease I or chymotrypsin family
Serine protease II or subtilisin family
Cysteine protease I or papain family
Cysteine protease II or clostripain family
Cysteine protease III or virus cysteine protease family
Aspartic protease or pepsin family
Metalloproteasse I or carboxypeptidase A family
Metalloprotease II or thermolysin family
Metalloprotease III or leucine aminopeptidase family

conserved residues, especially glycines, may be found in proteins evolved independently but having some similar structural elements such as β-turns or disulfide bridges. Even the catalytic sites of serine and cysteine proteases may exhibit some structural similarities as they act on similar substrates and operate by similar mechanisms, i.e., through acyl-enzyme formation. Such a poor similarity between two proteins, however, does not justify the assumption of evolutionary relationship.

II. EVOLUTION OF PROTEASES

As compared with the system of *Enzyme Nomenclature* discussed in Section I, studies of the structural features of proteases may reveal the more meaningful evolutionary relationship of these proteins. Comparison of the amino acid sequences of homologous proteins provides information about their evolutionary distances and renders it possible to construct a phylogenic tree that shows the evolutionary relationship of a group of proteins. Establishing homologies in protein sequences and constructing evolutionary trees have an extensive literature which has recently been reviewed.[5,8]

If amino acid sequences are to be compared, the first task is to align them optimally. This may be difficult in the case of evolutionarily distant proteins having different chain lengths and extensive point mutations. Homologous proteases, e.g., the pancreatic serine proteases, always have some short stretches in the amino acid sequences which can be aligned unambiguously in all proteins. Such stretches can be found around the catalytically competent residues such as the serine and histidine residues of serine proteases. These sections conserved strongly during evolution can serve as anchor points. Half cystines may also serve the same purpose since disulfide bridges are not readily changed during evolution. Several procedures have been devised for determining the optimum match between two anchor points.[5,8] Matched identities and similar residues are scored, cysteines usually with a weighted factor, whereas most scoring systems impose a "gap penalty" for each deletion or insertion. All these factors can be taken into account to compute an alignment score referring to the evolutionary distance of the two proteins compared.

Useful information about phylogenic relations can also be obtained in a rather simple way, namely by comparing the similarities of gaps or disulfide bridges.[3] Insertions and deletions of residues in sequences of homologous proteins occur less frequently than amino acid replacements. Similarly, loss or acquisition of a disulfide bond is also infrequent. Hence, distant evolutionary relationships can be tested by considering only gap events or disulfide changes. An estimation of evolutionary relationship of mammalian serine proteases based on the gaps occurring in the optimally aligned sequences is illustrated in Figure 1.[9]

The changes of proteases can be traced through their evolution. For instance, mammalian serine proteases are closely related to a protease isolated from the hepatopancreas of the crayfish (*Astacus fluviatilis*), an invertebrate species.[10] Moreover, this relationship goes back

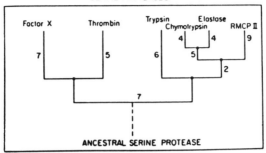

THE EVOLUTIONARY RELATIONSHIP OF MAMMALIAN
SERINE PROTEASES

FIGURE 1. The evolutionary relationship of mammalian serine
proteases as determined by the number of gaps arising from
optimally aligned sequences of the enzymes. Numbers indicate
the gaps occurring when the structures are compared. The sum
of the numbers between two proteases indicates their evolutionary distance. Full circles indicate points of divergence. (From
Woodbury, R. G. and Neurath, H., *FEBS Lett.*, 114, 189, 1980.
With permission.)

to the microbial enzymes such as those isolated from *Streptomyces griseus*.[11] Similarly, a
high degree of sequence identity has been observed between the plant cysteine protease
papain and the mammalian lysosomal enzymes, cathepsins B and H.[12] However, the tertiary
structure within a family of enzymes is even more conserved than the primary structure.
Evolution and the three-dimensional structure of proteins have recently been reviewed.[13]

Besides the above divergent evolution, proteases offer a good example of convergent
evolution as mentioned in the preceding section concerning the chymotrypsin and subtilisin
families. Furthermore, the evolution of specificity is best illustrated by proteases as will be
discussed later in Section V (and Chapter 3, Section III).

Another interesting feature is that in the course of evolutionary development, the protease
molecules which originally served simple digestive functions acquired more complex regulatory functions as well. In this process, they increased their specificity toward select
peptide bonds without destroying the protein molecule. Many biological processes, for
example, blood coagulation and fibrinolysis, are regulated by such limited proteolysis. The
proteases of the blood coagulation and fibrinolytic systems have acquired large noncatalytic
domains attached to the N-terminus of the trypsin-homolog serine protease modules. The
function of the noncatalytic domains is to ensure specific binding of the proteases or their
zymogens to the target macromolecules, whereby the cascades of blood coagulation and
fibrinolysis can be controlled.

The simple serine proteases, such as the pancreatic enzymes and the glandular kallikreins,
are synthesized as small precursors having a signal peptide and an activation peptide attached
to the N-terminus of the very protease.[14-16] On activation of the zymogen, the cleavage
occurs at the C-terminus of the activation peptide. The complex regulatory proteases, on
the other hand, contain very large segments between the signal and the activation peptides.[17-22] For example, in the enzymes of blood coagulation and fibrinolysis these large
segments are divided into several types of domains like kringle domains,[23] vitamin K-
dependent calcium-binding domains, finger domains, and growth factor domains.[24] These
structures as linked to the appropriate proteases are depicted in Figure 2. A new type of
domain has most recently been discovered in Factor XI and prekallikrein.[19]

It is an intriguing question as to at which points on the evolutionary pathway the regulatory
modules were inserted between the signal and the activation peptides. A hypothetical sequence of these events is illustrated in Figure 3.[24] An interesting feature of the dendrogram

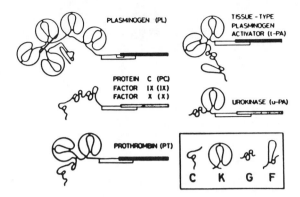

FIGURE 2. Structures of the proteases of blood coagulation and fibrinolysis. The cross-hatched bars represent the protease regions homologous to trypsin. The activation peptide is illustrated by a thin line. After activation, the peptide remains attached to the protease through a disulfide bond also shown by a thin line. The inset shows the different modules of the nonprotease regions: vitamin K-dependent calcium-binding module (C), kringle module (K), growth factor module (G), and finger module (F). (From Patthy, L., *Cell,* 41, 657, 1985. With permission.)

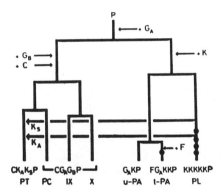

FIGURE 3. Evolutionary tree of the proteases shown in Figure 2. Abbreviations in part are the same as in Figure 2. Others are as follows. P, protease module; G_A and G_B, A- and B-type growth factor module, respectively; K_S and K_A, kringle S and A, respectively. • Represents an internal gene duplication. The vertical axis of the tree has a time dimension. (From Patthy, L., *Cell,* 41, 657, 1985. With permission.)

(Figure 3) is that the kringle structures characteristic of plasminogen are also found in the relatively distant prothrombin. This suggests that the kringles from the former were transferred to the latter. An analysis of the relationship of the sequences of both prothrombin and plasminogen kringles indicates that the ancestor of kringle S was transferred to prothrombin before the multiplication and divergence of the kringles of plasminogen. On the other hand, the closer relationship between kringle A of prothrombin and the kringles of plasminogen suggests that the transfer of this module (kringle A) occurred when the plas-

Table 4
BIOLOGICAL PROCESSES CONTROLLED BY
PROTEASES

Process	Example
Zymogen activation	Pancreatic protease zymogens
Blood coagulation	Prothrombin, factors IX-XII
Fibrinolysis	Plasminogen, plasminogen proactivator
Complement activation	C1r, C1s, C2, B, D, I
Polyprotein processing	Picornaviruses
Hormone processing	Proinsulin, proopiocortin
Production of active peptides	Angiotensinogen
Transmembrane processes	Secretory preproteins
Fertilization	Proacrosin
Development	Procoonase

minogen kringles multiplied and diverged. Figure 3 also implies losses of modules present only in ancestral proteases, for instance, loss of G_A from ancestral plasminogen (G_AKP) or losses of G_A and G_B from an ancestral prothrombin (CG_AG_BP).

III. PROTEASES AND REGULATION

We have seen in the preceding section that some regulatory proteases acquired a high degree of specialization in the course of evolution. Indeed, proteases are involved in a multitude of biological processes. Representative examples of such reactions are shown in Table 4.

The protease action in most regulatory processes is restricted to peptide bonds located at specific sites of specific proteins. The importance of this limited proteolysis was first recognized in the activation of pancreatic zymogens. Trypsin is the common activator of the pancreatic zymogens or proenzymes like chymotrypsinogen, proelastase, procarboxypeptidase, and even trypsinogen.[25] Specifically, the activation of trypsinogen is initiated in the duodenum by enteropeptidase (enterokinase), a large regulatory protein produced by the mucosa of the duodenum. It consists of two polypeptide chains (115 kdalton and 35 kdalton) linked by a disulfide bond. The catalytic function is associated with the small subunit.[26] This enzyme cleaves a particular Lys-Ile bond in trypsinogen at a rate about three orders of magnitude higher than the rate of autoactivation of trypsinogen.[27] In the action of pig enteropeptidase on bovine trypsinogen a hexapeptide, Val-(Asp)$_4$-Lys is released from the zymogen molecule.[28] The release of the hexapeptide leads to conformational changes at the active site of the trypsin molecule, which will be discussed in Chapter 3, Section III. The peculiar sequence of four aspartic acid residues in tandem was also found in several mammalian trypsinogens.[3] Hence, it was suggested that this series of acidic residues constitutes a specific recognition site for enteropeptidase.[27] It was, therefore, of an unexpected finding that the activation peptide of human cationic trypsinogen is merely a dipeptide (Asp-Lys).[29] In accordance with this result, human enteropeptidase, which is different from the bovine and porcine enzymes, activates human trypsinogen much more readily than bovine trypsinogen. Conversely, bovine trypsinogen is activated by porcine enteropeptidase much more rapidly than human trypsinogen. Thus, the two functionally related proteins, trypsinogen and enteropeptidase, may have evolved in a parallel manner in the human lineage.[29]

The activation peptides of the other pancreatic zymogens are larger than that of trypsinogen. The activation peptide of procarboxypeptidase A contains about 100 residues. It has been claimed that after isolation, this peptide is a potent inhibitor of carboxypeptidase A.[30,31]

The activation of chymotrypsinogen has been extensively reviewed.[32-34] This zymogen is

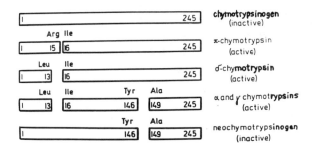

FIGURE 4. Enzyme forms produced on activation of chymotrypsinogen.

converted to the fully active π-chymotrypsin by cleavage of the peptide bond between Arg 15 and Ile 16 (Figure 4). However, the activation peptide of chymotrypsinogen is not released from the protein, but it remains connected with the active enzyme through a disulfide bridge from the N-terminal residue of the zymogen. Interestingly, the activation of chymotrypsinogen is complicated by further proteolysis by trypsin and chymotrysin. As shown in Figure 4, various active and inactive species can be formed depending on the reaction conditions. It is seen that α- and γ-chymotrypsins exhibit the same primary structure which, however, can assume distinct crystallization forms. The α-form is obtained at pH 4.0, whereas the γ-form is crystallized at pH 5.6.[35]

Zymogen activation is of special interest when it occurs in a series of reactions. In such an enzymic cascade, the activated form of one zymogen catalyzes the activation of the next zymogen. The numerous steps yield a large amplification, assuring an extremely efficient control mechanism. Blood coagulation and fibrinolysis are typical examples of cascade reactions. Clotting[36-38a] requires the interplay of two reaction series called intrinsic and extrinsic pathways. All the proteins and cofactors necessary for the former pathway are encountered in the blood plasma. The extrinsic pathway, on the other hand, also requires tissue factors. The two routes converge at a common pathway as illustrated in Figure 5. Clotting factors are designated by Roman numerals, the letter *a* showing that the factor is in the active form. Not all factors are proteases. Factors V and VIII, for example, can be regarded as modifier proteins. Protease C, which is not included in Figure 5, is a relatively novel component that inactivates Factor V and VIII.[39,40] It is seen from Figure 5 that the intrinsic pathway is initiated by the contact of Factor XII (Hageman factor) and kallikrein with an abnormal surface, whereas the extrinsic pathway begins with a trauma to the blood vessels which releases a lipoprotein called tissue factor. A complex of tissue factor and Factor VII activates Factor X.

Another important cascade system is operative in the activation of complement which causes lysis of cellular antigens.[41-43] The first component of complement (C1) is a macromolecular assembly of two distinct entities, C1q and $C1r_2C1s_2$. In the so-called classical pathway of complement activation, the binding of C1q to antibody-antigen aggregates causes conversion of the proenzymic $C1r_2C1s_2$ to the activated complex $C\bar{1}r_2C\bar{1}s_2$ (Figure 6). The structure of C1q seems to be very peculiar, inasmuch as it consists of six globular heads attached to six collagenous stalks, resembling a bunch of tulips.[43-45] The subcomponents C1r and C1s are both single polypeptide chain proteins of about 85 kdaltons; on activation, each gives rise to two polypeptide chains connected with a disulfide bridge.[46 – 46b] It should be mentioned that other serine proteases of the complement system are factor B,[41] factor D,[41] and factor I.[46c]

It is an interesting recent finding that C3 and C4, when activated, form covalent bonds with adjacent polysaccharides or proteins. The binding occurs by an acyltransfer mechanism resulting in the formation of an ester or amide bond.[47,48] An internal thiolester, which can

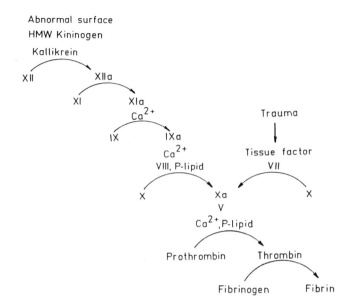

FIGURE 5.   A simplified scheme for blood clotting. The intrinsic pathway (left) and the extrinsic pathway (right) meet at Factor Xa and continue in the common pathway to produce fibrin clot.

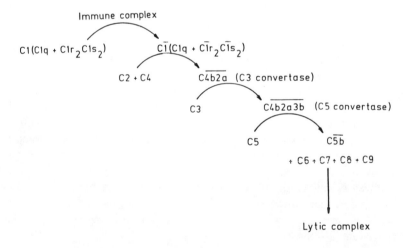

FIGURE 6.   A simplified scheme of the classical pathway of complement activation. The classical pathway involves only three proteases, C̄1r, C̄1s, and C2. The C̄1 cleaves both C2 and C4. The activated C2 and C4b form complexes inside the convertases. The smaller cleavage product, C4a (not shown in the figure), is an anaphilatoxin. The proteolytic cascade leads to a nonenzymic series of protein interactions responsible for the cytolytic activity of the complement system.

account for the covalent bond formation, has been demonstrated in C3 and C4.[49-52] This is similar to that found for α_2-macroglobulin (Section IV.D).[53,54] Indeed, C3, C4, and α_2-macroglobulin are evolution-related proteins.[55]

Another complex regulatory mechanism is associated with the blood pressure homeostasis. This involves two reciprocally interrelated protease systems: the renin-angiotensin and the kallikrein-kinin systems (Figure 7).[56-58] They produce peptide antagonists, namely angiotensin II, and kinins (kallidin and bradykinin) which increase and decrease, respectively,

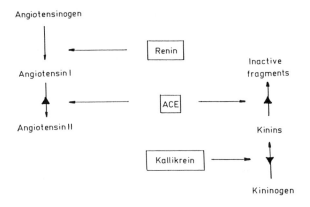

FIGURE 7. Interrelationship of the renin-angiotensin and the kallikrein-kinin systems. Proteases are shown in boxes. ACE stands for antiotensin-converting enzyme, ▲ and ▼ represents blood pressure increase and decrease, respectively.

the blood pressure. Figure 7 shows the three most important proteases implicated in regulation. They represent three different types of proteases: (1) kallikrein is a serine protease that produces the vasodilator kinins,[57] (2) renin is a highly specific aspartic protease which splits off a decapeptide (angiotensin I) from the N-terminus of the α-globulin angiotensinogen,[56] and (3) angiotensin-converting enzyme is a metalloprotease that exhibits peptidyl-dipeptidase activity.[56] It plays a dual role in the control of blood pressure by forming angiotensin II and eliminating bradykinin (Figure 7), with both processes leading to the elevation of blood pressure. The three enzymes will be discussed in the appropriate chapters (Chapters 3, 5, and 6).

The mammalian proteases discussed above and secretory proteins in general display an N-terminal extension, the so-called signal peptide or leader sequence, which is also cleaved by limited proteolysis. This peptide segment is required for transfer of the nascent pre-protein or pre-proenzyme across the membrane of the endoplasmic reticulum.[59-61] The transfer can occur after completion of the polypeptide chain by a post-translational mechanism or the growing polypeptide chain is translocated through the membrane by a cotranslational mechanism. The signal peptidase, a membrane-bound protease, removes the signal peptide as it enters the lumen.[62] It hydrolyzes a bond (Gly or Ala or Ser)–Xaa following a sequence of hydrophobic residues. Signal peptides are known to contain a hydrophobic segment with a minimal length of about nine residues. Microsomal signal peptidase has recently been purified as a complex of six polypeptide chains.[63]

Limited proteolysis is also essential for processing protein and peptide hormones.[61,64] For instance, insulin consisting of two polypeptide chains is synthesized as the single chain pre-proinsulin.[61] As the first event of processing, the leader sequence of the polypeptide chain is split off upon traversing the membrane of the endoplasmic reticulum. Then proteolysis at two definite peptide bonds (Figure 8) cleaves out a peptide segment, which gives rise to a connecting C-peptide and active insulin consisting of an A chain of 21 residues and a B chain of 30 residues, which are covalently joined by two disulfide bridges.[65] The connecting peptide contains a pair of basic amino acid residues at each end: Arg-Arg at the N-terminus and Lys-Arg at the C-terminus. It is has been suggested that a cathepsin B-related enzyme and a carboxypeptidase activate proinsulin.[66,67]

Pairs of basic residues are also encountered in other prohormones at the junctions with the residues which will form the active hormones, e.g., in procalcitonin,[68,69] prosomato-statin,[70,71] proglucagon,[72] and proopiomelanocortin or simply proopiocortin.[73,74] The latter

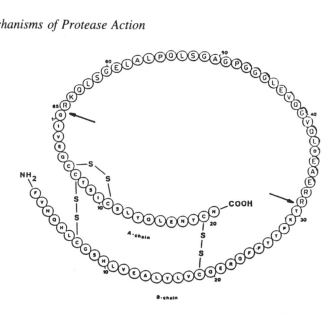

FIGURE 8.   Amino acid sequence of human proinsulin. The arrows show the points of cleavage liberating the C-peptide.

is the source of several potent peptide hormones (Figure 9) such as β-lipotropin, melanocyte-stimulating hormones (α- and β-MSH), and corticotropin or adrenocorticotropic hormone (ACTH). It is of particular interest that the larger hormones derived from the precursor protein may be further hydrolyzed to yield smaller hormones. Thus, β-endorphin is a fragment of β-lipotropin, and α-MSH is a portion of corticotropin. β-MSH is part of γ-lipotropin, whereas the sequence of γ-MSH is included in the 16-kdalton fragment. Although the opiate peptide Met-enkephalin might be produced from β-endorphin, recent evidence indicates the existence of separate precursors containing multiple copies of Leu- and Met-enkephalins.[75,76] It is worthy of note that not all pairs of basic amino acids are always cleaved and different tissues exhibit different cleavage patterns, which is of regulatory importance.

Not only hormones, but also virus proteins are produced as polyproteins. For example, poliovirus and other picornaviruses (see Reference 77) code for a >200-kdalton precursor protein. Cleavages of the polyprotein occur at highly specific sites which are achieved predominantly by a viral protease. This enzyme is also produced from the polyprotein, probably by autocatalytic cleavage.[78,79] The viral protease is presumably a cysteine protease not related to the papain family (see Chapter 4).

Besides the highly specific limited proteolysis we have discussed so far, regulation of intracellular protein degradation also constitutes an intriguing problem. In mammalian cells, there are two principal mechanisms of protein breakdown, the lysosomal and the nonlyso-somal pathways, which have been reviewed in detail.[80,81] It has been known for a long time that lysosomal autophagy plays the major role especially under conditions of nutritional deprivation, but the mechanisms that regulate the formation of autophagic vacuoles remain to be established. Some insight into the nonlysosomal mechanisms has recently been gained. The most interesting characteristics of these systems is that they are ATP-dependent. Since proteolysis is an exergonic process, energy is probably required for control. One of the most extensively studied ATP-dependent proteolytic systems is that found in reticulocytes, which synthesize predominantly hemoglobin. An important component of this proteolytic system is a heat-stable polypeptide, ubiquitin. As its name implies, the ubiquitin molecule is widespread, occurring in all eukaryotes thus far examined. It is composed of 76 amino acids, the C-terminal dipeptide being Gly-Gly. Its three-dimensional structure has recently been determined.[82,82a] In the ubiquitin proteolytic pathway, the polypeptide is covalently linked

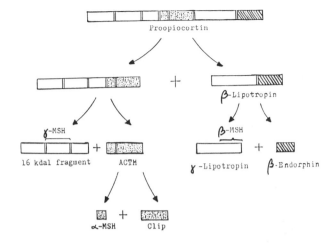

FIGURE 9. A scheme for the structure and processing of proopiocortin.
Paired basic amino acids are marked with double vertical lines.

to proteins destined for catabolism. It is generally believed that the binding of several ubiquitin molecules produces a conformational change in the target protein, thereby making it susceptible to attack by specific proteases. Equation 1 shows the activation of ubiquitin. In this process, the glycine carboxyl terminus of ubiquitin probably forms an acyl-adenylate with ATP which is followed by a thiolester formation in the reaction between the acyl-adenylate and a thiolenzyme. As shown by Equation 2, several activated ubiquitin molecules react with the protein to be degraded. This is accomplished through unusual isopeptide linkages involving the ϵ-amino groups of lysine residues in the target protein and the glycine carboxyl group of ubiquitin. Once the protein is tagged with ubiquitin, it is hydrolyzed rapidly by an as yet uncharacterized proteolytic system.

$$\text{Ub-Gly-COOH} + \text{HS-E} \xrightarrow{\quad\text{ATP}\quad\text{AMP} + \text{PP}_i\quad} \text{Ub-Gly-COS-E} \qquad (1)$$

$$n(\text{Ub-Gly-COS-E}) + \text{protein} \rightarrow (\text{Ub-Gly})_n\text{-protein} \qquad (2)$$

An enzyme that cleaves small amines, including lysine, from the carboxyl terminus of ubiquitin derivatives may serve for the regeneration of this factor from the products of proteolysis of ubiquitin-protein conjugates.[83]

The ubiquitin pathway, first detected in reticulocytes, has been demonstrated to occur in other cell types as well.[84-86]

The direct involvement of ATP in the function of a protease has recently been found in protein breakdown in *Escherichia coli*. This enzyme, protease La, exhibits an inherent ATPase activity.[87] It is a 450-kdalton tetramer composed of identical subunits. Addition of protein substrates to this enzyme enhances ATP hydrolysis, while inhibition of proteolytic activity leads to a fall in ATP hydrolysis. Protease La is capable of hydrolyzing also small peptides, such as glutaryl-Ala-Ala-Phe-methoxynaphthylamine,[88] but this process requires only ATP binding not hydrolysis. Thus, the nature of ATP requirement differs for the hydrolyses of peptides and proteins.[89]

IV. PROTEIN PROTEASE INHIBITORS

The regulatory role of proteases is intimately coupled with their inhibitors. These molecules are present in numerous tissues of all living creatures to prevent unwanted proteolysis. However, their true physiological function has been elucidated only in a few instances. For example, several lines of evidence suggest that in familial emphysema α_1-proteinase inhibitor is congenitally deficient which leads to destruction of connective tissue of the pulmonary alveoli by leukocyte elastase.[90,91] The deficiency can result from a single amino acid substitution in the molecule of about 50 kdaltons.[92]

A. Mechanism of Action

Inhibitors to proteases of all four mechanistic groups have been found. The best known examples will be discussed in the following chapters pertinent to the individual protease groups. Here we consider some problems of protein protease inhibitors in general. Of the four groups, inhibitors to serine proteases represent the only category that has been extensively studied.[93] As a result, two basic mechanisms of inhibition have been revealed: the reversible and the irreversible mechanisms.

The reversible mechanism has been clearly established for a number of serine protease inhibitors.[93] It is also called the standard mechanism. However, the former term (reversible) proposed in this book may be preferred since it points to the essential difference between the two alternative mechanisms whereby all serine protease inhibitors studied to date exert their effects. An inhibitor obeying the reversible mechanism has a reactive site, a specific peptide bond, which can combine with the active site of the target protease to form a proteolytically inactive complex. In the complex, the reactive peptide bond is hydrolyzed by the enzyme. Equation 3 shows the simplest form of complex formation.

$$E + I \rightleftharpoons C \rightleftharpoons E + I^* \tag{3}$$

where E is the protease, I and I* are the native (reactive site peptide bond intact) and modified (reactive site peptide bond cleaved) inhibitors, respectively, and C is the enzyme-inhibitor complex. The bond cleavage very probably proceeds through an acyl-enzyme intermediate. The reversibility of the above mechanism implies that the protease-inhibitor complex must be the same substance whether it is formed from the native or the modified inhibitor. This has been confirmed experimentally by studying the complex formation with both the native (I) and modified (I*) inhibitor. Such an experiment requires a free modified inhibitor that can be isolated at low pH where the protease-inhibitor complex dissociates.[94] It should be noted that in the case of the reversible mechanism, the hydrolysis of the reactive site peptide bond does not proceed to completion, but an equilibrium near unity is established between the forms with intact and cleaved peptide bonds.

The irreversible mechanism resembles the reversible one in that the complex formation involves a specific interaction between the active site of the protease and the reactive site of its inhibitor. Moreover, in the complex the reactive site peptide bond is hydrolyzed by the enzyme. However, the modified inhibitor becomes inactive and cannot recombine with the protease (Equation 4).[90] This may be rationalized in terms of the three-dimensional structure of the cleaved α_1-proteinase inhibitor released after complex formation.[95] In the cleaved structure, the reactive center is sprung with Met 358 at one end of the molecule separated by 6.9 nm from Ser 359 at the other. It appears then that the cleaved inhibitor is stable, whereas the native inhibitor is strained at the reactive site bond.

$$E + I \rightleftharpoons EI \rightarrow EI^* \rightarrow E + I^* \tag{4}$$

Another peculiarity of the inhibitors obeying the irreversible mechanism is that their complexes with the target enzymes resist dissociation by urea and sodium dodecylsulfate. This implies a covalent bond formation between the protease and its inhibitor which is most probably the result of the formation of an acyl-enzyme derivative.[96,97]

Table 5
BEST CHARACTERIZED FAMILIES OF
SERINE PROTEASE INHIBITORS OBEYING
THE REVERSIBLE MECHANISM

Family	Abbreviation
1. Bovine pancreatic trypsin inhibitor (Kunitz)	BPTI or PTI
2. Pancreatic secretory trypsin inhibitor (Kazal)	PSTI
3. Soybean trypsin inhibitor (Kunitz)	STI
4. Soybean protease inhibitor (Bowman-Birk)	—
5. *Streptomyces* subtilisin inhibitor	SSI

In summary, both types of serine protease inhibitors have a substrate-like reactive site which combines with the active site of the target enzyme to form a relatively stable complex. The similarity of the reactive site to the substrate is so great that the protease can hydrolyze the reactive site peptide bond. It is of an as yet unknown regulatory importance why some of the inhibitors are converted into an inactive form on reacting with the enzyme, whereas others remain active after liberation from the complex.

B. Inhibitors Obeying the Reversible Mechanism

Several families of serine protease inhibitors operate through the reversible mechanism. Table 5 includes the best characterized families, which are to be discussed below.

The pancreatic trypsin inhibitor (Kunitz) family has been named after the first inhibitor obtained in crystalline form.[98] The inhibitor is a small (58 amino acid residues),[99] unusually stable protein, whose three-dimensional structure is known (see Chapter 3, Section III.C). Hence, it has been the subject of extensive studies by protein physicochemists. However, the physiological function of this inhibitor is not yet known. Furthermore, it seems to be present only in bovids and caprids.[100] A number of other inhibitors belong to this family, some of them possessing multiple reactive sites on a single polypeptide chain. An interesting example is inter-α-trypsin inhibitor, a large glycoprotein molecule of 180 kdaltons, which is composed of a heavy chain and a light chain. The latter contains two tandem domains homologous to bovine pancreatic trypsin inhibitor.[93,101,101a] Kunitz-type inhibitors are also found in the garden snail, in venoms of several snakes, and in the egg white of Red Sea turtle.[93] Only one domain of this latter inhibitor, chelonianin, is homologous to the Kunitz inhibitor. This inhibits trypsin, while the other that inhibits subtilisin represents a new inhibitor family, which also includes antileukoprotease, an acid-stable protease inhibitor of human mucus secretions.[102]

Another family, which is distinguished by its unrelated primary structure, involves a large number of inhibitors and has been named after the pancreatic secretory trypsin inhibitor (Kazal).[103] In contrast to the pancreatic trypsin inhibitor (Kunitz), the Kazal-type inhibitor occurs in all vertebrates examined.[93] Inhibitors from a number of species were sequenced,[93] and the tertiary structure of the porcine Kazal inhibitors has been determined (see Chapter 3, Section III.C). Two tandem Kazal-type domains connected with a short peptide chain constitute the major protease inhibitor of the dog submandibular glands.[104] One domain inhibits trypsin and the other is complexed with chymotrypsin, elastase, and subtilisin. Three tandem Kazal domains were observed in some ovomucoids, the major glycoprotein inhibitors of avian egg whites (Figure 10), although single- and double-headed inhibitors were also found in certain species.[93,105] The carbohydrate-free third domain of chicken or Japanese quail egg white ovomucoid, each composed of 56 amino acid residues, can be obtained with limited proteolysis of the intact ovomucoid with staphylococcal protease.[105,106] This third domain could be crystallized and was analyzed by the X-ray diffraction method (see Chapter

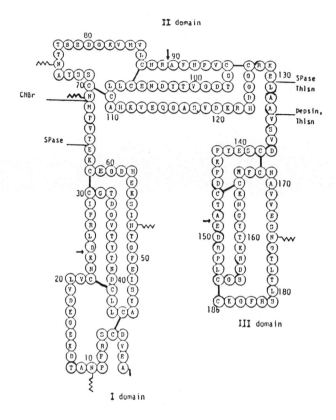

FIGURE 10.   Schematic representation of the structure of chicken egg white ovomucoid. Indicated in the figure are sugar attachment sites (sawtoothed lines), the reactive site locations (arrows), cleavage site for cyanogen bromide (CNBr), limited proteolysis sites for staphylococcal proteinase (SPase), thermolysin (Thlsn), and pepsin used to produce isolated structural domains of the molecule. (From Ogino, T., Croll, D. H., Kato, I., and Markley, J. L., *Biochemistry,* 21, 3452, 1982. With permission.)

3, Section III.C). Besides ovomucoid, avian egg whites contain another inhibitor, ovoinhibitor, which is composed of six Kazal-type domains.[93,107] The ovoinhibitor was also isolated from chicken plasma.[108]

Similar, to pancreas, soybean contains at least two families of serine protease inhibitors: the Kunitz and the Bowman-Birk families. Inhibitors of legumes have received a great deal of attention, mainly because of their antinutritional effects.[109] Soybean trypsin inhibitor (Kunitz) was the first plant protease inhibitor to be thoroughly investigated.[93,109] Most of the work that established the reversible mechanism was done with this inhibitor.[93,110] It is a large (181 residues), single-headed inhibitor that contains two disulfide bridges. Its amino acid sequence[111] and steric structure have been determined (see Chapter 3, Section III.C). The inhibitors of the Bowman-Birk family are much smaller (70 residues) and each contains seven disulfide bridges.[109] They are typical double-headed inhibitors capable of inhibiting two proteases at the same time with considerable variation in specificity. The amino acid sequences for several members of this family have been determined: soybean C-II,[112] soybean D-II,[113] lima bean,[114] garden bean,[115] and chick pea.[116]

Several members of the squash trypsin inhibitor family have also been sequenced.[117] They represent the smallest inhibitors (29 to 32 amino acid residues, 3 disulfide bonds) with Arg-Ile or Lys-Ile reactive site peptide bonds.

Table 6
SERINE PROTEASE INHIBITORS IN HUMAN BLOOD
PLASMA[90,93]

Inhibitor	Concentration mg/100 mℓ	Molecular mass[a] (kdal)	Abbreviation
α_1-Proteinase	290	53	α_1PI
α_1-Antichymotrypsin	50	68	α_1Achy
Antithrombin III	29	65	AT III
Heparin cofactor II		65	HC II
C$\overline{\text{I}}$-inhibitor	24	100	C$\overline{\text{I}}$-Inh
α_2-Antiplasmin	7	70	α_2AP
Inter-α-trypsin	50	180	IαI
α_2-Macroglobulin[b]	260	725	α_2M

[a] With carbohydrate.
[b] General protease inhibitor.

Among the serine protease inhibitors produced by various microorganisms, *Streptomyces* subtilisin inhibitor is best characterized.[118-120] Even the tertiary structure of the molecule has been determined (see Chapter 3, Section III.C). Its polypeptide chain of 113 amino acid residues forms two domains, only one reacting with subtilisin. This reactive site is very similar to that of the Kazal inhibitor, but the remainder of the molecule is unrelated. It remains to be established whether the two families are products of convergent or divergent evolution.

In contrast to the inhibitors discussed above, members of the potato inhibitor I family do not contain stabilizing disulfide bridges. The three-dimensional structures of two inhibitors of this family, notably chymotrypsin inhibitor 2 from barley seeds[121] and eglin[122,123] from the leech *Hirudo medicinalis* have most recently been determined. It has not been clearly established as yet whether the members of the potato inhibitor I family operate through the reversible mechanism.

C. Inhibitors Obeying Irreversible Mechanism

Most of the plasma inhibitors listed in Table 6 (α_1-proteinase inhibitor, α_1-antichymotrypsin, antithrombin III, heparin cofactor II, C$\overline{\text{I}}$-inhibitor, and α_2-antiplasmin) operate through the irreversible mechanism. As discussed above, inter-α-trypsin inhibitor contains two pancreatic trypsin inhibitor (Kunitz)-type domains. α_2-Macroglobulin is a general proteinase inhibitor that can inhibit all four mechanistic groups of proteases.

α_1-Proteinase inhibitor, α_1-antichymotrypsin inhibitor, and antithrombin III are the best studied members of the same inhibitor family,[90] recently called serpin family[91] since they are primarily serine protease inhibitors. This may not be an appropriate term because there are several other families of SERine Protease INhibitors as well. The reactive site peptide bond of the three inhibitors is Met-Ser, Leu-Ser, and Arg-Ser, respectively, the N-terminal residue of the peptide bond controlling the specificity of the individual inhibitor. It is interesting that the unrelated plant protease inhibitors of the Bowman-Birk family exhibit highly homologous amino acid sequences near the reactive site bond.[91] We have already mentioned that α_1-proteinase inhibitor is most specific toward leukocyte elastase and plays an important role in the defense mechanism of the lung against proteolytic attack. The original name of α_1-proteinase inhibitor, which is still in use, was α_1-antitrypsin because of its ability to inhibit trypsin. It turned out, however, that this inhibitor is far more effective in controlling the activity of several other serine proteases including elastase and chymotrypsin.

The presence of methionine at the reactive site of α_1-proteinase inhibitor raised the question about the effects of oxidizing agents on inhibitory activity. As it has recently been reviewed,[90,91] a number of studies dealt with this problem. The results indicated a significant reduction in the rate constant of association between the inhibitor and leukocyte elastase after oxidation.[124] By oxidizing methionyl residues, cigarette smoke significantly reduced the activity of α_1-proteinase inhibitor,[125,126] which could be deleterious to the protection of lung tissue, especially for individuals with abnormally low inhibitor concentration in serum.

α_1-Antichymotrypsin is a major acute phase protein, whose concentration increases markedly under traumatic conditions. It is a specific inhibitor of chymotrypsin-like proteases such as cathepsin G, which may be its primary target enzyme, mast cell chymases, and, of course, chymotrypsin.

Antithrombin III reacts with various serine proteases, including thrombin, at a relatively slow rate. However, its rate of association with thrombin is greatly increased in the presence of heparin. In fact, the coagulation process is most readily inhibited by a specific interaction between thrombin and antithrombin III.[38,90]

Heparin cofactor II is a relatively new inhibitor of thrombin which requires glycosaminoglycans for an efficient action. It has a molecular mass of 65 kdaltons and functionally resembles antithrombin III.[127] However, it does not inhibit trypsin and trypsin-like proteases including Factor Xa, Factor IXa, and plasmin, but inhibits chymotrypsin in accordance with the Leu-Ser reactive site bond.[128,129]

Nucleotide and amino acid sequencing have established the homology for α_1-proteinase inhibitor,[130] α_1-antichymotrypsin,[131] antithrombin III,[132] heparin cofactor II,[127] α_2-antiplasmin,[132a] and $\overline{C1}$ inhibitor.[133] It is of particular interest that the noninhibitory proteins, ovalbumin,[134,135] and protein Z,[136] a major barley endosperm protein, are also homologous with the above plasma inhibitors. The homology extends to the reactive sites as well. The Met_{372}-Ser_{373} bond of protein Z and the Ala_{372}-Ser_{373} bond of ovalbumin, which correspond to the reactive site of the plasma inhibitors, are both sensitive to proteolytic cleavage.[136] Indeed, the transformation of ovalbumin into plakalbumin in the presence of subtilisin has long been known.[137] The initial cleavage of this process involves the hydrolysis of the Ala_{372}-Ser_{373} bond.[136]

Finally, the protease inhibitors of the interstitial fluid should be mentioned. These have recently been discovered and named as protease nexins.[138,139] It was found that cultured cells release protease inhibitors which selectively form covalent linkages with certain serine proteases present in the medium. These protease-protease nexin complexes then bind to the cells and are internalized and degraded.

Protease nexin I or simply protease nexin was purified from culture medium of human fibroblasts.[140] It appears to be evolutionarily related to antithrombin III and operates through the irreversible mechanism. It inhibits trypsin-like serine proteases: trypsin (k_{ass} = 4.2 × 10^6 M^{-1} sec^{-1}), thrombin (k_{ass} = 6.0 × 10^5 M^{-1} sec^{-1}), urokinase (k_{ass} = = 1.5 × 10^5 M^{-1} sec^{-1}), and plasmin (k_{ass} = 1.3 × 10^5 M^{-1} sec^{-1}).[140] In the presence of heparin, protease nexin inhibits thrombin at a nearly diffusion-controlled rate. Functional and structural similiarities between protease nexin I and C1 inhibitor have been shown.[140a]

D. α_2-Macroglobulin

Whereas the serine protease inhibitors, acting through either the reversible or the irreversible mechanism, react with the active site of the target protease which results in complete inhibition, α_2-macroglobulin only inactivates proteases towards large protein substrates.[90,141] The activity towards small synthetic substrates remains unimpeded. Clearly, in the α_2-macroglobulin complex, the active site of the protease is accessible even for small protein-protease inhibitors such as pancreatic trypsin inhibitor (Kunitz).[142] The other remarkable feature of α_2-macroglobulin is its ability to inactivate proteases belonging to all four mechanistic groups.[90,141] This requires a special mechanism of inhibition by α_2-macroglobulin.

```
                                    E,PA
          T,SGT,PA      E    ┌──────┐  CS   CT   SP
          └────────┘    │  10│      ↓   │    │    │
                        ↓    ↓      ↓   ↓    ↓    ↓
  -His-Gly-Pro-Glu-Gly-Leu-Arg-Val-Gly-Phe-Tyr-Glu-Ser-Asp-Val-
```

```
                              S,SGB
          T,PL,TH,TL,S,SGT    ┌──┐        PA
          └──────────────┘  20│  │        │                    30
                             ↓  ↓        ↓
  -Met-Gly-Arg-Gly-His-Ala-Arg-Leu-Val-His-Val-Glu-Glu-Pro-His-
```

```
          TL  CS              CS
          │   │               │         40
          ↓   ↓               ↓
  -Thr-Glu-Thr-Val-Arg-Lys-Tyr-Phe-Pro-Glu-Thr-Trp-Ile-Trp-Asp-
```

FIGURE 11.   The amino acid sequence of the "bait region" of α_2- macro-globulin. The cleavage sites for different proteases are shown by arrows: T, bovine trypsin; SGT, *Streptomyces griseus* trypsin; PA, papain; E, pancreatic elastase; CS, calf chymosin; CT, chymotrypsin; SP, *Staphylococcus aureus* V8 protease; PL, plasmin; TH, thrombin; TL, thermolysin ; S, subtilisin Novo; and SGB, *Streptomyces griseus* protease B. (From Mortensen, S. B., Sottrup-Jensen, L., Hansen, H. F., Petersen T. E., and Magnusson, S., *FEBS Lett.*, 135, 295, 1981. With permission.)

Based on the observation that the protease hydrolyzes some particularly susceptible peptide bond in the inhibitor,[143] it was proposed that α_2-macroglobulin operates through a "trap" mechanism.[144] Specifically, the peptide bond cleavage in the inhibitor triggers a conformational change which entraps the enzyme. Conformational changes have been shown by an increase in the rate of migration of complexes in nondenaturing polyacrylamide gels relative to the migration of the native inhibitor.[145,146] The high molecular mass (Table 6) and the tetrameric structure of the inhibitor are consistent with its unique inhibitory mechanism. The inhibitor contains four identical polypeptide chains, linked in pairs by a single disulfide bond. The amino acid sequence of the subunit (1451 amino acid residues) has recently been reported.[147,147a]

The various proteases cleave near the middle of the α_2-macroglobulin subunit giving rise to two polypeptides of 85 and 95 kdaltons.[143,145] This finding suggests that it should be a short polypeptide segment, referred to as the "bait region", which displays specificity towards most endopeptidases except those that are very large or highly selective (C$\bar{1}$r, C$\bar{1}$s, Factor XIIa, urokinase, and renin). Proteolytic cleavage sites within this specific region are depicted in Figure 11.[90,148,149]

The hydrolysis of the "bait region" leads to the appearance of a free thiol group, (Cys 949), while the protease originally entrapped noncovalently becomes covalently attached to the inhibitor.[54,150,151] The thiol group liberated is assumed to be derived from a thiolester with a γ-glutamyl residue (Glu 952) of the native inhibitor. This thiolester acylates an amino group on the protease complexed with the inhibitor. The thiolester bond can also be cleaved by small primary amines. The amine incorporates into α_2-macroglobulin by forming a γ-glutamyl derivative.[150,151] In the human inhibitor, this cleavage results in a conformational change similar to that induced by proteases[152-154] and leads to inactivation of the inhibitor.[145,155] In contrast, cleavage of the thiolester in bovine α_2-macroglobulin by methylamine results in a limited conformational change and only decreases but not abolishes the activity of the inhibitor.[154,156,157] Thus, an intact thiolester bond is not an absolute requirement for the inhibitory action. However, the conformational change on formation of the enzyme-inhibitor complex appears to be essential for clearing the reacted α_2-macroglobulin from the circulation. Specifically, only the complexes and not the native inhibitor can be recognized by the cellular receptors, and then internalized by endocytosis.[141]

α_2-Macroglobulin occurs not only in the blood, it is also encountered in egg white as ovostatin (ovomacroglobulin).[158]

V. THE MAIN FEATURES OF PROTEASE SPECIFICITY

Specificity of a protease is expressed toward a portion of a polypeptide substrate which contains the peptide bond to be cleaved preferentially. This portion may be a single peptide bond or an extended polypeptide region. In the former case, it is relatively easy to recognize specificity. The hydrolysis by trypsin at basic amino acid residues demonstrates a well-known example. On the other hand, in the case of pancreatic elastase, a characteristic amino acid residue is not encountered around the scissile peptide bond. It is usually stated that elastase is specific for an alanine residue providing the carbonyl group to the sensitive peptide bond. However, a simple alanine derivative is a very poor substrate of elastase. Additional amino acid residues linked to the amino group of the alanine derivative increase by several orders of magnitude the rate of hydrolysis. Consequently, elastase is specific for a short polypeptide backbone consisting of 4 to 5 residues rather than for an alanine residue. Nonetheless, the alanine residue is favored over bulky residues because the latter exclude themselves from the catalytic site.

Table 7 shows the specificity of some of the best-known proteases. It is immediately apparent that the specificity of serine and cysteine proteases is associated mainly with the acyl group portion of a substrate RCO–X, where RCO– is the acyl group and X is the leaving group in the formation of an acyl-enzyme intermediate. On the other hand, the specificity of aspartic proteases and metalloproteases extends to both sides of the scissile peptide bond. In some cases, amino acids other than those flanking the sensitive peptide bond account for the specificity. For example, the specificity of papain is determined by the second rather then the first amino acid residue to the bond to be cleaved. This second residue should possess a large hydrophobic side chain (Phe, Leu, Val), while the first residue may be varied widely. Among the enzymes listed in Table 7, clostripain seems to be one of the most bond-specific proteases. Its preference for arginyl bonds is virtually absolute.

In the case of polypeptide substrates, the interaction between the enzyme and substrate can extend over several amino acid residues. For convenience of discussion, the notation proposed by Schechter and Berger[172] is widely used to designate the amino acid residues flanking the scissile bond. This is depicted in Figure 12, where P_n stands for an amino acid residue of the polypeptide substrate, and S_n is the subsite on the enzyme surface that binds the P_n amino acid residue. In the case of trypsin, for instance, P_1 is an Arg or Lys residue; in the case of papain, P_2 is a large hydrophobic residue. These residues determine the P_1–P'_1 bond cleavage in the respective enzyme.

It is customary to distinguish between primary and secondary specificities. Here again, there is a considerable confusion in the literature. The case of trypsin or chymotrypsin is relatively simple and straightforward. In these instances, the P_1 residue is of primary importance in controlling the specificity, and the primary specificity is associated with the P_1 residue. However, the primary specificity is often attributed to the P_1 residue even though it does not control specificity at all, as in the case of papain. Contrary to the repeatedly occurring belief, in papain catalysis, the P_2 rather than the P_1 residue of the substrate accounts for the primary specificity. In the pepsin-catalyzed hydrolysis, two residues, P_1 and P'_1, are

Table 7
CHARACTERISTIC SPECIFICITIES
OF SOME PROTEASES

Enzyme	Preferential cleavage	Ref.
Serine proteases		
Trypsin	Arg$\perp$	159
	Lys	
Staphylococcus	Glu$\perp$	160, 161
aureus V8	Asp	
protease		
Post-proline	Pro$\perp$	162, 163
endopeptidase		
Cysteine proteases		
Papain	Phe–Xaa$\perp$	164
	Leu	
	Val	
Clostripain	Arg$\perp$	165, 166
Aspartic proteases		
Pepsin	Phe$\perp$Trp	167
	Tyr Phe	
	Leu Tyr	
Sorghum carboxyl	Glu$\perp$	168
proteinase	Asp	
Metalloproteases		
Thermolysin	$\perp$Val	169
	Leu	
	Phe	
Achromobacter	$\perp$Gly–Pro	170, 171
iophagus	Ala	
collagenase		

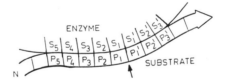

FIGURE 12. Schematic representation of the binding site of proteolytic enzymes. The amino acid residues extending from the sensitive bond indicated by an arrow toward the amino terminus are denoted as P_1, P_2, ... P_n (the acyl group side), and those extending from the scissile bond toward the carboxyl terminus are denoted as P'_1, P'_2, ... P'_n (the leaving group side). The corresponding binding sites on the enzyme are S_1, S_2, ... S_n and S'_1, S'_2, ... S'_n.

associated with the primary specificity. Consequently, the primary specificity of a protease does not necessarily find expression in the P_1 residue.[173,174]

In the case of polypeptide substrates, residues other than those associated with the primary specificity may also interact with the enzyme to enhance specificity. When the primary specificity is strongly expressed, as in the catalysis by trypsin, the contribution by the secondary interactions is less important. However, the secondary specificity becomes more important with proteases exhibiting lower primary specificity, e.g., chymotrypsin and subtilisin. In some instances, as in the elastase catalysis, it is hard to distinguish between primary and secondary specificities because virtually all residues interacting with the enzyme are important. In short, the primary specificity mainly accounts for the selection of the bond to be cleaved (a qualitative feature), while the secondary specificity promotes the cleavage of the selected bond (a quantitative feature).[174]

Besides primary and secondary specificities, several other types of specificities have been proposed. In contrast to the flexible substrates, native proteins may contain a relatively rigid segment of the main chain that is complementary to the extended binding site of a certain protease. This can be of importance to certain zymogen activations and other specific regulatory processes controlled by limited proteolysis.[175] However, it should be kept in mind that limited proteolysis generally occurs at the most flexible part of a protein (Chapter 3, Section III). Wright denoted the assumed specificity for a particular three-dimensional conformation around the susceptible bond of the substrate as conformational specificity.[175] The same term was also applied to the *cis/trans* isomeric specificity.[176] To avoid confusion about the meaning of conformational specificity, it seems to be more appropriate to use the term "three-dimensional specificity" towards substrates of definite steric structure and the term "*cis/trans* isomeric specificity" towards substrates containing a proline residue.[174]

Besides the substrate specificity discussed above, biological specificity should also be mentioned. This can be mediated, for example, by specialized modules associated with the protease (Figure 2), by the ubiquitin system (Section III), or simply by compartmentation.

The specificity of an enzyme may be most straightforwardly studied by kinetic investigation using various peptide substrates. This approach renders it possible to estimate the contribution to the reaction by the different parts of the substrate. For example, for mapping the primary substrate specificities of serine proteases, a series of tripeptide thiobenzyl esters were employed.[177] The general formula for the series is *t*-butyloxycarbonyl-Ala-Ala-Xaa-SBzl, where Xaa represents various amino acids. Thiobenzyl esters, used in a coupled reaction with Ellman's reagent

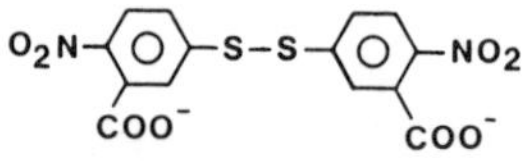

as chromogen, are sensitive substrates of chymotrypsin and subtilisin[178] and the trypsin-like enzymes.[179]

As a measure of specificity, k_{cat}/K_m is generally employed. This rate constant is equal to the proteolytic coefficient C used in earlier studies.[180] The k_{cat}/K_m has the advantage over other rate constants, e.g., k_{cat}, that it is not affected by nonproductive binding.[181] It is directly related to the free-energy difference ($\Delta G^{\neq}$) between the transition state ($ES^{\neq}$) and the enzyme and the free unbound substrate (E + S) as shown by Equations 5 and 6.

$$E + S \overset{\Delta G^{\neq}}{\rightleftharpoons} ES^{\neq} \tag{5}$$

$$k_{cat}/K_m = A \exp(-\Delta G^{\neq}/RT) \tag{6}$$

VI. SPECIFICITY OF PROTEASES IN PEPTIDE-BOND SYNTHESIS

The specificity of proteases is reflected not only in hydrolytic reactions but also the synthesis of the peptide bond. At first sight it may be surprising that proteases, which mediate hydrolytic reactions to completion, can also catalyze the reverse reaction to synthesize peptides in high yield. In fact, the unfavorable equilibrium for peptide-bond synthesis in aqueous solution is to a great extent a consequence of the energy required to convert the ionized forms of the reactants ($RCOO^-$ and RNH_3^+) to the reacting nonionized forms ($RCOOH$ and RNH_2).[182] Thus, the favorable equilibrium for the hydrolysis of the peptide bond is readily explicable in terms of the energy gained on dissociation of the split products. Hence, addition of organic solvents to the reaction mixture may substantially promote the synthesis both by changing the pK_a of the reactants, in particular by raising the pK_a of the carboxylic acid component and by lowering the concentration of water needed for the hydrolysis. Furthermore, the equilibrium may be shifted toward synthesis by precipitation of the synthesized peptide or by its specific transfer to some organic phase.

The use of proteases as catalysts for peptide-bond synthesis has recently been extensively reviewed[183-186] thanks to a revival of interest in this field. Although protease-catalyzed peptide synthesis to date does not have the versatility of the chemical methods, the enzyme-catalyzed synthesis of a particular peptide bond, such as in semisynthesis of enzymes, protein hormones, or other biologically important proteins, seems to be an invaluable technique. For example, trypsin specifically catalyzes both the hydrolysis and the resynthesis of the bond between Arg 63 and Ile 64 in soybean trypsin inhibitor.[93,187,188] Moreover, Arg 63 can be removed from the modified inhibitor by carboxypeptidase B, and in the presence of lysine, the Lys 63 analog of the inhibitor is synthesized. In an analogous manner, a tryptophyl residue can be substituted for Arg 63, which gives rise to an inhibitor that preferentially inhibits chymotrypsin.[188]

Another example, which is of practical significance, is the conversion of porcine insulin into human insulin. The former contains an alanine (Ala 30) instead of threonine as the C-terminal residue of the B chain. The conversion was performed by the removal of Ala 30 by carboxypeptidase A followed by the coupling of the new C-terminus, Lys 29, with an excess of tertiary butylester of threonine which was affected by the action of trypsin.[189] The protecting tertiary butyloxy group was removed with trifluoroacetic acid in the presence of anisole. Desalanine insulin may also be prepared from porcine insulin by means of *Achromobacter* protease, which is specific in its action for lysyl bonds.[190] In the presence of organic cosolvent, this enzyme also catalyzes the above coupling reaction with tertiary butylester of threonine.[191] Owing to its high specificity for lysine residues, *Achromobacter* protease does not cleave the Arg 22–Gly 23 bond in the B chain, which is a problem in the trypsin-catalyzed reaction. In fact, desoctapeptide insulin, which is devoid of the C-terminal residues 23 to 30, can readily be prepared by means of trypsin. The coupling of porcine desoctapeptide insulin and a synthetic octapeptide could also be affected by trypsin to provide human insulin[192] or its analogs mutated in the synthetic portion.

Proteases of all four mechanistic groups have been used for peptide synthesis.[184] The specificities of the enzymes found in the hydrolytic reactions are also reflected in the synthetic processes. Thus, in the chymotrypsin-catalyzed peptide-bond synthesis, the reactants contain the preferred amino acid residues tryptophan, tryrosine, phenylalanine, and leucine at the P_1 position.[193] It is noteworthy that the stereospecificity for the P_1' residue is much less stringent than that for the P_1 residue. This is indicated by the observation that D-Leu-NH_2 is a good reactant in the chymotrypsin-catalyzed peptide-bond synthesis.[193] Various substituted anilide groups and their amino acid derivatives may also interact with the S_1' and S_2' subsites of trypsin, respectively, owing to some hydrophobic character of these binding sites.[194] In contrast, the metalloprotease, thermolysin, which has a strong preference for a hydrophobic P_1' residue, is entirely specific for the L-amino acid residues at the P_1' position.[195]

The specificity of papain for hydrophobic residues at the P_2 position could be exploited in the syntheses of enkephalins.[196] The condensation reaction leading to the Met-enkephalin derivative is shown by Equation 7. The phenylhydrazide group was removed from the product by oxidation with $FeCl_3$. The use of the *O*-benzyl derivative was essential, possibly because of the greater solubility of a product lacking this group.

$$Boc-Tyr(Bzl)-Gly-OH \ + \ H-Gly-Phe-Met-NHNHPh \rightarrow$$

$$Boc-Tyr(Bzl)-Gly-Gly-Phe-Met-NHNHPh \qquad (7)$$

An interesting feature of the papain-catalyzed synthesis is the formation of oligopeptide esters from amino acid esters.[197,198] This may be exemplified by the condensation of H-Tyr-OMe to the product $H-(Tyr)_n-OMe$, where n is about 7. The observation of polycondensation raises the possibility of unwanted side reactions in the papain-catalyzed peptide synthesis. This complication, however does not appear to occur in the reactions with phenylhydrazide derivatives (Equation 7).

For the synthesis of hydrocabon soluble peptides, the use of reverse micelles has recently been proposed.[199] The water pool, which may be viewed as a microreactor, is stabilized by the surfactant sodium *bis*-(2-ethylhexyl) sulfosuccinate [1]. The enzymic reaction proceeds in the water pool with the continuous transfer of the product into the bulk hydrocarbon, e.g., isooctane. Another technique takes advantage of chymotrypsin modified with poly-ethyleneglycol. This derivative is capable of catalyzing peptide synthesis in benzene,[200] thereby obviating the problem of product hydrolysis associated with the usual procedures carried out in water or water-containing solvent.

COOR
|
CH_2
|
$CH-SO_3Na$ R = 2-ethylhexyl
|
COOR

[1]

Peptide-bond synthesis is exploited also by nature. It is of utmost interest that in the initial translational product of concanavalin A, the order of the N- and C-terminal half molecules is reversed. Proteolytic cleavage and subsequent resynthesis of the correct polypeptide chain are accomplished by some unknown protease(s) which are specific for Asn as the P_1 residue.[201]

VII. BINDING AND CATALYSIS

Two important phenomena are associated with enzymic catalysis: rate enhancement and specificity. It is evident that the rate enhancement is brought about to a great extent by the chemical catalysis (see Chapter 1) involving functional groups of the enzyme and the sensitive bond of the substrate. The specificity is made possible by the favorable binding energy between the enzyme and its specific substrates. It is not immediately apparent as to whether this binding energy contributes to specificity only or to the chemical catalysis as well. Indeed, in the early "lock and key" model which implied a rigid binding site, the specificity is entirely separated from the very catalytic process. However, strong evidence indicates that binding interaction between the nonreacting parts of the substrate and the enzyme is used to lower the activation energy of the reaction. Notably, as shown in Table 8, pepsin hydrolyzes the tetrapeptide Z-(Ala)$_2$-Phe-Phe-OP4P (OP4P = 3-(4-pyridyl) propyl-*1*-oxy)

Table 8
KINETIC PARAMETERS OF PEPSIN-CATALYZED REACTIONS[a]

Substrate	k_{cat} (sec^{-1})	K_m (mM)	Relative k_{cat}/K_m
Z-Phe-Phe-OP4P[b]	0.7	0.2	1
Z-(Ala)$_2$-Phe-Phe-OP4P[b]	282	0.04	2010

[a] pH 3.0 at 37°C.
[b] OP4P stands for 3-(4-pyridyl)propyl-1-oxy.

at a rate substantially higher than the hydrolysis rate of the dipeptide Z-Phe-Phe-OP4P.[202] The large catalytic efficiency (k_{cat}/K_m) is apparently associated with the secondary interactions by the Ala-Ala dipeptide unit in the P_2-P_3 positions. However, the relative binding affinity (K_m, which is probably the same as K_s in the present case), did not change appreciably, indicating that a portion of the binding energy has been utilized to decrease the activation energy of the catalytic process.

Utilization of binding energy in the catalysis has been reviewed in detail.[203,204] It is clear from the above example (Table 8) that the nonreacting part of a specific substrate, which provides additional binding energy compared to a nonspecific substrate, may result in little or no increase or even in decrease in the observed binding energy. This energy only represents what is left over after utilization for conformational change, strain, and any other mechanisms of destabilization of the reactant state, and what remained after the loss of entropy on the formation of the enzyme-substrate complex. Furthermore, the binding energy may also be utilized in the stabilization of the transition state.

An important way of utilization of binding energy is a compensation for the unfavorable entropy change that accompanies formation of the enzyme-substrate and transition-state complexes. The complete restriction of a medium-sized substrate results in a decrease in entropy of about -150 J K^{-1} mol^{-1}. This implies and unfavorable change of 10^{-8}-fold in rate or equilibrium at 25°C.[203,204] In contrast to a specific substrate, a nonspecific substrate may have insufficient binding energy to compensate for the entropy loss required for reaching the transition state. This results in a reduced value of k_{cat}/K_m.

A stronger binding of the substrate in the transition state than in the reactant state is essential to stabilize the activated form of the substrate in its passage to product. In effect, stabilization implies a lower energy for the transition state leading to a more efficient catalysis. On the other hand, stabilization of the enzyme-substrate or the enzyme-product complex would inhibit rather than promote catalysis.

This basic theory of enzyme catalysis was proposed many years ago by Pauling,[205] and it was discussed from different aspects in several reviews.[206-209] An important corollary of the transition-state theory is that a stable compound resembling the transition-state species rather than the substrate should exhibit an affinity for the enzyme much higher than the substrate or product does. Indeed, such stable substances, called transition-state analogs, proved to be unusally effective inhibitors. In addition to their potential use as antimetabolites, transition-state analogs can provide a useful indication of the mechanism on which their design was based. Inhibitors of this kind are of great mechanistic value since the transition state is not readily accessible for study by any physical or chemical method. Transition-state analog inhibitors, like peptide aldehydes for serine and cysteine proteases or pepstatin for aspartic proteases, will be discussed in the following chapters concerning the individual protease groups.

VIII. KINETICS AND PROTEASE MECHANISM

A. Basic Equations

Much useful information about protease mechanisms has been obtained from kinetic investigations. Generally, proteases obey the Michaelis-Menten equation (Equation 8) expressing the variation of rate (v) with substrate (S) concentration when $[S] >> [E_0]$ if $[E_0]$ represents the total enxyme concentration (free plus bound substrate).

$$v = \frac{V[S]}{[S] + K_m} \tag{8}$$

In Equation 8, V stands for the maximal rate that is obtained when the enzyme is saturated with substrate and K_m represents the Michaelis constant, a pseudo equilibrium constant, which is operationally defined as the substrate concentration at half-maximum rate. Equation 8 can be accounted for in terms of a two-step mechanism (Equation 9) involving an equilibrium and the decomposition of the enzyme-substrate complex (ES) to enzyme and product (P).

$$E + S \underset{k_{-1}}{\overset{k_1}{\rightleftharpoons}} ES \xrightarrow{k_{cat}} E + P \tag{9}$$

where k_{cat}, the overall rate constant is equal to $V/[E_0]$. The equilibrium assumption for Equation 9 can only be made if $k_{cat} << k_{-1}$. In this case K_m will approximate the dissociation constant for the enzyme-substrate complex (k_{-1}/k_1). For the usual case, the more general steady-state assumption should be adopted where the formation and decomposition of the ES complex are equal, i.e., $d[ES]/dt = 0$. In this case $K_m = (k_{-1} + k_{cat})/k_1$.

In the catalysis by serine and cysteine proteases, Equation 9 becomes somewhat more complicated because another intermediate, an acyl-enzyme (ES'), is also encountered on the reaction path (Equation 10).

$$E + S \underset{k_{-1}}{\overset{k_1}{\rightleftharpoons}} ES \underset{P_1}{\overset{k_2}{\rightleftharpoons}} ES' \underset{P_2}{\overset{k_3}{\rightleftharpoons}} E \tag{10}$$

whcre P_1 is the amine or alcohol moiety of the substrate, P_2 is the acid part, and k_2 and k_3 stand for the acylation and deacylation rate constants, respectively. The steady-state kinetic constants can be expressed in terms of the Michaelis-Menten parameters (k_{cat} and K_m) as shown by Equations 11 and 12.[210]

$$k_{cat} = \frac{V}{[E_0]} = \frac{k_2 k_3}{k_2 + k_3} \tag{11}$$

$$K_m = K_s \frac{k_3}{k_2 + k_3} \tag{12}$$

It should be noted that K_s of Equation 12 is not k_{-1}/k_1 but $(k_{-1} + k_2)/k_1$. When $k_2 >> k_3$, Equations 11 and 12 simplify to Equations 13 and 14, respectively.

$$k_{cat} = k_3 \tag{13}$$

$$K_m = \frac{k_3}{k_2} \tag{14}$$

When $k_3 \gg k_2$, the steady-state kinetic parameters are represented by Equations 15 and 16.

$$k_{cat} = k_2 \tag{15}$$

$$K_m = K_s \tag{16}$$

Dividing Equation 11 by Equation 12, we obtain Equation 17. The k_{cat}/K_m or k_2/K_s is the second-order acylation rate constant, also called the specificity constant.[211]

$$\frac{k_{cat}}{K_m} = \frac{k_2}{K_s} \tag{17}$$

The parameters $k_{cat} = V/[E_0]$ and K_m can be determined from a linear form of Equation 8. Several linear plots have been introduced, the most widely used being the Lineweaver-Burk plot (Equation 18).[212]

$$\frac{1}{v} = \frac{1}{V} + \frac{K_m}{V} \cdot \frac{1}{[S]} \tag{18}$$

A plot of $1/v$ vs. $1/[S]$ yields a straight line of slope K_m/V and an ordinate intercept $1/V$. The Eadie-Hofstee method involves plotting $v/[S]$ vs. v,[213] and the Hanes plot method uses plotting $[S]/v$ vs. $[S]$.[214] A more recent linear plot introduced by Eisenthal and Cornish-Bowden[215] is primarily designed for computer calculations. All these linear plotting methods require difficult statistical weighting processes.[216] The widespread adoption of computer methods now rely on the fitting of a rectangular hyperbola to the plot of v vs. $[S]$ without any linear transformation.[217,218]

B. Estimation of k_2, k_3, and K_s

Generally, the Michaelis parameters, k_{cat} and K_m, can be readily determined. Poor solubility of the substrate, of course, may pose a problem in some cases. It is more difficult, however, to estimate, the first-order acylation constant (k_2), the deacylation constant (k_3), and K_s; the latter is equal to $(k_{-1} + k_2)/k_1$. Under special conditions, when the values of k_2 and k_3 differ by more than an order of magnitude, some of these constants may be obtained from Equations 13, 15, and 16. When k_2 and k_3 are of the same order of magnitude, they can be determined from steady-state kinetics in the presence of an added nucleophile, such as methanol or 1,4-butanediol,[219-221] or more generally from pre-steady-state kinetics.[222]

The added nucleophile method is applicable to those proteases that form an acyl-enzyme intermediate according to Equation 10. The nucleophile competes with water and gives rise to an alternative product P_3 as shown by Equation 19.

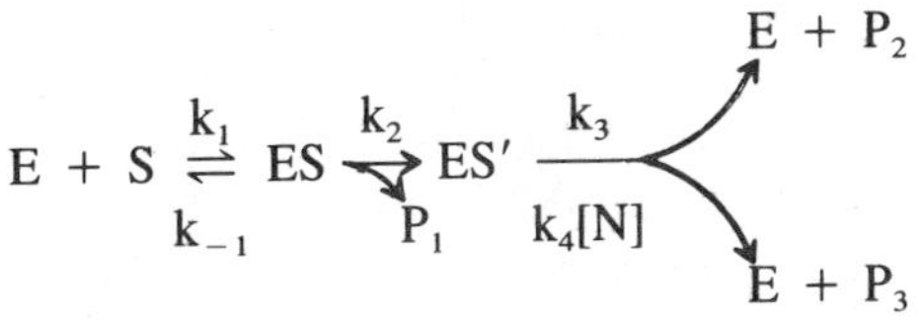

$$E + S \underset{k_{-1}}{\overset{k_1}{\rightleftharpoons}} ES \underset{P_1}{\overset{k_2}{\rightharpoondown}} ES' \overset{k_3}{\underset{k_4[N]}{<}} \begin{array}{c} E + P_2 \\ \\ E + P_3 \end{array} \tag{19}$$

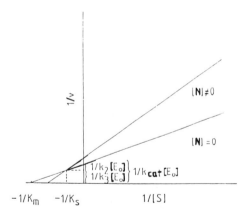

FIGURE 13.   Determination of k_2, k_3, and K_s from
a reciprocal plot obtained in the presence and absence
of a nucleophile (N).

where [N] is the concentration of the added alcohol as a nucleophile, k_4 is the rate constant
of alcoholysis, and k_3 is actually a composite constant, $k_3 = K_3'[H_2O]$. The hydrolysis of
an ester substrate may be followed by monitoring P_2, the acidic product in a pH state. The
other product, P_3, is an ester, which does not interfere with the determination of P_2. The
formation of P_2 in the double reciprocal plot is shown by Equation 20.

$$\frac{1}{v} = \frac{K_s(k_3 + k_4[N])}{k_2 k_3 [S][E_0]} + \frac{k_2 + k_3 + k_4[N]}{k_2 k_3 [E_0]} \tag{20}$$

In the absence of a nucleophile, Equation 20 simplifies to the Lineweaver-Burk equation
(Equation 18). However, when comparing Equation 20 with Equation 18, one should consider
that $v = k_{cat}/K_m = k_2/K_s$. Figure 13 shows that the lines calculated from measurements
carried out in the presence and in the absence of nucleophile intersect at $1/[S] = -1/K_s$.
From the ordinate intercepts we can obtain the individual rate constants. It is evident from
Figure 13 that in the interest of accuracy, the value of k_2/k_3 should be in the range of 0.1
to 10. Another restriction of using this method is that the nucleophile should not be bound
to any of the enzyme species as this will alter the mechanism of Equation 19.

A major advantage of the determination of k_2, k_3, and K_s under steady-state conditions
is that we do not need to resort to fast reaction techniques. The transient phase during which
the steady-state level becomes established is usually too fast to be monitored in steady-state
measurements. Yet, in the pre-steady-state portion of the reaction, the individual steps
characterized by k_1, k_{-1}, k_2, or k_3 may be directly observed.[222] This requires special techniques, such as stopped flow and temperature jump methods. The pre-steady state for a
simple reaction (Equation 9) is illustrated in Figure 14. For Equation 10 the pre-steady state
becomes more complicated because there are two intermediates on the reaction path.

C. The Use of pH-Rate Profiles

The pH-dependence studies can furnish useful mechanistic information as to the participation of ionizing groups in the catalysis. In fact, the involvement of a histidine residue in
the catalysis by chymotrypsin was indicated by such studies.[182] Of course, this information
is not always straightforward because the dissociation of enzymic groups may seriously be
affected by their environment. Nonetheless, the observation of an ionizing group in the
catalysis is always an important starting point for further studies.

The pK_a, which is characteristic of the ionizing group implicated in catalysis can be

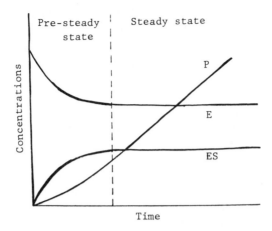

FIGURE 14.   A schematic representation of the pre-steady
state for Equation 9 when $[S_0] >> [E_0]$.

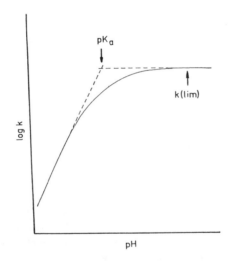

FIGURE 15.   Dependence of the rate constant,
k, on one basic ionizing group.

determined from measuring a certain rate constant, e.g., k_2, k_3, or k_{cat}/K_m, over a pH range
where the enzyme is stable. Equations 21 and 22 show the simple case, i.e., sigmoid pH-
rate constant profiles, when the ionizable group functions catalytically in the basic and acidic
form, respectively.

$$k = \frac{k(lim)}{1 + [H^+]/K_a} \tag{21}$$

$$k = \frac{k(lim)}{1 + K_a/[H^+]} \tag{22}$$

where k(lim) stands for the pH-independent, maximum rate constant, and K_a is the apparent
ionization constant of the group involved in the reaction (Figure 15). For instance, deacylation
of chymotrypsin obeys Equation 21 showing the participation of a base of $pK_a = 7$. On the

other hand, acylation of chymotrypsin depends on the ionization of two groups, one acid and one base. This gives rise to a bell-shaped pH dependence as described by Equation 23.

$$k = \frac{k(\lim)}{1 + [H^+]/K_1 + K_2/[H^+]} \tag{23}$$

where K_1 and K_2 represent the ionization constant of the base and acid, respectively. In the acylation of chymotrypsin, pK_1 is about 7 and pK_2 is 8.5 to 9.0 (see Chapter 3).

A meaningful conclusion can only be drawn from pH-dependence studies involving well-defined rate constants, e.g., k_{cat}/K_m, k_2, or k_3. It is of great mechanistic value that the pH dependencies of k_{cat}/K_m, k_2, and k_3 reflect the pK_a of the ionizing group important in the free enzyme, enzyme-substrate complex, and acyl-enzyme, respectively.[223] Obviously, the substrate should not dissociate over the pH range studied. Furthermore, care is required when using this approach because the pH dependencies of the rate constants may be adversely affected by other factors as well.[223-225] For example, with most specific substrates the diffusion-controlled binding and the chemical reaction may proceed at commensurable rates, one being rate determining at low pH and the other at high pH with a consequent distortion of the ionization curve.

D. Active Site Titrations

Determination of rate constants requires knowledge of the enzyme concentration in the reaction mixture. In the case of apparently pure enzymes such as crystallized enzymes, the total protein concentration is often used instead of the active enzyme concentration. The two quantities, however, are not necessarily the same. The enzyme may contain inert protein, and proteases frequently have substantial amounts of degradation products as a consequence of autodigestion. Commercial crystalline papain, for example, usually contains less than 60% active and activatable (Chapter 4, Section I.A) enzyme.

Determination of the operational molarity of proteases may be achieved by active site titration.[226,227] This implies a stoichiometric reaction between the enzyme and the titrant. The conversion of enzyme, E, according to Equation 10 into acyl-enzyme, ES′, and product one, P_1, can be stoichiometric if the decomposition of the acyl-enzyme is much slower than its formation. In effect, a pre-steady-state acylation and a steady-state deacylation may be observed, the former often being too fast to measure. In most cases, liberation of the P_1 product is monitored spectrophotometrically or spectrofluorimetrically. In some other cases, e.g., with *N-trans*-cinnamoylimidazole as titrant (Figure 16), the decrease in the substrate concentration is measured. Extrapolation to the zero time from the steady-state reaction gives rise to the absorbence change that corresponds to the amount of substrate consumed or the product liberated in the stoichiometric reaction. This is equivalent to the concentration of the active enzyme if $[S_0] \gg K_m$. If the condition $[S_0] \gg K_m$ does not hold, a series of experiments at constants $[E_0]$ but varying $[S_0]$ enables one to circumvent this problem.[226,227]

A titrant for a particular protease has to be sufficiently similar to a specific substrate to ensure a rapid acylation of the enzyme, but its structure should be altered in such a way that deacylation is substantially reduced. Several substrates of this kind have been designed for serine proteases.[226,227,229,230] For example, a useful titrant for chymotrypsin is *p*-nitrophenyl acetyl-azaphenyl-alaninate[231] [2]. In this phenylalanine analog the Cα-H group has been replaced by a nitrogen atom. More specific aza-peptide *p*-nitrophenyl esters have recently been proposed for titration of chymotrypsin-like enzymes[232] and elastase.[233] For trypsin titration, *N*-methyl-lysine analogs[234] [3] and *p*-guanidinobenzoate derivatives.[229,235,236] can be employed. The fluorimetric reagents, like 4-methylumbelliferyl-*p*-guanidinobenzoate[235] [4] and fluorescein mono-*p*-guanidinobenzoate[236] [5], are the most sensitive titrants.

73

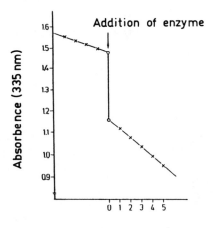

FIGURE 16. Titration of subtilisin with *N-trans*-cinnamoyl-imidazole. (From Polgár, L., *Acta Biochim. Biophys. Acad. Sci. Hung.*, 3, 397, 1968. With permission.)

[2]

[3]

[4]

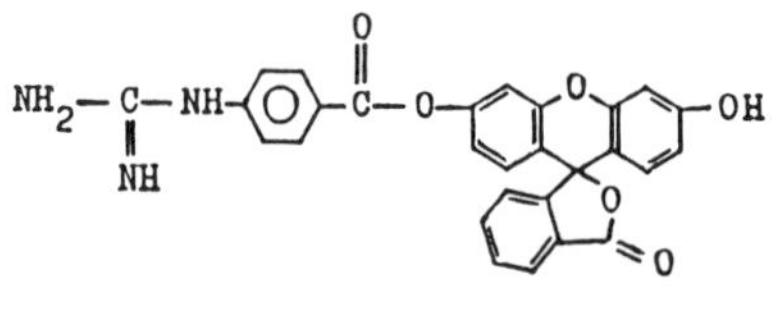

[5]

Interestingly, in the case of cysteine proteases, substrates or substrate analogs are rarely used as titrants.[237] These enzymes, however, by virtue of their highly reactive cysteine residue can readily react with chromogen disulfide compounds such as 5,5'-dithiobis(2-nitrobenzoate)[238] (Nbs_2) or 2,2'-dipyridyl disulfide[239] (PDS). The former is widely employed for the determination of thiol compounds in the pH range of 7 to 8 and may be used also for the active site titration of pure cysteine enzymes containing only one thiol group, i.e., the active site thiol group. PDS at low pH (pH 4) is specific for the active site thiol of most cysteine proteases of the papain family. The reaction is not affected by other enzymic thiols or low molecular mass thiol compounds present at not too high concentrations. The liberation of 2-thiopyridone on titration (Equation 24) can be monitored spectrophotometrically at 343 nm.

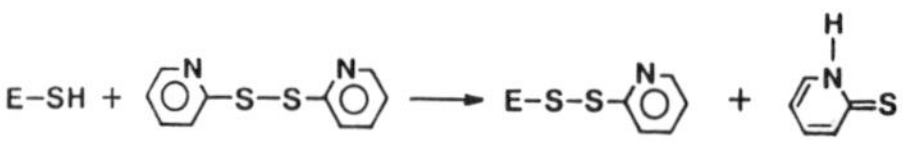

$$(24)$$

Epoxide inhibitors represent a different kind of titrant for cysteine proteases. The first epoxide inhibitor, L-*trans*-epoxysuccinyl-leucylamido(4-guanidino)butane (E-64), was isolated from a culture extract of *Aspergillus japonicus* TPR-64.[240] Its structure [6] is composed of three molecular species: L-*trans*-epoxysuccinyl acid, L-leucine, and agmatine.[241] E-64 and its several synthetic analogs, such as Ep-475 [L-*trans*-epoxysuccinyl-leucylamido(3-methyl)butane], react specifically and irreversibly with the active site thiol group of most cysteine proteases of the papain family.[242] The decrease in enzymic activity is measured on each addition of the inhibitor. Although this kind of titration is not a continuous assay, it requires usually less enzyme than the conventional spectrophotometric titration procedures.

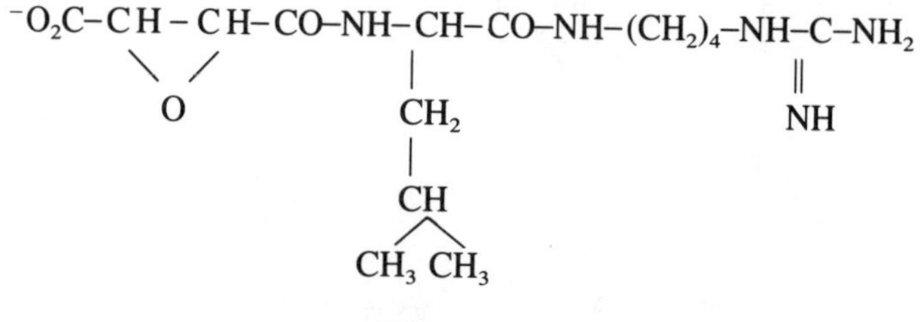

[6]

Finally, it may be noted that rate assays are often used for the determination of enzyme concentration. This method, in contrast to active site titrations, cannot give the absolute enzyme concentration without a known standard. However, the relationship between titration and rate assay may be established and then the accurate enzyme concentration can be estimated by using the rate assay under strictly controlled conditions (temperature, pH, ionic strength). The rate assay is of particular significance when the amount of enzyme available is limited.

IX. STRUCTURAL BASES OF PROTEASE MECHANISM

Although kinetic investigations can provide valuable mechanistic information, the kinetic mechanism does not bear any stereochemical feature needed for the understanding of enzymic catalysis. Therefore, the first X-ray crystallographic studies providing a picture of the three-dimensional structure of chymotrypsin[243] proved to be a breakthrough in our knowledge of protease mechanisms. The results confirmed much previous data pertinent to the catalytic groups, eliminated suggestions not compatible with the active site structure, and revealed new catalytically important groups whose presence could not be suspected from any other studies. Furthermore, X-ray crystallography explored the mode of substrate binding and the possible structures of the catalytic intermediates. Subsequently, the structures of a number of proteases have been solved, some of them at very high resolution (see the following chapters). Among the known structures, members of all four mechanistic groups of proteases are represented. Although knowledge of the structure is essential for understanding the mechanism; in itself, it is insufficient. Carboxypeptidase A may serve as an example. It was one of the first protease structures to be solved,[244] yet elucidation of one of its most important mechanistic features, i.e., whether or not a covalent intermediate is on the reaction path, has awaited until most recently (Chapter 6, Section V).

One of the major problems associated with the X-ray diffraction method is that taking a picture of a productive catalytic intermediate is extremely difficult. The most common approach toward achieving this result has been the application of the difference Fourier technique to enzyme complexes formed with substrate analogs, products, inhibitors, and very poor substrates that can be considered as inhibitors rather than substrates. One problem associated with this approach is that the geometry obtained for poor substrates and inhibitors is not expected to be entirely relevant to that of the natural substrates. Furthermore, the difference Fourier technique, which implies the diffusion of a ligand into the crystal, may not reveal the conformational changes in the enzyme structure that would otherwise occur in solution if conformational changes were not prevented by possible stabilizing effects of the crystal lattice. Evaluation of the difference map is complicated by the water and solvent molecules bound to the protein. Where these molecules are displaced by the ligand, in that region of the map the difference in electron density reflects the difference between the ligand and the solvent displaced and does not show the ligand clearly. This problem, however, can be obviated by refinement of the enzyme-ligand complex independently of the free enzyme.

The question about the relationship between protein structures in crystals and in solution has often been raised. Indeed, X-ray crystallography does not give direct information about the solution form of the enzyme which may be different from the crystalline form. This is conceivable since a crystal structure, different from the solution structure, can be stabilized by the crystal lattice. The difference may be significant at the protein surface where the side chains of the neighboring protein molecules interact. That the structures in the two phases are not likely to be too much different follows from the appreciable catalytic activity observed with some proteases.[245,246] Many of the problems and perspectives of mechanistic X-ray crystallography, with particular emphasis on sub-zero temperature studies, have been discussed in detail.[247] Also, the determination of flexible parts of the protein structure, as reflected by disordered regions (absence of strong electron-density) on the map, has been reviewed.[248] Such flexibility may have functional significance, e.g., in zymogen activation[249] or ligand binding.[250]

Progress in X-ray crystallography has been enormous and appears to continue. For example, if polychromatic X-rays derived from synchrotron radiation are used, they generate a Laue diffraction pattern from single crystals that can be obtained in <1 sec.[251] Increases in X-ray intensity could reduce exposure time to about 10 msec. This time-resolved X-ray crystallography may provide pictures about catalytic intermediates. However, transient intermediates cannot be crystallized, nor is the diffusion into the interior of the crystal fast enough for data collection. This problem hopefully will be somehow overcome, for instance, by generating the substrate *in situ*. In favorable cases, this can be achieved by photoactivation of a stable structure in the crystal. Laser flash photolysis of blocked substrates, such as "caged" ATP [252] or cAMP,[253] have already been used with various biological systems. Even amino acid derivatives are known which contain photosensitive protecting groups.[254] Thus, to pursue this line appears to be promising.

Neutron scattering studies can expand the structural information obtained from X-ray diffraction measurements. Through neutron diffraction analysis of single protein crystals, it is possible to determine the positions of hydrogen atoms that are involved in the catalytic function and cannot be detected by X-ray crystallography. Indeed, neutron scattering proved to be a valuable technique in the mechanistic studies of serine proteases.[255]

As it is clear from the above discussion, X-ray diffraction results should not be considered in isolation, but rather in conjunction with other techniques. NMR studies are of particular importance in this respect since they can provide information about the solution structure of enzymes and render it possible to investigate certain mechanistically competent individual groups.[256-258] The NMR studies substantially contributed, for example, to understanding proton-transfer reactions in serine protease catalysis[259,260] and to ruling out an assumed covalent tetrahedral intermediate between trypsin and pancreatic trypsin inhibitor [261,262] which was suggested to occur on the basis of X-ray diffraction measurements.[263]

The latest exceedingly promising approach to establishing the relationship between structure and function of proteases is site-directed mutagenesis.[264,266] The first site-specific mutagenesis was accomplished 20 years ago.[267] This involved the chemical conversion of the active site serine of subtilisin into a cysteine residue, which showed the importance of the precise geometry at the active site of a serine protease (Chapter 4, Section V.B). By means of the recent general technique of site-directed mutagenesis, any specific amino acid residue at or near the active site can be replaced by another residue, and its possible involvement in catalysis can be examined. Important results concerning the mechanisms of trypsin and carboxypeptidase A have already been obtained from site-directed mutagenesis.[264-266]

REFERENCES

1. Nomenclature Committee of the International Union of Biochemistry, *Enzyme Nomenclature,* Academic Press, New York, 1984, 330.
2. **Barrett, A. J. and McDonald, J. K.,** Nomenclature: a possible solution to the "peptidase anomaly", *Biochem. J.,* 231, 807, 1985.
3. **De Haën, C., Neurath, H., and Teller, D. C.,** The phylogeny of trypsin-related serine proteases and their zymogens. New methods for the investigation of distant evolutionary relationships, *J. Mol. Biol.,* 92, 225, 1975.
4. **Neurath, H.,** Evolution of proteolytic enzymes, *Science,* 224, 350, 1984.
5. **Dayhoff, M. O., Barker, W. C., and Hunt, L. T.,** Establishing homologies in protein sequences, *Methods Enzymol.,* 91, 524, 1983.
6. **Doolittle, R. F.,** Similar amino acid sequences: change or common ancestry?, *Science,* 214, 149, 1981.
7. **Gorbalenya, A. E., Blinov, V. M., and Donchenko, A. P.,** Poliovirus-encoded proteinase 3C: a possible evolutionary link between cellular serine and cysteine proteinase families, *FEBS Lett.,* 194, 253, 1986.
8. **Doolittle, R. F.,** Protein evolution, in *The Proteins,* Vol. 4, 3rd ed., Neurath, H. and Hill, R. L., Eds., Academic Press, New York, 1984, chap. 6.
9. **Woodbury, R. G. and Neurath, H.,** Structure, specificity and localization of the serine proteases of connective tissue, *FEBS Lett.,* 114, 189, 1980.

10. **Titani, K., Sasagawa, T., Woodbury, R. G., Ericsson, L. H., Dörsam, H., Kraemer, M., Neurath, H., and Zwilling, R.,** Amino acid sequence of crayfish *(Astacus fluviatilis)* trypsin I$_f$, *Biochemistry,* 22, 1459, 1983.

11. **James, M. N. G.,** An X-ray crystallographic approach to enzyme structure and function, *Can. J. Biochem.,* 58, 251, 1980.

12. **Takio, K., Towatari, T., Katunuma, N., Teller, D. C., and Titani, K.,** Homology of amino acid sequences of rat liver cathepsins B and H with that of papain, *Proc. Natl. Acad. Sci. U.S.A.,* 80, 3666, 1983.

13. **Bajaj, M. and Blundell, T.,** Evolution and the tertiary structure of proteins, *Annu. Rev. Biophys. Bioeng.,* 13, 453, 1984.

14. **MacDonald, R. J., Swift, G. H., Quinto, C., Swain, W., Pictet, R. L., Nikovits, W., and Rutter, W. J.,** Primary structure of two distinct rat pancreatic preproelastases determined by sequence analysis of the complete cloned messenger ribonucleic acid sequences, *Biochemistry,* 21, 1453, 1982.

15. **Mason, A. J., Evans, B. A., Cox, D. R., Shine, J., and Richards, R. I.,** Structure of mouse kallikrein gene family suggests a role in specific processing of biologically active peptides, *Nature (London),* 303, 300, 1983.

16. **Ashley, P. L. and MacDonald, R. J.,** Kallikrein-related mRNAs of the rat submaxillary gland: nucleotide sequences of four distinct types including tonin, *Biochemistry,* 24, 4512, 1985.

17. **Degen, S. J. F., MacGillivray, R. T. A., and Davie, E. W.,** Characterization of the complementary deoxyribonucleic acid and gene coding for human prothrombin, *Biochemistry,* 22, 2087, 1983.

18. **MacGillivray, R. T. A. and Davie, E. W.,** Characterization of bovine prothrombin mRNA and its transition product, *Biochemistry,* 23, 1626, 1984.

19. **Fujikawa, K., Chung, D. W., Hendrickson, L. E., and Davie, E. W.,** Amino acid sequence of human factor XI, a blood coagulation factor with four tandem repeats that are highly homologous with plasma prekallikrein, *Biochemistry,* 25, 2417, 1986.

20. **Kurachi, K. and Davie, E. W.,** Isolation and characterization of a cDNA coding for human factor IX, *Proc. Natl. Acad. Sci. U.S.A.,* 79, 6461, 1982.

21. **Fung, M. R., Hay, C. W., and MacGillivray, R. T. A.,** Characterization of an almost full-length cDNA coding for human blood coagulation factor X, *Proc. Natl. Acad. Sci. U.S.A.,* 82, 3591, 1985.

22. **Que, B. G. and Davie, E. W.,** Characterization of a cDNA coding for human factor XII (Hageman factor), *Biochemistry,* 25, 1525, 1986.

23. **Park, C. H. and Tulinsky, A.,** Three-dimensional structure of the kringle sequence: structure of prothrombin fragment 1, *Biochemistry,* 25, 3977, 1986.

24. **Patthy, L.,** Evolution of proteases of blood coagulation and fibrinolysis by assembly from modules, *Cell,* 41, 657, 1985.

25. **Neurath, H. and Walsh, K. A.,** Role of proteolytic enzymes in biological regulation (a review), *Proc. Natl. Acad. Sci. U.S.A.,* 73, 3825, 1976.

26. **Light, A. and Fonseca, P.,** The preparation and properties of the catalytic subunit of bovine enterokinase, *J. Biol. Chem.,* 259, 13195, 1984.

27. **Maroux, S., Baratti, J., and Desnuelle, P.,** Purification and specificity of porcine enterokinase, *J. Biol. Chem.,* 246, 5031, 1971.

28. **Davie, E. W. and Neurath, H.,** Identification of a peptide released during autocatalytic activation of trypsinogen, *J. Biol. Chem.,* 212, 515, 1955.

29. **Brodrick, J. W., Largman, C., Hsiang, M. W., Johnson, J. H., and Geokas, M. C.,** Structural basis for the specific activation of human cationic trypsinogen by human enteropeptidase, *J. Biol. Chem.,* 253, 2737, 1978.

30. **San Segundo, B., Martinez, M. C., Vilanova, M., Cuchillo, C. M., and Aviles, F. X.,** The severed activation segment of porcine pancreatic procarboxypeptidase A is a powerful inhibitor of the active enzyme. Isolation and characterisation of the activation peptide, *Biochim. Biophys. Acta,* 707, 74, 1982.

31. **Avilés, F. X., San Segundo, B., Vilanova, M., Cuchillo, C. M., and Turner, C.,** The activation segment of procarboxypeptidase A from porcine pancreas constitutes a folded structural domain, *FEBS Lett.,* 149, 257, 1982.

32. **Desnuelle, P.,** Chymotrypsin, in *The Enzymes,* Vol. 4, 2nd ed., Boyer, P. D., Lardy, H., and Myrbäck, K., Eds., Academic Press, New York, 1960, 93.

33. **Hess, G. P.,** Chymotrypsin — chemical properties and catalysis, in *The Enzymes,* Vol. 3, 3rd ed., Boyer, P. D., Ed., Academic Press, New York, 1971, 213.

34. **Kraut, J.,** Chymotrypsinogen: X-ray structure, in *The Enzymes,* Vol. 3, 3rd ed., Boyer, P. D., Ed., Academic Press, New York, 1971, 165.

35. **Corey, R. B., Battfay, O., Brueckner, D. A., and Mark, F. G.,** Preliminary X-ray diffraction studies of crystal forms of free and inhibited chymotrypsin, *Biochim. Biophys. Acta,* 94, 535, 1965.

36. **Davie, E. W. and Fujikawa, K.,** Basic mechanisms in blood coagulation, *Annu. Rev. Biochem.,* 44, 799, 1975.

37. **Davie, E. W., Fujikawa, K., Kurachi, K., and Kisiel, W.,** The role of serine proteases in the blood coagulation cascade, *Adv. Enzymol. Relat. Areas Mol. Biol.,* 48, 277, 1979.
38. **Jackson, C. M. and Nemerson, Y.,** Blood coagulation, *Annu. Rev. Biochem.,* 49, 765, 1980.
38a. **Mann, K. G.,** The assembly of blood clotting complexes on membranes, *Trends Biochem. Sci.,* 12, 229, 1987.
39. **Kisiel, W., Canfield, W. M., Ericsson, L. H., and Davie, E. W.,** Anticoagulant properties of bovine plasma protein C following activation by thrombin, *Biochemistry,* 16, 5824, 1977.
40. **Vehar, G. A. and Davie, E. W.,** Preparation and properties of bovine factor VIII (antihemophilic factor), *Biochemistry,* 19, 401, 1980.
41. **Reid, K. B. M. and Porter, R. R.,** The proteolytic activation systems of complement, *Annu. Rev. Biochem.,* 50, 433, 1981.
42. **Reid, K. B. M.,** Proteins involved in the activation and control of the two pathways of human complement, *Biochem. Soc. Trans.,* 11, 1, 1983.
43. **Schumaker, V. N., Zavodszky, P., and Poon, P. H.,** Activation of the first component of complement, *Annu. Rev. Immunol.,* 5, 21, 1987.
44. **Hanson, D. C., Siegel, R. C., and Schumaker, V. N.,** Segmental flexibility of the C1q subcomponent of human complement and its possible role in the immune response, *J. Biol. Chem.,* 260, 3576, 1985.
45. **Kilchherr, E., Hofmann, H., Steigemann, W., and Engel, J.,** Structural model of the collagen-like region of C1q comprising the kink region and the fibre-like packing of the six triple helices, *J. Mol. Biol.,* 186, 403, 1985.
46. **Sim, R. B.,** The human complement system serine proteases C$\bar{1}$r and C$\bar{1}$s and their proenzymes, *Methods Enzymol.,* 80, 26, 1981.
46a. **Leytus, S. P., Kurachi, K., Sakariassen, K. S., and Davie, E. W.,** Nucleotide sequence of the cDNA coding for human complement C1r, *Biochemistry,* 25, 4855, 1986.
46b. **Mackinnon, C. M., Carter, P. E., Smyth, S. J., Dunbar, B., and Fothergill, J. E.,** Molecular cloning of cDNA for human complement component C1s, *Eur. J. Biochem.,* 169, 547, 1987.
46c. **Goldberger, G., Bruns, G. A. P., Rits, M., Edge, M. D., and Kwiatlowski, D. J.,** Human complement factor I: analysis of cDNA-derived primary structure and assignment of its gene to chromosome 4, *J. Biol. Chem.,* 262, 10065, 1987.
47. **Law, S. K., Lichtenberg, N. A., and Levine, R. P.,** Evidence for an ester linkage between the labile binding site of C3b and receptive surfaces, *J. Immunol.,* 123, 1388, 1979.
48. **Gadd, K. J. and Reid, K. B. M.,** The binding of complement component C3 to antibody-antigen aggregates after activation of the alternative pathway in human serum, *Biochem. J.,* 195, 471, 1981.
49. **Howard, J. B.,** Methylamine reaction and denaturation-dependent fragmentation of complement component 3: comparison with α_2-macroglobulin, *J. Biol. Chem.,* 255, 7082, 1980.
50. **Campbell, R. D., Gagnon, J., and Porter, R. R.,** Amino acid sequence around the thiol and reactive acyl groups of human complement component C4, *Biochem. J.,* 199, 359, 1981.
51. **Pangburn, M. K. and Müller-Eberhard, H. J.,** Relation of a putative thiolester bond in C3 to activation of the alternative pathway and the binding of C3b to biological targets of complement, *J. Exp. Med.,* 152, 1102, 1980.
52. **Thomas, M. L., Janatova, J., Gray, W. R., and Tack, B. F.,** Third component of human complement: localization of the internal thiolester bond, *Proc. Natl. Acad. Sci. U.S.A.,* 79, 1054, 1982.
53. **Swenson, R. P. and Howard, J. B.,** Amino acid sequence of the tryptic peptide containing the alkylamine-reactive site from human α_2-macroglobulin. Identification of γ-glutamylmethylamide, *J. Biol. Chem.,* 255, 8087, 1980.
54. **Sottrup-Jensen, L., Hansen, H. F., Mortensen, S. B., Petersen, T. E., and Magnusson, S.,** Sequence location of the reactive thiol ester in human α_2-macroglobulin, *FEBS Lett.,* 123, 145, 1981.
55. **Sottrup-Jensen, L., Stepanik, T. M., Kristensen, T., Lønbald, P. B., Jones, C. M., Wierzbicki, D. M., Magnusson, S., Domdey, H., Wetsel, R. A., Lundwall, Å., Tack, B. F., and Fey, G. H.,** Common evolutionary origin of α_2-macroglobulin and complement components of C3 and C4, *Proc. Natl. Acad. Sci. U.S.A.,* 82, 9, 1985.
56. **Ondetti, M. A. and Cushman, D. W.,** Enzymes of the renin-angiotensin system and their inhibitors, *Annu. Rev. Biochem.,* 51, 283, 1982.
57. **Schachter, M.,** Kallikreins (kininogenases) — a group of serine proteases with bioregulatory actions, *Pharmacol. Rev.,* 31, 1, 1980.
58. **Kato, H., Nagasawa, S., and Iwanaga, S.,** HMW and LMW kininogens, *Methods Enzymol.,* 80, 172, 1981.
59. **Blobel, G. and Dobberstein, B.,** Transfer of proteins across membranes, *J. Cell Biol.,* 67, 835, 1975.
60. **Kreil, G.,** Transfer of proteins across membranes, *Annu. Rev. Biochem.,* 50, 317, 1981.
61. **Docherty, K. and Steiner, D. F.,** Post-translational proteolysis in polypeptide hormone biosynthesis, *Annu. Rev. Physiol.,* 44, 625, 1982.

62. **Lively, M. O. and Walsh, K. A.,** Hen oviduct signal peptidase is an integral membrane protein, *J. Biol. Chem.,* 258, 9488, 1983.
63. **Evans, E. A., Gilmore, R., and Blobel, G.,** Purification of microsomal signal peptidase as a complex, *Proc. Natl. Acad. Sci. U.S.A.,* 83, 581, 1986.
64. **Herbert, E. and Uhler, M.,** Biosynthesis of polyprotein precursors to regulatory proteins, *Cell,* 30, 1, 1982.
65. **Kitabchi, A. E., Duckworth, W. C., Stentz, F. B., and Yu, S.,** Properties of proinsulin and related polypeptides, *Crit. Rev. Biochem.,* 1, 59, 1972.
66. **Docherty, K., Caroll, R., and Steiner, D. F.,** Identification of a 31,500 molecular weight islet cell protease as cathepsin B, *Proc. Natl. Acad. Sci. U.S.A.,* 80, 3245, 1983.
67. **Docherty, K., Hutton, J. C., and Steiner, D. F.,** Cathepsin B related protease in insulin secretory granule, *J. Biol. Chem.,* 259, 6041, 1984.
68. **Amara, S. G., David, D. N., Rosenfeld, M. G., Ross, B. A., and Evans, R. M.,** Characterization of rat calcitonin mRNA, *Proc. Natl. Acad. Sci. U.S.A.,* 77, 4444, 1980.
69. **Jacobs, J. W., Goodman, R. H., Chin, W. W., Dee, P. C., Bell, N. H., Potts, J. T., Jr., and Habener, J. F.,** Calcitonin messenger RNA encodes multiple polypeptides in a single precursor, *Science,* 213, 457, 1981.
70. **Hobart, P., Crawford, R., Shen, L. P., Pictet, R., and Rutter, W. J.,** Cloning and sequence analysis of cDNAs encoding two distinct somatostatin precursors found in the endocrine pancreas of anglerfish, *Nature (London),* 288, 137, 1980.
71. **Goodman, R. H., Jacobs, J. W., Chin, W. W., Lund, P. K., Dee, P. C., and Habener, J. F.,** Nucleotide sequence and cloned structural gene coding for a precursor of pancreatic somatostatin, *Proc. Natl. Acad. Sci. U.S.A.,* 77, 5869, 1980.
72. **Lund, P. K., Goodman, R. H., Montminy, M. R., Dee, P. C., and Habener, J. F.,** Anglerfish islet pre-proglucagon II. Nucleotide and corresponding amino acid sequence of the cDNA, *J. Biol. Chem.,* 258, 3280, 1983.
73. **Nakanishi, S., Inoue, A., Kita, T., Nakamura, M., Chang, A. C. Y., Cohen, S. N., and Numa, S.,** Nucleotide sequence of cloned cDNA for bovine corticotropin-β-lipotropin precursor, *Nature (London),* 278, 423, 1979.
74. **Drouin, J. and Goodman, H. M.,** Most of the coding region of rat ACTH β-LPH precursor gene lacks intervening sequences, *Nature (London),* 288, 610, 1980.
75. **Stern, A. S., Jones, B. N., Shively, J. E., Stein, S., and Udenfriend, S.,** Two adrenal opioid polypeptides: proposed intermediates in the processing of proenkephalin, *Proc. Natl. Acad. Sci. U.S.A.,* 78, 1962, 1981.
76. **Comb, M., Seeburg, P. H., Adelman, J., Eiden, L., and Herbert, E.,** Primary structure of human Met- and Leu-enkephalin precursor and its mRNA, *Nature (London),* 295, 663, 1982.
77. **Korant, B. D., Lonberg-Holm, K., and LaColla, P.,** Picornaviruses and togaviruses: targets for design of antivirals, in *Targets for the Design of Antiviral Agents,* de Clercq, E. and Walker, R. T., Eds., Plenum Press, New York, 1984, 61.
78. **Hanecak, R., Semler, B. L., Ariga, H., Anderson, C. W., and Wimmer, E.,** Expression of a cloned gene segment of poliovirus in *E. coli:* evidence for auotcatalytic production of the viral proteinase, *Cell,* 37, 1063, 1984.
79. **Ivanoff, L. A., Towatari, T., Ray, J., Korant, B. D., and Petteway, S. R., Jr.,** Expression and site-specific mutagenesis of the poliovirus 3C protease in *Escherichia coli, Proc. Natl. Acad. Sci. U.S.A.,* 83, 5392, 1986.
80. **Hershko, A. and Ciechanover, A.,** Mechanisms of intracellular protein breakdown, *Annu. Rev. Biochem.,* 51, 335, 1982.
81. **Mayer, R. J. and Doherty, F.,** Intracellular protein catabolism: state of the art, *FEBS Lett.,* 198, 181, 1986.
82. **Vijay-Kumar, S., Bugg, C. E., Wilkinson, K. D., and Cook, W. J.,** Three-dimensional structure of ubiquitin at 2.8 Å resolution, *Proc. Natl. Acad. Sci. U.S.A.,* 82, 3582, 1985.
82a. **Vijay-Kumar, S., Bugg, C. E., Wilkinson, K. D. Viersta, R. D., Hatfield, P. M., and Cook, W. J-.,** Comparison of the three-dimensional structures of human, yeast, and oat ubiquitin, *J. Biol. Chem.,* 262, 6396, 1987.
83. **Pickart, C. M. and Rose, I. A.,** Ubiquitin carboxyl-terminal hydrolase acts on ubiquitin carboxyl-terminal amides, *J. Biol. Chem.,* 260, 7903, 1985.
84. **Özkaynak, E., Finley, D., and Vashavsky, A.,** The yeast ubiquitin gene: head-to-tail repeats encoding a polyubiquitin precursor protein, *Nature (London),* 312, 663, 1984.
85. **Haas, A. L., Murphy, K. E., and Bright, P. M.,** The inactivation of ubiquitin accounts for the inability to demonstrate ATP, ubiquitin-dependent proteolysis in liver extracts, *J. Biol. Chem.,* 260, 4694, 1985.
86. **Vierstra, R. D., Langan, S. M., and Haas, A. L.,** Purification and initial characterization of ubiquitin from the higher plant, *Avena sativa, J. Biol. Chem.,* 260, 12015, 1985.

87. **Waxman, L. and Goldberg, A. L.**, Protease La from *Escherichia coli* hydrolyses ATP and proteins in a linked fashion, *Proc. Natl. Acad. Sci. U.S.A.*, 79, 4883, 1982.

88. **Waxman, L. and Goldberg, A. L.**, Protease La, the *lon* gene product, cleaves specific fluorogenic peptides in an ATP-dependent reaction, *J. Biol. Chem.*, 260, 12022, 1985.

89. **Goldberg, A. L. and Waxman, L.**, The role of ATP hydrolysis in the breakdown of proteins and peptides by protease La from *Escherichia coli*, *J. Biol. Chem.*, 260, 12029, 1985.

90. **Travis, J. and Salvesen, G. S.**, Human plasma proteinase inhibitors, *Annu. Rev. Biochem.*, 52, 655, 1983.

91. **Carrell, R. and Travis, J.**, α_1-Antitrypsin and the serpins: variation and countervariation, *Trends Biochem. Sci.*, 10, 20, 1985.

92. **Kidd, V. J., Wallace, R. B., Itakura, K., and Woo, S. L. C.**, α_1-Antitrypsin deficiency detection by direct analysis of the mutation in the gene, *Nature (London)*, 304, 230, 1983.

93. **Laskowski, M., Jr. and Kato, I.**, Protein inhibitors of proteinases, *Annu. Rev. Biochem.*, 49, 593, 1980.

94. **Finkenstadt, W. R. and Laskowski, M., Jr.**, Resynthesis by trypsin of the cleaved peptide bond in modified soybean trypsin inhibitor, *J. Biol. Chem.*, 242, 771, 1967.

95. **Loebermann, H., Tokuoka, R., Deisenhofer, J., and Huber, R.**, Human α_1- proteinase inhibitor. Crystal structure analysis of two crystal modifications, molecular model and preliminary analysis of the implications for function, *J. Mol. Biol.*, 177, 531, 1984.

96. **Moroi, M. and Yamasaki, M.**, Mechanism of interaction of bovine trypsin with human α_1-antitrypsin, *Biochim. Biophys. Acta*, 359, 130, 1974.

97. **Owen, W. G.**, Evidence for the formation of an ester between thrombin and heparin cofactor, *Biochim. Biophys. Acta*, 405, 380, 1975.

98. **Kunitz, M. and Northrop, J. H.**, Isolation from beef pancreas of crystalline trypsinogen, trypsin, a trypsin inhibitor and an inhibitor-trypsin compound, *J. Gen. Physiol.*, 19, 991, 1936.

99. **Kassel, B. and Laskowski, M., Sr.**, The basic trypsin inhibitor of bovine pancreas. V. The disulfide linkages, *Biochem. Biophys. Res. Commun.*, 20, 463, 1965.

100. **Werle, E.**, Über einen Hemmkörper für Kallikrein und Trypsin in der Rinderlunge, *Hoppe-Seyler's Z. Physiol. Chem.*, 338, 228, 1964.

101. **Wachter, E. and Hochstrasser, K.**, Kunitz-type proteinase inhibitors derived by limited proteolysis of the inter-α-trypsin inhibitor. IV. The amino acid sequence of the human urinary trypsin inhibitor isolated by affinity chromatography, *Hoppe-Seyler's Z. Physiol. Chem.*, 362, 1351, 1981.

101a. **Salier, J. P., Diarra-Mehrpour, M., Sesboue, R., Bourguignon, J., Benarous, R., Ohkubo, I., Kurachi, S., Kurachi, K., and Martin, J. P.**, Isolation and characterization of cDNAs encoding the heavy chain of human inter-α-trypsin inhibitor (IαTI): unambiguous evidence for multipolypeptide chain structure of IαTI, *Proc. Natl. Acad. Sci. U.S.A.*, 84, 8272, 1987.

102. **Seemüller, U., Arnhold, M., Fritz, H., Wiedenmann, K., Machleidt, W., Heinzel, R., Appelhans, H., Gassen, H.-G., and Lottspeich, F.**, The acid-stable proteinase inhibitor of human mucous secretions (HUSI-I, antileucoprotease). Complete amino acid sequence as revealed by protein and cDNA sequencing and structural homology to whey proteins and Red Sea turtle proteinase inhibitor, *FEBS Lett.*, 199, 43, 1986.

103. **Kazal, L. A., Spicer, D. S., and Brahinsky, R. A.**, Isolation of a crystalline trypsin inhibitor-anticoagulant protein from pancreas, *J. Am. Chem. Soc.*, 70, 3034, 1948.

104. **Hochstrasser, K. and Fritz, H.**, Die Aminosäuresequenz des doppelköpfigen Proteinasen-Inhibitors aus der Glandula submandibularis des Hundes. Strukturelle Homologie zu den sekretorischem Trypsin-Inhibitoren der Bauchspeicheldrüse, *Hoppe-Seyler's Z. Physiol. Chem.*, 356, 1659, 1975.

105. **Ogino, T., Croll, D. H., Kato, I., and Markley, J. L.**, Properties of conserved amino acid residues in tandem homologous protein domains, hydrogen-1 nuclear magnetic resonance studies of the histidines of chicken ovomucoid, *Biochemistry*, 21, 3452, 1982.

106. **Bogard, W. C., Jr., Kato, I., and Laskowski, M., Jr.**, A Ser^{162}/Gly^{162} polymorphism in Japanese quail ovomucoid, *J. Biol. Chem.*, 255, 6569, 1980.

107. **Shechter, Y., Burstein, Y., and Gertler, A.**, Effect of oxidation of methionine residues in chicken ovoinhibitor on its inhibitory activities against trypsin, chymotrypsin, and elastase, *Biochemistry*, 16, 992, 1977.

108. **Barrett, A. J.**, Chicken α_2-proteinase inhibitor: a serum protein homologous with ovoinhibitor of egg white, *Biochim. Biophys. Acta*, 371, 52, 1974.

109. **Steiner, R. F. and Frattali, V.**, Purification and properties of soybean protein inhibitors of proteolytic enzymes, *J. Agric. Food Chem.*, 17, 513, 1969.

110. **Ozawa, K. and Laskowski, M., Jr.**, The reactive site of trypsin inhibitors, *J. Biol. Chem.*, 241, 3955, 1966.

111. **Koide, T. and Ikenaka, T.**, Studies on soybean trypsin inhibitors. III. Amino-acid sequence of the carboxyl-terminal region and the complete amino-acid sequence of soybean trypsin inhibitor (Kunitz), *Eur. J. Biochem.*, 32, 417, 1973.

112. **Odani, S. and Ikenaka, T.**, Studies on soybean trypsin inhibitors. XI. Complete amino acid sequence of a soybean trypsin-chymotrypsin-elastase inhibitor, C-II, *J. Biochem. (Tokyo)*, 82, 1523, 1977.

113. **Odani, S. and Ikenaka, T.**, Studies on soybean trypsin inhibitors. XII. Linear sequences of two soybean double-headed trypsin inhibitors, D-II and E-I. *J. Biochem. (Tokyo)*, 83, 737, 1978.

114. **Tan, C. G. L. and Stevens, F. C.**, Amino acid sequence of lima bean protease inhibitor component IV. II. Isolation and sequence determination of the chymotryptic peptides and the complete amino acid sequence, *Eur. J. Biochem.*, 18, 515, 1971.

115. **Wilson, K. A. and Laskowski, M., Sr.**, The partial amino acid sequence of trypsin inhibitor II from garden bean, *Phaseolus vulgaris*, with location of the trypsin and elastase-reactive sites, *J. Biol. Chem.*, 250, 4261, 1975.

116. **Belew, M. and Eaker, D.**, The trypsin and chymotrypsin inhibitors in chick peas *(Cicer arietinum* L.). Identification of the trypsin-reactive site, partial-amino-acid sequence and further physico-chemical properties of the major inhibitor, *Eur. J. Biochem.*, 62, 499, 1976.

117. **Wieczorek, M., Otlewski, J., Cook, J., Parks, K., Leluk, J., Wilimowska-Pelc, A., Polanowski, A., Wilusz, T., and Laskowski, M., Jr.**, The squash family of serine proteinase inhibitors. Amino acid sequences and association equilibrium constants of inhibitors from squash, summer squash, zucchini, and cucumber seeds, *Biochem. Biophys. Res. Commun.*, 126, 646, 1985.

118. **Inouye, K., Tonomura, B., and Hiromi, K.**, The interaction of a tyrosyl residue and carboxyl groups in the specific interaction between *Streptomyces* subtilisin inhibitor and subtilisin BPN', *J. Biochem. (Tokyo)*, 85, 1115, 1979.

119. **Inouye, K., Tonomura, B., Hiromi, K., Fujiwara, K., and Tsuru, D.**, Further studies on the interaction between a protein proteinase inhibitor, *Streptomyces* subtilisin inhibitor, and thiolsubtilisin BPN', *J. Biochem. (Tokyo)*, 85, 1127, 1979.

120. **Inouye, K., Tonomura, B., and Hiromi, K.**, The effect of sodium dodecyl sulfate on the structure and function of a protein proteinase inhibitor, *Streptomyces* subtilisin inhibitor, *Arch. Biochem. Biophys.*, 192, 260, 1979.

121. **McPhalen, C. A., Svendesen, I., Jonassen, I., and James, M. N. G.**, Crystal and molecular structure of chymotrypsin inhibitor 2 from barley seeds in complex with subtilisin Novo, *Proc. Natl. Acad. Sci. U.S.A.*, 82, 7242, 1985.

122. **McPhalen, C. A., Schnebli, H. P., and James, M. N. G.**, Crystal and molecular structure of the inhibitor eglin from leeches in complex with subtilisin Carlsberg, *FEBS Lett.*, 188, 55, 1985.

123. **Bode, W., Papamokos, E., Musil, D., Seemueller, U., and Fritz, H.**, Refined 1.2 Å crystal structure of the inhibitor eglin c. Molecular structure of eglin and its detailed interaction with subtilisin, *EMBO J.*, 5, 813, 1986.

124. **Beatty, K., Bieth, J., and Travis, J.**, Kinetics of association of serine proteinases with native and oxidized α-1-proteinase inhibitor and α-1-antichymotrypsin, *J. Biol. Chem.*, 255, 3931, 1980.

125. **Carp, H., Miller, F., Hoidal, J. R., and Janoff, A.**, Potential mechanism of emphysema: α_1-proteinase inhibitor recovered from lungs of cigarette smokers contains oxidized methionine and has decreased elastase inhibitory capacity, *Proc. Natl. Acad. Sci. U.S.A.*, 79, 2041, 1982.

126. **Beatty, K., Robertie, P., Senior, R. M., and Travis, J.**, Determination of oxidized alpha-1-proteinase inhibitor in serum, *J. Lab. Clin. Med.*, 100, 186, 1982.

127. **Tollefsen, D. M., Majerus, D. W., and Blank, M. K.**, Heparin cofactor II. Purification and properties of a heparin-dependent inhibitor of thrombin in human plasma, *J. Biol. Chem.*, 257, 2162, 1982.

128. **Church, F. C., Noyes, C. M., and Griffith, M. J.**, Inhibition of chymotrypsin by heparin cofactor II, *Proc. Natl. Acad. Sci. U.S.A.*, 82, 6431, 1985.

129. **Griffith, M. J., Noyes, C. M., Tyndall, J. A., and Church, F. C.**, Structural evidence for leucine at the reactive site of heparin cofactor II, *Biochemistry*, 24, 6777, 1985.

130. **Kurachi, K., Chandra, T., Degen, S. J. F., White, T. T., Marchioro, T. L., Woo, S. L. C., and Davie, E. W.**, Cloning and sequence of cDNA coding for α_1-antitrypsin, *Proc. Natl. Acad. Sci. U.S.A.*, 78, 6826, 1981.

131. **Chandra, T., Stackhouse, R., Kidd, V. J., Robson, K. J. H., and Woo, S. L. C.**, Sequence homology between human α_1-antichymotrypsin, α_1-antitrypsin, and antithrombin III, *Biochemistry*, 22, 5055, 1983.

132. **Bock, S. C., Wion, K. L., Vehar, G. A., and Lawn, R. M.**, Cloning and expression of the cDNA for human antithrombin III, *Nucleic Acids*, 10, 8113, 1982.

132a. **Holmes, W. E., Nelles, L., Lijnen, H. R., and Collen, D.**, Primary structure of human α_2-antiplasmin, a serine protease inhibitor (serpin), *J. Biol. Chem.*, 262, 1659, 1987.

133. **Bock, S. C., Skriver, K., Nielsen, E., Thøgersen, H.-C., Wiman, B., Donaldson, V. H., Eddy, R. L., Marrinan, J., Radziejewska, E., Huber, R., Shows, T. B., and Magnusson, S.**, Human C1 inhibitor: primary structures, cDNA cloning, and chromosomal localization, *Biochemistry*, 25, 4292, 1986.

134. **McReynolds, L., O'Malley, B. W., Nisbet, A. D., Fothergill, J. E., Givol, D., Fields, S., Robertson, M., and Brownlee, G. G.**, Sequence of chicken ovalbumin mRNA, *Nature (London)*, 273, 723, 1978.

135. **Hunt, L. T. and Dayhoff, M. O.**, A surprising new protein family containing ovalbumin, antithrombin III, and alpha₁-proteinase inhibitor, *Biochem. Biophys. Res. Commun.*, 95, 864, 1980.

136. **Hejgaard, J., Rasmussen, S. K., Brandt, A., and Svendsen, I.** Sequence homology between barley endosperm protein Z and protease inhibitors of the α_1-antitrypsin family, *FEBS Lett.*, 180, 89, 1985.

137. **Ottensen, M.**, The transformation of ovalbumin into plakalbumin. A case of limited proteolysis, *C. R. Trav. Lab. Carlsberg Ser. Chim.*, 30, 211, 1958.

138. **Baker, J. B., Low, D. A., Simmer, R. L., and Cunningham, D. D.**, Protease-nexin: a cellular component that links thrombin and plasminogen activator and mediates their binding to cells, *Cell*, 21, 37, 1980.

139. **Knauer, D. J. and Cunningham, D. D.**, Protease nexins: cell-secreted proteins which regulate extracellular proteases, *Trends Biochem. Sci.*, 9, 231, 1984.

140. **Scott, R. W., Bergman, B. L., Bajpai, A., Hersh, R. T., Rodriguez, H., Jones, B. N., Barreda, C., Watts, S., and Baker, J. B.**, Protease nexin. Properties and modified purification procedure, *J. Biol. Chem.*, 260, 7029, 1985.

140a. **Van Nostrand, W. E., McKay, L. D., Baker, J. B., and Cunningham, D. D.**, Functional and structural similarities between protease nexin I and C1 inhibitor, *J. Biol. Chem.*, 263, 3979, 1988.

141. **Van Leuven, F.**, Human α_2-macroglobulin: structure and function, *Trends Biochem. Sci.*, 7, 185, 1982.

142. **Eddeland, A. and Ohlsson, K.**, The elimination in dogs of trypsin-α-macroglobulin complexes inactivated by the Kazal or the Kunitz inhibitor, *Z. Physiol. Chem.*, 359, 379, 1978.

143. **Harpel, P. C.**, Studies on human plasma α-2-macroglobulin-enzyme interactions, *J. Exp. Med.*, 138, 508, 1973.

144. **Barrett, A. J. and Starkey, P. M.**, The interaction of α_2-macroglobulin with proteinases. Characteristics and specificity of the reaction, and a hypothesis concerning its molecular mechanism, *Biochem. J.*, 133, 709, 1973.

145. **Barrett, A. J., Brown, M. A., and Sayers, C. A.**, The electrophoretically "slow" and "fast" forms of the α_2-macroglobulin molecule, *Biochem. J.*, 181, 401, 1979.

146. **Nelles, L. P., Hall, P. K., and Roberts, R. C.**, Human alpha-2- macroglobulin. Studies on the electrophoretic heterogeneity, *Biochim. Biophys. Acta*, 623, 46, 1980.

147. **Sottrup-Jensen, L., Stepanik, T. M., Kristensen, T., Wierzbicki, D. M., Jones, C. M., Lonblad, P. B., Magnuson, S., and Petersen, T. E.**, Primary structure of human α_2-macroglobulin. V. The complete structure, *J. Biol. Chem.*, 259, 8318, 1984.

147a. **Jensen, P. E. H. and Sottrup-Jensen, L.**, Primary structure of human α_2-macroglobulin. Complete disulfide bridge assignment and localization of two interchain bridged in the dimeric proteinase binding unit, *J. Biol. Chem.*, 261, 15863, 1986.

148. **Sottrup-Jensen, L., Lonblad, P. B., Stepanik, T. M., Petersen, T. E., Magnusson, S., and Jörnvall, H.**, Primary structure of the "bait" region for proteinases in α_2-macroglobulin, *FEBS Lett.*, 127, 167, 1981.

149. **Mortensen, S. B., Sottrup-Jensen, L., Hansen, H. F., Petersen, T. E., and Magnusson, S.**, Primary and secondary cleavage sites in the bait region of α_2-macroglobulin, *FEBS Lett.*, 135, 295, 1981.

150. **Howard, J. B.**, Reactive site in human α_2-macroglobulin: circumstantial evidence for a thiolester, *Proc. Natl. Acad. Sci. U.S.A.*, 78, 2235, 1981.

151. **Salvesen, G. S., Sayers, C. A., and Barrett, A. J.**, Further characterization of the covalent linking reaction of α_2-macroglobulin, *Biochem. J.*, 195, 453, 1981.

152. **Björk, I. and Fish, W. W.**, Evidence for similar conformational changes in α_2- macroglobulin on reaction with primary amines or proteolytic enzymes, *Biochem. J.*, 207, 347, 1982.

153. **Gonias, S. L., Reynolds, J. A., and Pizzo, S. V.**, Physical properties of human α_2-macroglobulin following reaction with methylamine and trypsin, *Biochim. Biophys. Acta.*, 705, 306, 1982.

154. **Dangott, L. J., Puett, D., and Cunningham, L. W.**, Conformational changes induced in human α_2-macroglobulin by protease and nucleophilic modification, *Biochemistry*, 22, 3647, 1983.

155. **Steinbuch, M., Pejaudier, L., Quentin, M., and Martin, V.**, Molecular alteration of α_2-macroglobulin by aliphatic amines, *Biochim. Biophys. Acta*, 154, 228, 1968.

156. **Dangott, L. J. and Cunningham, L. W.**, Residual α_2-macroglobulin in fetal calf serum and properties of its complex with thrombin, *Biochem. Biophys. Res. Commun.*, 107, 1243, 1982.

157. **Björk, I., Lindblom, T., and Lindahl, P.**, Changes of the proteinase binding properties and conformation of bovine α_2-macroglobulin on cleavage of the thiol ester bonds by methylamine, *Biochemistry*, 24, 2653, 1985.

158. **Nagase, H., Harris, E. D., Jr., and Brew, K.**, Evidence for a thiol ester in duck ovostatin (ovomacroglobulin) in egg white, *J. Biol. Chem.*, 261, 1421, 1986.

159. **Keil, B.**, Trypsin, in *The Enzymes*, Vol. 3, 3rd ed., Boyer, P. D., Ed., Academic Press, New York, 1971, chap. 8.

160. **Drapeau, G. R.**, Protease from *Staphylococcus aureus*, *Methods Enzymol.*, 45, 469, 1976.

161. **Wittman-Liebold, B. and Marzinzig, E.**, Primary structure of protein L28 from the large subunit of *Escherichia coli* ribosomes, *FEBS Lett.*, 81, 214, 1977.

162. **Walter, R.,** Partial purification and characterization of post-proline cleaving enzyme: enzymatic inactivation of neurohypophyseal hormones by kidney preparations of various species, *Biochim. Biophys. Acta,* 422, 138, 1976.
163. **Lin, L.-N. and Brandts, J. F.,** Evidence showing that a proline-specific endopeptidase has an absolute requirement for a trans peptide bond immediately preceding the active bond, *Biochemistry,* 22, 4480, 1983.
164. **Glazer, A. N. and Smith, E. L.,** Papain and other plant sulfhydryl proteolytic enzymes, in *The Enzymes,* Vol. 3, 3rd ed., Boyer, P. D., Ed., Academic Press, New York, 1971, chap. 14.
165. **Mitchell, W. M. and Harrington, W. F.,** Clostripain, in *The Enzymes,* Vol. 3, 3rd ed., Boyer, P. D., Ed., Academic Press, New York, 1971, chap. 19.
166. **Gilles, A.-M., Imhoff, J.-M., and Keil, B.,** α-Clostripain. Chemical characterization, activity, and thiol content of the highly active form of clostripain, *J. Biol. Chem.,* 254, 1462, 1979.
167. **Fruton, J. S.,** Pepsin, in *The Enzymes,* Vol. 3, 3rd ed., Boyer, P. D., Ed., Academic Press, New York, 1971, chap. 4.
168. **Garg, G. K. and Virupaksha, T. K.,** Acid protease from germinated sorghum. II. Substrate specificity with synthetic peptides and ribonuclease A, *Eur. J. Biochem.,* 17, 13, 1970.
169. **Matsubara, H. and Feder, J.,** Other bacterial, mold, and yeast proteases, in *The Enzymes,* Vol. 3, 3rd ed., Boyer, P. D., Ed., Academic Press, New York, 1971, chap. 20.
170. **Keil, B.,** Some newly characterized collagenases from procaryotes and lower eucaryotes, *Mol. Cell. Biochem.,* 23, 87, 1979.
171. **Lecroisey, A. and Keil, B.,** Differences in the degradation of native collagen by two microbial collagenases, *Biochem. J.,* 179, 53, 1979.
172. **Schechter, I. and Berger, A.,** On the size of the active site in proteases. I. Papain, *Biochem. Biophys. Res. Commun.,* 27, 157, 1967.
173. **Fruton, J. S.,** Proteinase-catalyzed synthesis of peptide bonds, *Adv. Enzymol. Relat. Areas Mol. Biol.,* 53, 239, 1982.
174. **Polgár, L.,** Structure and function of serine proteinases, in *New Comprehensive Biochemistry,* Vol. 16, Neuberger, A. and Brocklehurst, K., Eds., Elsevier, Amsterdam, 1987, chap. 3.
175. **Wright, H. T.,** Secondary and conformational specificities of trypsin and chymotrypsin, *Eur. J. Biochem.,* 73, 567, 1977.
176. **Fischer, G., Bang, H., Berger, E., and Schellenberger, A.,** Conformational specificity of chymotrypsin toward proline-containing substrates, *Biochim. Biophys. Acta,* 791, 87, 1984.
177. **Harper, J. W., Cook, R. R., Roberts, C. J., McLaughlin, B. J., and Powers, J. C.,** Active site mapping of the serine proteases human leukocyte elastase, cathepsin G, porcine pancreatic elastase, rat mast cell proteases I and II, bovine chymotrypsin Aα, and *Staphylococcus aureus* protease V-8 using tripeptide thiobenzyl ester substrates, *Biochemistry,* 23, 2995, 1984.
178. **Farmer, D. A. and Hageman, J. H.,** Use of N-benzoyl-L-tyrosine thiobenzyl ester as a protease substrate. Hydrolysis by α-chymotrypsin and subtilisin BPN′, *J. Biol. Chem.,* 250, 7366, 1975.
179. **Green, G. D. J. and Shaw, E.,** Thiobenzyl benzyloxycarbonyl-L-lysinate, substrate for a sensitive colorimetric assay for trypsin-like enzymes, *Anal. Biochem.,* 93, 223, 1979.
180. **Irwing, G. W., Jr., Fruton, J. S., and Bergmann, M.,** Kinetics of proteinase action. Application to specificity problems, *J. Biol. Chem.,* 138, 231, 1941.
181. **Bender, M. L. and Kézdy, F. J.,** Mechanism of action of proteolytic enzymes, *Annu. Rev. Biochem.,* 34, 49, 1965.
182. **Carpenter, F. H.,** The free energy change in hydrolytic reactions: the nonionized compound convention, *J. Am. Chem. Soc.,* 82, 1111, 1960.
183. **Chaiken, I. M.,** Semisynthetic peptides and proteins, *Crit. Rev. Biochem.,* 11, 255, 1981.
184. **Fruton, J. S.,** Proteinase-catalyzed synthesis of peptide bonds, *Adv. Enzymol. Relat. Areas Mol. Biol.,* 53, 239, 1982.
185. **Konopinska, D. and Muzalewski, F.,** Proteolytic enzymes in peptide synthesis, *Mol. Cell. Biochem.,* 51, 165, 1983.
186. **Jakubke, H.-D., Kuhl, P., and Könnecke, A.,** Grundprinzipien der proteasekatalysierten Knüpfung der Peptidbindung, *Angew. Chem.,* 97, 79, 1985.
187. **Finkelstadt, W. R., Hamid, M. A., Mattis, J. A., Schrode, J., Sealock, R. W., Wang, D., and Laskowski, M., Jr.,** Kinetics and thermodynamics of the interaction of proteinases with protein inhibitors, in *Proteinase Inhibitors,* Fritz, H., Tschesche, H., Greene, L. J., and Trusheit, E., Eds., Springer-Verlag, Berlin, 1974, 389.
188. **Kowalski, D., Leary, T. R., McKee, R. W., Sealock, R. W., Wang, D., and Laskowski, M., Jr.,** Replacements, insertions, and modifications of the amino acid residues in reactive site of soybean trypsin inhibitor (Kunitz), in *Protease Inhibitors,* Fritz, H., Tschesche, H., Greene, L. J., and Trusheit, E., Eds., Springer-Verlag, Berlin, 1974, 311.
189. **Morihara, K., Oka, T., and Tsuzuki, H.,** Semi-synthesis of human insulin by trypsin-catalyzed replacement of Ala-B30 by Thr in porcine insulin, *Nature (London),* 280, 412, 1979.

190. **Masaki, T., Nakamura, K., Isono, M., and Soejima, M.,** A new proteolytic enzyme from *Achromobacter lyticus* M 497-1, *Agric. Biol. Chem.*, 42, 1443, 1978.

191. **Morihara, K., Oka, T., Tsuzuki, H., Tochino, Y., and Kanaya, T.,** *Achromobacter* protease I-catalyzed conversion of porcine insulin into human insulin, *Biochem. Biophys. Res. Commun.*, 92, 396, 1980.

192. **Inouye, K., Watanabe, K., Morihara, K., Tochino, Y., Kanaya, T., Emura, J., and Sakakibara, S.,** Enzyme-assisted semisynthesis of human insulin, *J. Am. Chem. Soc.*, 101, 751, 1979.

193. **Oka, T., and Morihara, K.,** Peptide bond synthesis catalyzed by α-chymotrypsin, *J. Biochem. (Tokyo)*, 84, 1277, 1978.

194. **Christova, E., Petkov, D. D., and Stoineva, I.,** S_2-P_2' interaction and the trypsin anilide hydrolysis, *Arch. Biochem. Biophys.*, 218, 626, 1982.

195. **Oka, T. and Morihara, K.,** Peptide bond synthesis by thermolysins, *J. Biochem. (Tokyo)*, 88, 807, 1980.

196. **Kullman, W.,** Proteases as catalysts for enzymic synthesis of opioid peptides, *J. Biol. Chem.*, 255, 8234, 1980.

197. **Sluyterman, L. A. AE. and Wijdenes, J.,** Sigmoidal progress curves in the polymerization of leucine methyl ester catalyzed by papain, *Biochim. Biophys. Acta*, 289, 194, 1972.

198. **Anderson, G. and Luisi, P. L.,** Papain-induced oligomerization of α-amino acid esters, *Helv. Chim. Acta*, 62, 488, 1979.

199. **Lüthi, P. and Luisi, P. L.,** Enzymatic synthesis of hydrocarbon-soluble peptides with reverse micelles, *J. Am. Chem. Soc.*, 106, 7285, 1984.

200. **Matsushima, A., Okada, M., and Inada, Y.,** Chymotrypsin modified with polyethylene glycol catalyzes peptide synthesis reaction in benzene, *FEBS Lett.*, 178, 275, 1984.

201. **Bowles, D. J., Marcus, S. E., Pappin, D. J. C., Findlay, J. B. C., Eliopoulos, E., Maycox, P. R., and Burgess, J.,** Posttranslational processing of concanavalin A precursors in jackbean cotyledons, *J. Cell Biol.*, 102, 1284, 1986.

202. **Sachdev, G. P. and Fruton, J. S.,** Secondary enzyme-substrate interactions and the specificity of pepsin, *Biochemistry*, 9, 4465, 1970.

203. **Jencks, W. P.,** Binding energy, specificity, and enzymic catalysis: the Circe effect, *Adv. Enzymol. Relat. Areas Mol. Biol.*, 43, 219, 1975.

204. **Page, M. I.,** The energetics and specificity of enzyme-substrate interactions, in *New Comprehensive Biochemistry*, Vol. 6, Page, M. I., Ed., Elsevier, Amsterdam, 1984, chap. 1.

205. **Pauling, L.,** Molecular architecture and biological reactions, *Chem. Eng. News*, 24, 1375, 1946.

206. **Wolfenden, R.,** Analog approaches to the structure of the transition state in enzyme reactions, *Acc. Chem. Res.*, 5, 10, 1972.

207. **Wolfenden, R.,** Transition state analog inhibitors and enzyme catalysis, *Annu. Rev. Biophys. Bioeng.*, 5, 271, 1976.

208. **Lienhard, G. E.,** Enzymatic catalysis and transition-state theory. Transition-state analogs show that catalysis is due to tighter binding of transition states than of substrates, *Science*, 180, 149, 1973.

209. **Page, M. I.,** Transition-states, standard states, and enzymic catalysis, *Int. J. Biochem.*, 11, 331, 1981.

210. **Gutfreund, H. and Sturtevant, J. M.,** The mechanism of the reaction of chymotrypsin with p-nitrophenyl acetate, *Biochem. J.*, 63, 656, 1956.

211. **Brot, F. E. and Bender, M. L.,** Use of the specificity constant of α-chymotrypsin, *J. Am. Chem. Soc.*, 91, 7187, 1969.

212. **Lineweaver, H. and Burk, D.,** The determination of enzyme dissociation constants, *J. Am. Chem. Soc.*, 56, 658, 1934.

213. **Eadie, G. S.,** The inhibition of cholinesterase by physostigmine and prostigmine, *J. Biol. Chem.*, 146, 85, 1942.

214. **Hanes, C. S.,** Studies on plant amylases. I. The effect of starch concentration upon the velocity of hydrolysis by the amylase of germinated barley, *Biochem. J.*, 26, 1406, 1932.

215. **Eisenthal, R. and Cornish-Bowden, A.,** The direct linear plot. A new graphical procedure for estimating enzyme kinetic parameters, *Biochem. J.*, 139, 715, 1974.

216. **Markus, M., Hess, B., Ottaway, J. H., and Cornish-Bowden, A.,** The analysis of kinetic data in biochemistry. A critical evaluation of methods, *FEBS Lett.*, 63, 225, 1976.

217. **Duggleby, R. G.,** A nonlinear regression program for small computers, *Anal. Biochem.*, 110, 9, 1981.

218. **Canela, E. I. and Franco, R.,** Enzyme kinetic studies from progress curves, *Biochem. J.*, 233, 599, 1986.

219. **Bender, M. L., Clement, G. E., Gunter, C. R., and Kézdy, F. J.,** The kintics of α-chymotrypsin reactions in the presence of added nucleophiles, *J. Am. Chem. Soc.*, 86, 3697, 1964.

220. **Berezin, I. V., Kazanskaya, N. F., and Klyosov, A. A.,** Determination of the individual rate constants of α-chymotrypsin-catalyzed hydrolysis with the added nucleophilic agent, 1,4-butanediol, *FEBS Lett.*, 15, 121, 1971.

221. **Hinberg, I. and Laidler, K. J.,** Steady-state kinetics of enzyme reactions in the presence of added nucleophiles, *Can. J. Biochem.*, 50, 1334, 1972.

222. **Bender, M. L., Kézdy, F. J., and Wedler, F. C.,** α-Chymotrypsin: enzyme concentration and kinetics, *J. Chem. Educ.,* 44, 84, 1967.
223. **Peller, L. and Alberty, R. A.,** Multiple intermediates in steady state enzyme kinetics. I. The mechanism involving a single substrate and product, *J. Am. Chem. Soc.,* 81, 5907, 1959.
224. **Knowles, J. R.,** The intrinsic pK_a-values of functional groups in enzymes: improper deductions from the pH-dependence of steady-state parameters, *Crit. Rev. Biochem.,* 4, 165, 1976.
225. **Polgár, L. and Halász, P.,** Evidence for multiple reactive forms of papain, *Eur. J. Biochem.,* 88, 513, 1978.
226. **Bender, M. L., Begue-Canton, M. L., Blakely, R. L., Brubacher, L. J., Feder, J., Gunter, C. R., Kézdy, F. J., Killheffer, J. V., Jr., Marshall, T. H., Miller, C. G., Roeske, R. W., and Stoops, J. K.,** The determination of the concentration of hydrolytic enzyme solutions: α-chymotrypsin, trypsin, papain, elastase, subtilisin, and acetylcholinesterase, *J. Am. Chem. Soc.,* 88, 5890, 1966.
227. **Kézdy, F. J. and Kaiser, E. T.,** Principles of active site titration of proteolytic enzymes, *Methods Enzymol.,* 19, 3, 1970.
228. **Polgár, L.,** Conversion of the serine residue at the active site of alcalase to a cysteine side chain, *Acta Biochim. Biophys. Acad. Sci. Hung.,* 3, 397, 1968.
229. **Chase, T., Jr. and Shaw, E.,** Titration of trypsin, plasmin, and thrombin with p-nitrophenyl p'-guanidinobenzoate HCl, *Methods Enzymol.,* 19, 20, 1970.
230. **Coleman, P. L., Latham, H. G., Jr., and Shaw, E. N.,** Some sensitive methods for the assay of trypsin-like enzymes, *Methods Enzymol.,* 45, 12, 1976.
231. **Elmore, D. T. and Smyth, J. J.,** A new method for determining the absolute molarity of solutions of trypsin and chymotrypsin by using p-nitrophenyl N^2-acetyl-N^1-benzylcarbazate, *Biochem. J.,* 107, 103, 1968.
232. **Gupton, B. F., Carroll, D. L., Tuhy, P. M., Kam, C.-M., and Powers, J. C.,** Reaction of azapeptides with chymotrypsin-like enzymes. New inhibitors and active site titrants for chymotrypsin A_α, subtiltisin BPN', subtiltisin Carlsberg, and human leukocyte cathepsin G, *J. Biol. Chem.,* 259, 4279, 1984.
233. **Powers, J. C., Boone, R., Carroll, D. L., Gupton, B. F., Kam, C.-M., Nishino, N., Sakamoto, M., and Tuhy, P. M.,** Reaction of azapeptides with human leukocyte elastase and porcine pancreatic elastase. New inhibitors and active site titrants, *J. Biol. Chem.,* 259, 4288, 1984.
234. **Elmore, D. T. and Smyth, J. J.,** The behaviour of trypsin towards α-N-methyl-α-N-toluene-p-sulphonyl-L-lysine β-naphthyl ester. A new method for determining the absolute molarity of solutions of trypsin, *Biochem. J.,* 107, 97, 1968.
235. **Jameson, G. W., Roberts, D. V., Adams, R. W., Kyle, W. S. A., and Elmore, D. T.,** Determination of the operational molarity of solutions of bovine α-chymotrypsin, trypsin, thrombin and factor Xa by spectrofluorimetric titration, *Biochem. J.,* 131, 107, 1973.
236. **Melhado, L. L., Peltz, S. W., Leytus, S. P., and Mangel, W. F.,** p-Guanidinobenzoic acid esters of fluorescein as active-site titrants of serine proteases, *J. Am. Chem. Soc.,* 104, 7299, 1982.
237. **Kézdy, F. J. and Kaiser, E. T.,** Active site titration of cysteine proteases, *Methods Enzymol.,* 45, 3, 1976.
238. **Ellman, G. L.,** Tissue sulfhydryl groups, *Arch. Biochem. Biophys.,* 82, 70, 1959.
239. **Brocklehurst, K.,** Two-protonic-state elctrophiles as probes of enzyme mechanism, *Methods Enzymol.,* 87, 427, 1982.
240. **Hanada, K., Tamai, M., Yamagishi, M., Ohmura, S., Sawada, J., and Tanaka, I.,** Isolation and characterization of E-64, a new thiol proteaso inhibitor, *Agric. Biol. Chem.,* 42, 523, 1978.
241. **Hanada, D., Tamai, M., Ohmura, S., Sawada, J., Seki, T., and Tanaka, I.,** Structure and synthesis of E-64, a new thiol protease inhibitor, *Agric. Biol. Chem.,* 42, 529, 1978.
242. **Barrett, A. J. and Kirschke, H.,** Cathepsin B, cathepsin H, and cathepsin L, *Methods Enzymol.,* 80, 535, 1981.
243. **Matthews, B. W., Sigler, P. B., Henderson, R., and Blow, D. M.,** Three-dimensional structure of tosyl-α-chymotrypsin, *Nature (London),* 214, 652, 1967.
244. **Reeke, G. N., Hartsuck, J. A., Ludwig, M. L., Quiocho, F. A., Steitz, T. A., and Lipscomb, W. N.,** The structure of carboxypeptidase A. VI. Some results at 2.0-Å resolution, and the complex with glycyl-tyrosine at 2.8 Å resolution, *Proc. Natl. Acad. Sci. U.S.A.,* 58, 2220, 1967.
245. **Sluyterman, L. A. AE. and de Graaf, M. J. M.,** The activity of papain in the crystalline state, *Biochim. Biophys. Acta,* 171, 277, 1969.
246. **Rossi, G. L. and Bernhard, S. A.,** Are the structure and function of an enzyme the same in aqueous solution and in the wet crystal? *J. Mol. Biol.,* 49, 85, 1970.
247. **Douzou, P. and Petsko, G. A.,** Proteins at work: "Stop-action" pictures at sub-zero temperature, *Adv. Protein Chem.,* 36, 245, 1984.
248. **Petsko, G. A. and Ringe, D.,** Fluctuations in protein structure from X-ray diffraction, *Annu. Rev. Biophys. Bioeng.,* 13, 331, 1984.

249. **Walter, J., Steigemann, W., Singh, T. P., Bartunik, H. D., Bode, W., and Huber, R.,** On the disordered activation domain in trypsinogen: chemical labeling and low temperature crystallography, *Acta Crystallogr. Sect. B,* 38, 1462, 1982.

250. **James, M. N. G. and Sielecki, A. R.,** Structure and refinement of penicillopepsin at 1.8 Å resolution, *J. Mol. Biol.,* 163, 299, 1983.

251. **Moffat, K., Szebenyi, D., and Bilderback, D.,** X-ray Laue diffraction from protein crystals, *Science,* 223, 1423, 1984.

252. **McRay, J. A., Herbette, L., Kihara, T., and Trentham, D. R.,** A new approach to time-resolved studies of ATP-requiring biological systems: laser flash photolysis of caged ATP, *Proc. Natl. Acad. Sci. U.S.A.,* 77, 7237, 1980.

253. **Nargeot, J., Nerbonne, J. M., Engels, J., and Lester, H. A.,** Time course of the increase in the myocardial slow inward current after a photochemically generated concentration jump of intracellular cAMP, *Proc. Natl. Acad. Sci. U.S.A.,* 80, 2395, 1983.

254. **Patchornik, A., Amit, B., and Woodward, R. B.,** Photosensitive protecting groups, *J. Am. Chem. Soc.,* 92, 6333, 1970.

255. **Kossiakoff, A. A. and Spencer, S. A.,** Direct determination of the protonation states of aspartic acid-102 and histidine-57 in the tetrahedral intermediate of the serine proteases: neutron structure of trypsin, *Biochemistry,* 20, 6462, 1981.

256. **Steitz, T. A. and Shulman, R. G.,** Crystallographic and NMR studies of the serine proteases, *Annu. Rev. Biophys. Bioeng.,* 11, 419, 1982.

257. **Markley, J.-L. and Ulrich, E.,** Detailed analysis of protein structure and function by NMR spectroscopy, *Annu. Rev. Biophys. Bioeng.,* 13, 493, 1984.

258. **Mackenzie, N. E., Malthouse, J. P. G., and Scott, A. I.,** Studying enzyme mechanism by ^{13}C nuclear magnetic resonance, *Science,* 225, 883, 1984.

259. **Bachovchin, W. W. and Roberts, J. D.,** Nitrogen-15 nuclear magnetic resonance spectroscopy. The state of histidine in the catalytic triad of α-lytic protease. Implications for the charge-relay mechanism of peptide-bond cleavage by serine proteases, *J. Am. Chem. Soc.,* 100, 8041, 1978.

260. **Jordan, F. and Polgár, L.,** Proton nuclear magnetic resonance evidence for the absence of a stable hydrogen bond between the active site aspartate and histidine residues of native subtilisins and for its presence in thiolsubtilisins, *Biochemistry,* 20, 6366, 1981.

261. **Baillargeon, M. W., Laskowski, M., Jr., Neves, D. E., Porubcan, M. A., Santini, R. E., and Markley, J. L.,** Soybean trypsin inhibitor (Kunitz) and its complex with trypsin. Carbon-13 nuclear magnetic resonance studies of the reactive site arginine, *Biochemistry,* 19, 5703, 1980.

262. **Richarz, R., Tschesche, H., and Wüthrich, K.,** Carbon-13 nuclear magnetic resonance studies of the selectively isotope-labeled reactive site peptide bond of the basic pancreatic trypsin inhibitor in the complex with trypsin, trypsinogen, and anhydrotrypsin, *Biochemistry,* 19, 5711, 1980.

263. **Rühlmann, A., Kukla, D., Schwager, P., Bartels, K., and Huber, R.,** Structure of the complex formed by bovine trypsin and bovine pancreatic trypsin inhibitor. Crystal structure determination and stereochemistry of the contact region, *J. Mol. Biol.,* 77, 417, 1973.

264. **Kaiser, E. T.,** Better enzymes by design, *Nature (London),* 313, 630, 1985.

265. **Craik, C. S., Largman, C., Fletcher, T., Roczniak, S., Barr, P. J., Fletterick, R., and Rutter, W. J.,** Redesigning trypsin: alteration of substrate specificity, *Science,* 228, 291, 1985.

266. **Shaw, W. V.,** Protein engineering. The design, synthesis, and characterization of factitious proteins, *Biochem. J.,* 246, 1, 1987.

267. **Polgár, L. and Bender, M. L.,** Simulated mutation at the active site of biologically active proteins, *Adv. Enzymol. Relat. Areas Mol. Biol.,* 33, 381, 1970.

Chapter 3

SERINE PROTEASES

I. HISTORICAL BACKGROUND

Serine proteases attract a growing interest as their vital role becomes apparent in even more biological processes (see Chapter 2, Section III). They have been most extensively studied among the proteolytic enzymes, and a vast literature is concerned with the diverse problems associated with this important group of enzymes. Here, we concentrate on the mechanistic aspects of serine proteases and attempt to give a picture of the mechanism of action that is based on the early modification and kinetic studies, and that emerges now in a clearer form, thanks to the advent of sophisticated techniques, such as X-ray crystallography, nuclear magnetic resonance (NMR) spectroscopy, and site-directed mutagenesis.

The basic features of the mechanism of action have been covered in a few prior reviews including those by Cunningham[1] and Bender and Kézdy.[2] The largest collection of references is in the review by Bender and Killheffer.[3] Specifically, chymotrypsin was reviewed by Hess;[4] trypsin by Keil;[5] elastase by Hartley and Shotton;[6] subtilisins and thiolsubtilisins by Markland and Smith,[7] Polgár and Bender,[8] and Philipp and Bender.[9] The more recent reviews were concerned with crystallographic studies and the structural aspects of catalysis.[10-19]

Most of the modification and kinetic studies that laid down the foundation of our knowledge about the mechanism of action were carried out with chymotrypsin. In the majority of instances, these results are applicable to other serine proteases as well. Therefore, in this section it seems to be appropriate to focus our attention on the basic studies performed with chymotrypsin and to refer to the interesting peculiarities of other enzymes.

A. The Active Site Serine

Chymotrypsin reacts stoichiometrically with diisopropylphosphofluoridate (DFP) to form an inactive diisopropylphosphoryl (DIP) enzyme.[20,21] Amino acid sequence studies have shown that a particular serine residue, later identified as Ser 195, becomes phosphorylated in the reaction.[22] The remarkable reactivity of serine 195 is illustrated by the fact that the other 27 serine side chains of chymotrypsin do not react with DFP.

Phenylmethanesulfonyl fluoride (PMSF) is another good inhibitor of chymotrypsin-like serine proteases.[23] Trypsin-like enzymes react more readily with (p-amidinophenyl)methanesulfonyl fluoride.[24] The sulfonyl fluoride inhibitors are of practical importance, being not as dangerous to work with as DFP. The latter reagent strongly inactivates acetyl-cholinesterase, an enzyme crucial for the transmission of nerve impulses at certain synapsis.[25]

The ester bond formed between the enzyme and the above phosphoryl or sulfonyl inhibitors is only relatively stable and reactivation of the protease may occur. Therefore, a one-step addition of the inhibitor usually cannot permanently suppress the proteolytic activity. The stability of the ester bond changes with the enzymes studied. It was also observed that in the case of diisopropylphosphoryl serine proteases, the inhibited enzyme can be converted into a monoisopropyl derivative.[26-30] The formation of this "aged" protein can be monitored by NMR technique.

Interestingly, not only phosphoryl and sulfonyl but also certain carboxylic acid derivatives can inhibit serine proteases. The active site titrant, p-nitrophenyl p'-guanidinobenzoate (Chapter 2, Section VIII.D), is such a reactant. Again, the inhibition is reversible and differs considerably from one enzyme to another. Of course, carboxylic acid esters and amides are usually substrates rather than inhibitors. They all react with the same serine residue to form an acyl-enzyme, which is a catalytic intermediate. The existence of the intermediate was

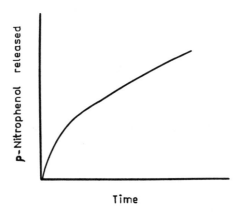

FIGURE 1. Schematic representation of the burst and steady-state phases in the hydrolysis of *p*-nitrophenyl acetate by chymotrypsin.

first demonstrated by kinetic studies on the hydrolysis of *p*-nitrophenyl acetate. Figure 1 shows that the kinetics of the hydrolysis is biphasic; there is a rapid burst of *p*-nitrophenol release, roughly equivalent to the concentration of active chymotrypsin, followed by a slow, steady-state liberation of *p*-nitrophenol.[31] This was rationalized in terms of a fast acyl-enzyme formation with a concomitant release of P_1 product Chapter 2, Equation 10) which is followed by the slow deacylation step yielding acetate ion (P_2) and free enzyme.[31]

The acyl-enzyme proved to be sufficiently stable to be isolated at low pH. Hence, the amino acid sequence could be determined around the acyl-enzyme forming serine residue which had been labeled with a ^{14}C-acetyl group.[32] As expected, the sequence (-Gly-Asp-*Ser*-Gly-Gly-Pro-) was identical with that found previously for DIP-chymotrypsin.[22]

The reaction pathway proposed for *p*-nitrophenyl acetate has also been confirmed for other nonspecific substrates such as cinnamoyl imidazole,[33] indoleacryloyl imidazole,[34] and methyl cinnamate.[35] In the hydrolysis of specific substrates, acyl-enzyme formation has also been demonstrated by both steady-state and pre-steady-state kinetic studies.[2,4] Thus, determination of similar k_{cat} values for a series of chymotrypsin-catalyzed hydrolysis of different esters of the same *N*-acetyl-L-amino acid strongly supported the existence of an acyl-enzyme intermediate.[36] It is seen from Table 1, that the methyl, ethyl, and *p*-nitrophenyl esters are hydrolyzed with similar k_{cat} while the K_m values are markedly different, although one would expect a much higher rate for the hydrolysis of the *p*-nitrophenyl ester possessing a good leaving group than for the reaction of the methyl or ethyl ester of the corresponding acid. This reactivity difference is indeed reflected in the relative alkaline hydrolysis rates of the substrates (Table 1). Consequently, the identical k_{cat} values obtained in the enzymic reactions indicate a common rate-determining step for all ester substrates shown in Table 1. Most likely this step pertains to the hydrolysis of the acyl-enzyme. This implies that k_{cat} reflects deacylation, i.e., $k_{cat} \approx k_3$ (see also Chapter 2, Section VIII). Thus, the higher reactivity of the *p*-nitrophenyl ester substrate is reflected only in k_2 which affects K_m.[36,37] On the other hand, the amide substrate shown in Table 1 has a much lower k_{cat} value, indicating rate-limiting acylation ($k_{cat} \approx k_2$). This is consistent with the relative stability of amides as compared with esters.

Further kinetic evidence for the formation of acyl-enzyme in the hydrolysis of specific ester substrates was provided by the determination of the individual acylation and deacylation rate constants, k_2 and k_3. This was performed under steady-state conditions by the added nucleophile method[38,39] (see Chapter 2, Section VIII). The values of k_2 and k_3 could also be determined under pre-steady-state conditions by fast kinetic methods. A notable approach

Table 1
THE CHYMOTRYPSIN-CATALYZED
HYDROLYSIS OF DERIVATIVES OF *N*-
ACETYL-L-PHENYLALANINE[36]

Derivative	k_{cat} (sec^{-1})	K_m (mM)	Relative k_{OH^-} of acetate
Amide	0.039	37	1
Ethyl ester	63.1	0.88	2,750
Methyl ester	57.5	1.50	5,500
p-Nitrophenyl ester	77	0.024	315,000

involved the synthesis of substrates possessing furylacryloyl or indoleacryloyl groups which permitted the monitoring of spectral changes upon acyl-enzyme formation.[40,41] Another approach utilized the spectral changes occurring (near 450 nm) when the competitive inhibitor, proflavin, was displaced from chymotrypsin by the substrate.[42-45]

The hydrolysis of specific amide substrates catalyzed by chymotrypsin is considerably less well characterized than the hydrolysis of esters. Nevertheless, from steady-state kinetic investigations[2,4,44,46] (Table 1) it appears that in contrast to ester hydrolysis, the formation of the acyl-enzyme is rate determining with amides ($k_2 < k_3$). However, in some cases the hydrolysis of specific peptide amides and *p*-nitroanilides appears to be controlled by deacylation and not by acylation.[47,48]

By capturing the acyl-enzyme with an alanine amide acceptor, its existence was clearly demonstrated in the chymotrypsin catalysis,[49] contrary to the suggestion that amide hydrolysis does not follow an acyl-enzyme mechanism.[50] Transpeptidation reactions catalyzed by chymotrypsin in the presence of $H_2^{18}O$ have also confirmed the acyl-enzyme mechanism for peptide substrates.[51]

B. The Active Site Histidine

The implication of a histidine residue in serine protease catalysis was first indicated by pH-dependence studies.[52,53] These investigations demonstrated that the k_{cat} for a specific substrate of chymotrypsin depends on the ionization of a group with an apparent pK_a of 7. This group is most probably identified as the imidazole moiety of a histidine residue, the only group on the enzyme with a pK_a around neutrality. Later studies have also shown that it is the basic form of the group of pK_a about 7 that is operative in both acylation and deacylation.[54-57]

In contrast to deacylation, which exhibits a sigmoid pH k_3 profile, the acylation of chymotrypsin shows a bell-shaped pH-rate profile, i.e., the reaction is dependent on two ionizing groups: a base of pK_a 7 and an acid of pK_a 8.5 to 9.[54-57] It is interesting that the group of higher pK_a is not directly involved in the catalysis. This group is the α-amino group of Ile 16 liberated on activation of chymotrypsinogen (see Chapter 2, Section III, and Chapter 3, Section II.B). The protonated α-amino group interacts with the carboxylate group of Asp 194, thereby ensuring the proper active site conformation for catalysis.[4,58] This is quite reasonable as Asp 194 is adjacent to the active site Ser 195. The unimportance of the protonated α-amino group in the chemical catalysis is in accordance with the absence of a bell-shaped pH-rate profile in the catalysis by subtilisin. This bacterial enzyme, which is devoid of a zymogen form characteristic of pancreatic serine proteases, exhibits a simple sigmoid pH-rate profile both in acylation and deacylation.[59,60]

Because of the influence of environmental and other factors on the pK_a value (see Chapter 2, Section VIII.C), unequivocal assignment of a direct catalytic role to histidine cannot be made on the basis of a pK_a alone. Therefore, chemical modification studies proved to be highly rewarding. Specifically, photooxidation experiments demonstrated that destruction of a single histidine residue of chymotrypsin led to inactivation of the enzyme.[61,62] Furthermore, it was shown by Shaw and Schoellman[63,64] that tosyl-L-phenylalanine chloromethyl ketone (TPCK) [1], a derivative of a specific substrate of chymotrypsin, alkylated the active site histidine specifically and irreversibly with the concomitant loss of enzymic activity. Numerous chloromethyl ketone derivatives of amino acids and peptides have been synthesized, which by virtue of a specific interaction with the binding site selectively inhibit the various serine proteases.[65] Another type of reagent, which selectively alkylates the active site histidine of chymotrypsin, is methyl *p*-nitrobenzene-sulfonate [2].[66-68]

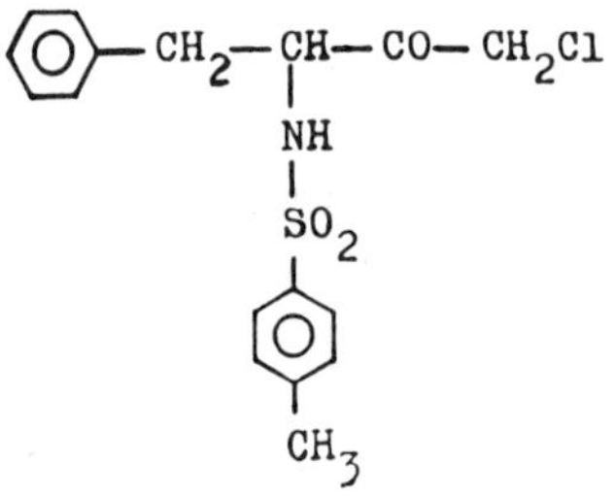

$$C_6H_5-CH_2-\underset{\underset{\underset{CH_3}{|}}{\overset{\displaystyle C_6H_4}{|}}}{\underset{|}{\overset{|}{CH}}}\,}\!\!\!\!-CO-CH_2Cl$$

[1]

$$NO_2-C_6H_4-\overset{\displaystyle O}{\underset{\displaystyle O}{\overset{\|}{\underset{\|}{S}}}}-OCH_3$$

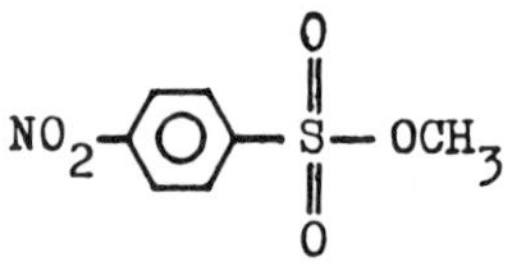

[2]

It was shown in Chapter 1 that imidazole catalyzes ester hydrolysis as both nucleophile and general base. Indeed, there was some speculation in the literature that chymotrypsin-catalyzed reactions proceeded through an *N*-acyl imidazole derivative, i.e., by nucleophilic catalysis.[69] No direct evidence for such an intermediate was found[70] although the intermediate, if it existed, should be detectable with kinetic or spectroscopic methods.[2] The lack of nucleophilic catalysis is also supported by chemical studies (Chapter 1, Section VI) showing that substrates with a poor leaving group, such as physiological substrates of proteases, are hydrolyzed through general base rather than nucleophilic catalysis. In fact, the kinetic isotope effects observed in deuterium oxide as solvent are consistent with general base catalysis both in acylation and deacylation.[55,71] Although deuterium isotope effects may also be consistent with the pre-equilibrium formation of *N*-acyl imidazole followed by rate-determining attack of water in deacylation, this mechanism is inconsistent with the fact that deacylation is dependent on a basic group of pK_a about 7.[2]

It was later suggested that in the case of activated substrates like nitrophenyl esters, the acylation process is facilitated by nucleophilic catalysis involving the active site histidine side chain.[72,73] Notwithstanding other difficulties,[3,74] nucleophilic catalysis by imidazole in acylation and general base catalysis in deacylation is incompatible with the principle of microscopic reversibility and symmetry arguments[2] to be discussed next.

C. The Basic Mechanism

As it was pointed out in the preceding section, both acylation and deacylation depend on the ionization of the same catalytic group. Furthermore, the rate constants for the two steps decrease to a similar extent in ^2H_2O. This implies a mechanistic equivalence of acylation and deacylation.[75,76] A similarity of the two mechanisms is also supported by isotope exchange reactions. Specifically, in the chymotrypsin-catalyzed exchange of ^{14}C-methyl ester substrates with methanol,[77] acylation and deacylation (methanolysis) represent the same catalytic process (Equation 1), inasmuch as the serine side chain and methanol are chemically equivalent.

$$^{14}CH_3OH$$

$$RCO-O^{14}CH_3 + E \rightleftharpoons RCO-O^{14}CH_3.E \longleftarrow$$

$$RCO\!-\!E$$

$$RCO\!-\!OCH_3 + E \rightleftharpoons RCO\!-\!OCH_3.E \longleftarrow \tag{1}$$

$$CH_3OH$$

The exchange reaction with the nonlabeled methanol can be monitored by determining the changes in the relative isotope content of the starting ester substrate in the course of the catalysis. As the reaction proceeds, the specific radioactivity of the substrate decreases due to its reformation from the nonlabeled methanol.[77]

The mechanistic equivalence of acylation and deacylation demands both of these reaction steps to proceed through an intermediate.[75,76,78] This follows from the principle of microscopic reversibility (Chapter 1, Section XIV) which requires that the microscopic reverse of general base catalysis must be general acid catalysis. Such a system, corresponding to acylation in one direction and deacylation in the other, is illustrated by Equation 2.

$$A \underset{\text{general acid catalysis}}{\overset{\text{general base catalysis}}{\rightleftharpoons}} I \underset{\text{general base catalysis}}{\overset{\text{general acid catalysis}}{\rightleftharpoons}} B \tag{2}$$

Equation 2 indicates that, in either direction, the catalytic process involves general base catalysis followed by general acid catalysis. If an intermediate (I) were not present on the reaction path, the reaction would be catalyzed by a general base in one direction and a general acid in the other direction, as prescribed by the principle of microscopic reversibility. However, different mechanisms in the forward and reverse directions would conflict with the dependence of acylation and deacylation reactions on the *same base* of pK_a 7. The postulation of an intermediate is also consistent with the mechanism established for simple organic reactions, such as the base-catalyzed hydrolysis and alcoholysis of esters. These reactions are known to proceed through tetrahedral addition intermediates (Chapter 1) which suggests that the enzymic reaction also does. Accordingly, the mechanism of action of serine proteases may be depicted as in Figure 2.

Figure 2 shows the basic features of the mechanism deduced from extensive studies on chymotrypsin. Both the formation and the hydrolysis of the acyl-enzyme, i.e., acylation and deacylation, proceed by the following two steps. First, a nucleophilic attack of the hydroxyl group of serine or water on the carbonyl carbon atom of the substrate. This process, leading to the formation of a tetrahedral intermediate, is catalyzed by a histidine residue as a general base. Second, breakdown of the tetrahedral intermediate, which results in the formation of an acyl-enzyme or free enzyme, is catalyzed by an imidazolium ion as a general acid. The negative ρ value (Chapter 1, Section VIII.B) obtained for the hydrolysis of anilide substrates by chymotrypsin is consistent with a proton transfer from the imidazolium ion to the leaving aniline.[79,80] However, it should be kept in mind that the specification of general

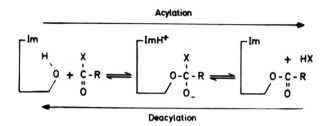

FIGURE 2.   Scheme of the mechanism of action of serine proteases. X stands for an OR′ or NHR′ group in acylation and for an OH group in deacylation.

acid catalysis by the imidazolium ion, as well as the formation of the tetrahedral adduct, is deduced principally from the combination of microscopic reversibility and the symmetry principle and not from direct experimental observation. Efforts expended on experimental demonstration of the tetrahedral intermediate will be discussed in Section VII.

The mechanism illustrated in Figure 2 satisfies all experimental data with the exception that it does not predict a bell-shaped pH-rate profile, as has been found for the deacylation of chymotrypsin with nucleophiles that ionize in the experimentally available pH region such as glycine hydroxamic acid, phenylacetohydroxamic acid, and isonitrosoacetone (Figure 3).[81-83] These deacylation reactions depend on two groups, one with the pK_a of the active site histidine and the other with the pK_a of the nucleophile ($pK_a = 8.3$ for isonitrosoacetone). In hydrolysis and methanolysis, the right hand leg of the bell-shaped curve of deacylations cannot be demonstrated because water and methanol do not ionize in the pH region available for enzymic reactions. On account of the higher reactivity of the nucleophile in its anionic form rather than in its un-ionized form, Figure 2 predicts a double sigmoid rather than a bell-shaped pH-rate profile observed in deacylation with isonitrosoacetone (Figure 3). To overcome this difficulty, a catalytic function must be given to the proton on the nucleophile.[2,76] This requirement is fulfilled only if the same proton participates both in the general base- and in the subsequent general acid-catalyzed steps.

In the absence of the special enzymic environment, the proton accepted from the nucleophile by the general base would immediately be released to a solvent water molecule. However, the proton can be preserved for the next catalytic step, i.e., for the transfer to the leaving group if the positive imidazolium ion is stabilized by an interaction with the negative tetrahedral intermediate.[84] This interaction implies the formation of an intimate ion-pair such as that which holds the proton between the entering and leaving atoms of the tetrahedral intermediate as depicted in Figure 4. In such a constellation of the groups which involves a bifurcated hydrogen bond, the proton can be transferred from the nucleophile through the imidazole to the leaving group without the possibility of being lost to the solvent. Thus, with the stipulation that the tetrahedral intermediate and the imidazolium ion must form an intimate ion-pair, the scheme in Figure 2 may correctly describe the basic features of the mechanism of action of serine proteases.[84]

II. THE THREE-DIMENSIONAL STRUCTURE

A. The Structure of the Catalytic Site

The extensive kinetic and modification studies on chymotrypsin, as discussed in the preceding section, led to the well-established mechanism depicted in Figure 2. This mechanism, however, does not bear any stereochemical feature that would permit a much deeper insight into the mechanism of action. It was, therefore, a breakthrough in the mechanistic investigation of serine proteases when the first information was reported on the steric structure

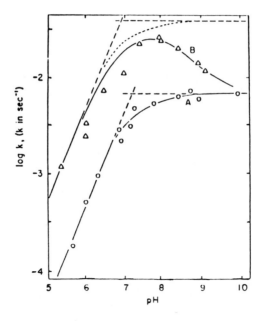

FIGURE 3. The pH dependence of the rate constants of deacylation of acetyl-chymotrypsin in the presence (Δ, curve B) and absence (o, curve A) of isonitroso-acetone. (From Wedler, F. C., Killian, F. L., and Bender, M. L., *Proc. Natl. Acad. Sci. U.S.A.*, 65, 1120, 1970. With permission.)

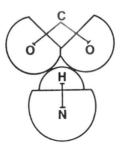

FIGURE 4. Space filling model representing an interaction between the tetrahedral intermediate and the imidazolium ion. The tetrahedral form of the substrate carbonyl carbon (C) is linked to the serine and the leaving oxygen atoms. Only the protonated NE2 of the imidazolium ion is shown. (From Polgár, L., *J. Theor. Biol.*, 31, 165, 1971. With permission.)

of chymotrypsin.[85] The data obtained by X-ray diffraction measurements have shown the serine and histidine residues in a proper position to function as implied by the above mechanism (Figure 2). Figure 5 shows the active site region of chymotrypsin, including OG (O$^\gamma$) of Ser 195 and NE2 (N$^{\epsilon 2}$) of His 57, which represent the nucleophile and the general base, respectively. The other nitrogen atom, ND1 (N$^{\delta 1}$), of the imidazole ring is hydrogen-bonded to OD2 of Asp 102. The OD1 of Asp 102 is also engaged in hydrogen bond formation, namely with the main chain nitrogen of His 57. Also, OD2 of Asp 102 forms a hydrogen bond with OG of Ser 214 (not shown in Figure 5). Asp 102 is shielded from the solvent by

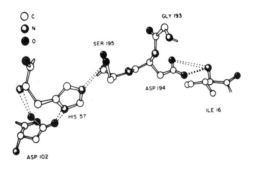

FIGURE 5.   The conformation of a few amino acids in the active center of chymotrypsin. Potential hydrogen bonds are illustrated with dashed lines. (From Blow, D. M. and Steitz, T. A., *Annu. Rev. Biochem.*, 39, 63, 1970. With permission.)

the side chains of Ala 55, Ala 56, His 57, Cys 58, Tyr 94, Ile 99, and Ser 214.[86] Two water molecules are, however, buried close to the carboxylate group.[87] The interaction between Ile 16 and Asp 194 (Figure 5) is generated on the activation of chymotrypsinogen (see Sections I.B and II.B).

On discovering the buried aspartate, Asp 102, Blow and his co-workers[86] speculated about its functional role. They suggested that electrons from Asp 102 are channelled through hydrogen bonds and the imidazole ring to Ser 195 whereby the OG of Ser 195 acquires a significant amount of negative charge. The Ser . . . His . . . Asp triad, which they called the charge relay system, would account for the enhanced nucleophilic reactivity of Ser 195.[86] Although this idea became very popular in the literature, later studies could not verify the charge relay mechanism as summarized in recent reviews.[17,18] Instead the catalytic triad appears to work as a charge stabilizing system (Section VI.A).

Difference electron-density maps of tosyl-α-chymotrypsin[88] and the complex formed between chymotrypsin and formyl-L-tryptophan[89] have demonstrated the nature of the specificity site and suggested how the scissile bond of a real substrate might be oriented.[89,90] It was shown that the carboxylate group of formyl-L-tryptophan is fixed in the vicinity of Ser 195 and His 57 by binding of the indolyl side chain in a pocket and by interaction of the -NH- of the formylamido group with the backbone carbonyl oxygen of Ser 214, which is an S_1-P_1 hydrogen bond (see Chapter 2, Section V for this notation). The hydrogen bonding capacity of the acylamido group has been demonstrated by kinetic specificity studies in the case of chymotrypsin,[91,92] as well as in the subtilisin catalysis.[92]

A further mechanistically important discovery by X-ray diffraction studies on chymotrypsin is the oxyanion-binding site, which consists of two peptide backbone -NH- groups belonging to Gly 193 and Ser 195, respectively.[90] These main chain groups may be implicated in stabilizing the tetrahedral intermediate by forming hydrogen bonds with its negatively charged oxygen atom. This catalytic subsite, as a general feature of serine proteases, was called oxyanion hole when a similar site was also observed in subtilisin.[93] The importance of the oxyanion binding site in the catalysis by serine proteases will be discussed in Section VI.B.

The structure of α-chymotrypsin has most recently been determined by X-ray diffraction at high resolution (0.167 and 0.168 nm).[94,95] This crystalline form of chymotrypsin contains dimers. The two independently refined molecules exhibited the same basic structure, but differences in the positions of the side chains were noted.[95] Comparison of these two molecules with the monomeric γ-chymotrypsin refined at 0.19 nm resolution also revealed differences in the positions of several side chains.[95,96] As mentioned in Chapter 2, Section III, γ-chymotrypsin is chemically identical with α-chymotrypsin.

Besides chymotrypsin, the structure of several other serine proteases has been solved at high resolution: notably, β-trypsin at 0.15 nm,[97,98] α-lytic protease at 0.17 nm,[99] *Streptomyces griseus* trypsin at 0.17 nm,[100] *Streptomyces griseus* protease A at 0.18 nm,[101] and porcine pancreatic kallikrein at 0.20 nm.[102] Also, structure determinations were performed at medium resolution for elastase at 0.25 nm,[103] subtilisin BPN' at 0.25 nm,[13,104,105] subtilisin Novo at 0.28 nm,[106] and *Streptomyces griseus* protease B at 0.28 nm.[107]

The catalytic triad occupies practically the same steric position in all serine proteases.[104] This is of particular interest for subtilisin and chymotrypsin having entirely unrelated amino acid sequences (Chapter 2, Section I). The sequential order is His 57, Asp 102, Ser 195 in chymotrypsin, and Asp 32, His 64, Ser 221 in subtilisin. The similarity goes beyond the triad inasmuch as a second hydrogen bond is provided to the aspartate side chain in both enzymes. In subtilisin, the hydrogen bond is provided by the buried side chain OH group of Ser 33 to OD2 of Asp 32, whereas in chymotrypsin the OH group of Ser 214 plays the corresponding role.

B. The Constellation of the Catalytic Residues in the Zymogen

The knowledge of the spatial arrangement of the catalytic triad in the practically inactive zymogen of chymotrypsin or trypsin may be an important clue to the understanding of the mechanism of action. Evidence that zymogen forms are not completely inactive was obtained for trypsinogen, which can catalyze its own activation.[108] A rough calculation suggested that the zymogen has approximately 10^{-5} of the activity of trypsin. Chymotrypsinogen also exhibits a very small but significant reactivity.[109-112] Comparison of the first X-ray structure of chymotrypsinogen[58,113] with that of chymotrypsin revealed no major difference in the structure of the catalytic triad. The activation cleavage of the peptide bond between Arg 15 and Ile 16 (Chapter 2, Section III, Figure 4), however, resulted in movements of segment 16 to 21 and segment 191 to 194. As a consequence, the Asp 194 side chain rotated to form an ion-pair with the α-ammonium group of Ile 16 (Figure 5). Furthermore, segment 189 to 192 became more exposed to the surface so that the S_1-specificity pocket and the oxyanion hole, which are incomplete in the zymogen, could be completed. A detailed description of the structure of chymotrypsinogen determined at 0.18 nm resolution has recently been reported.[114] Comparison with the high resolution structure of α-chymotrypsin[96] confirmed the earlier finding[113] that the inadequate formation of the specificity pocket and the absence of the oxyanion hole are the main reasons for the extremely low catalytic power of chymotrypsinogen. It was also concluded that the autolysis loop containing residues 142 to 153 exhibits considerable flexibility, which explains why the second cleavage in the polypeptide chain (Chapter 2, Section III, Figure 4) occurs in this region.

Essentially the same structural changes could be observed when trypsinogen[15,115,116] was compared to trypsin. In trypsinogen a so-called activation domain has been distinguished,[15,116] which comprises the amino terminal segment Val 10 to Gly 19 (chymotrypsin numbering), and three internal chain segments Gly 142 to Pro 152, Gly 184 to Gly 193, and Gly 216 to Asn 223. For these segments no significant electron density is seen on the Fourier map of trypsinogen even at low temperature (103 K), which is consistent with predominantly static disorder of the activation domain.[117] Upon the activation cleavage, which results in the release of the N-terminal hexapeptide, the activation domain becomes clearly visible on the electron-density map indicating a well-defined structure for the active enzyme. This may be essential for the catalysis. The active structure also involves a small pocket for the new N-terminal dipeptide, Ile 16-Val 17.[15,118,119] Interestingly, in the complex formed between trypsinogen and the pancreatic trypsin inhibitor (Kunitz), the zymogen adopts a rigid, trypsin-like conformation for the activation domain.[15] The Ile-Val binding site is also formed in the complex, but remains empty since Ile 16 is blocked by the activation hexapeptide in the zymogen. The binary complex can strongly bind exogenous Ile-Val dipeptide, thereby forming a ternary complex. The small inhibitor *p*-guanidobenzoate also causes a similar ordering of trypsinogen as the pancreatic trypsin inhibitor does, but only in the presence of the Ile-Val dipeptide.[118,119]

The self-protecting mechanism of zymogen formation is absent in the extracellular bacterial enzymes such as *Streptomyces griseus* proteases A and B or α-lytic protease, which belong to the chymotrypsin family (Chapter 2, Section II). Instead of Ile 16, the buried Arg 138 forms the ion-pair with Asp 194.[16,99] When the bacterial enzymes were compared to chymotrypsin, a major structural difference was found at the N-terminal parts. James[16] has suggested that evolution to the precursor system necessitated the generation of a longer stretch of N-terminal polypeptide chain in the mammalian enzymes to allow Ile 16 to approach Asp 194.

III. SUBSTRATE BINDING

A. The S$_1$ Binding Site

While the catalytic groups implicated in the peptide-bond cleavage exhibit virtually identical conformation in the various serine proteases, the binding sites have been modified in the course of evolution giving rise to enzymes of different specificity. The primary specificity of chymotrypsin is directed by the P$_1$ amino acid residue that gives its carbonyl group to the sensitive peptide bond. Therefore, simple amino acid derivatives, instead of peptides, are generally suitable for most studies with chymotrypsin. In this enzyme the S$_1$-specificity site is mediated by a well-defined pocket which can accept various aromatic groups. A difference Fourier map has shown that the indolyl group of formyl-L-tryptophan is located in the pocket which is close to Ser 195.[89] The two sides of the pocket were found to be composed of the peptide bonds of residues 214 to 216 and 190 to 192, respectively. At the bottom of the pocket Ser 189 was situated.

Trypsin-like enzymes (kallikrein, plasmin B-chain, thrombin B- chain, urokinase, factor IX$_a$, factor X$_a$, and nerve growth factor γ-subunit) are specific toward basic residues, arginine and lysine, and this is explained by the existence of an aspartate in place of Ser 189 at the bottom of the pocket. The positioning of the lysine and arginine side chains in the pocket of trypsin were studied by X-ray diffraction measurements with the use of the complexes of trypsin formed with the pancreatic trypsin inhibitor (Kunitz)[120] and benzamidine.[98,115,121] The side chain of Lys 15 of the pancreatic trypsin inhibitor is inserted into the specificity pocket on the formation of the complex.[120] Interestingly, the lysine amino group does not bind directly to the aspartate carboxyl group, but by utilizing its entire hydrogen-bonding capability, it donates hydrogen bonds to the carbonyl oxygen of Ser 190 and two water molecules, one of which is bonded to Asp 189. On the other hand, benzamidine directly forms hydrogen bonds to the carboxylate of Asp 189 as well as to the OG of Ser 190, to the carbonyl oxygen of Gly 219, and to a water molecule. Similar to benzamidine, the longer arginine side chain is expected to extend deeper into the binding pocket than the lysine side chain and may form a cyclic network of direct hydrogen bonds with Asp 189.

The three-dimensional structure of pig pancreas kallikrein has also been determined.[102] Kallikreins liberate vasodepressor peptides from kininogen (Chapter 2, Section III). Several other endopeptidases, such as the γ-component of mouse nerve growth factor, are closely related to pancreatic kallikrein. The specificity pocket of kallikrein is significantly enlarged compared to that of trypsin due to a longer peptide segment between residues 217 and 220, and to the unique outward orientation of the carbonyl group of *cis*-Pro 219. The Ser 226 (the corresponding residue is glycine in trypsin and chymotrypsin) partially covers Asp 189 at the bottom of the pocket. Furthermore, Tyr 99, which is a leucine residue in trypsin, protrudes into the binding site and interferes with the binding of peptide substrates. These structural differences explain the greater specificity of kallikrein with respect to trypsin. In other words, kallikrein does not split the peptide bonds at basic residues equally well because secondary specificities, especially at the S_2 site, modify its selectivity.[102] Most unusually, in its natural substrate, kininogen, kallikrein cleaves a Met-Lys bond, probably due to the favorable binding by P_2, P_1' and, in particular, P_2' residues[122] shown in form [3]. The other cleavage site (Arg-Ser), which is normal, is also illustrated in the partial amino acid sequence [3].

P_2 P_1 P_1' P_2' P_1 P_1'
-Ser-Leu–Met-Lys-Arg-Pro-Pro-Gly-Phe-Ser-Pro-Phe-Arg-Ser-Val-Gln-

[3]

The deep substrate binding pocket observed in chymotrypsin, trypsin, and kallikrein is occluded by the bulky side chains of Val 216 and Thr 226 in elastase.[103,123] These residues are replaced by glycine in trypsin and chymotrypsin. (In kallikrein residue 226 is serine.) Consequently, only relatively small aliphatic side chains, such as the methyl group of alanine, can be accomodated by the shallow cavity. The microbial α-lytic protease has a similar specificity to elastase, but it is achieved in a different way. Here, residue 216 is glycine, but the side chains of other residues (a methionine and a valine) are directed into the binding pocket.[124] *Streptomyces griseus* protease A has a larger binding pocket that can accomodate a phenylalanine side chain.[125]

Besides studying the natural variations in the binding pocket, a recent investigation employed site-specific mutagenesis to alter the specificity site of trypsin.[126] Glycine residues at positions 216 and 226 in the binding cavity were replaced by alanine residues resulting in three trypsin mutants. The specificity rate constants (k_{cat}/K_m) measured with specific tripeptide substrates were reduced for all three mutants. For the Ala 226 mutant, and in particular for the double mutant, the catalytic activity was virtually extinguished. The observed reductions in rate were different for the lysine and arginine substrates.

Another interesting specificity study has most recently been reported.[127] The Asp 189 residue at the bottom of the specificity pocket of rat trypsin has been replaced by Lys. This mutation could result in reversing the specificity of trypsin toward substrates containing acidic residues. As expected, the mutant enzyme was not able to hydrolyze arginyl and lysyl substrates, but displayed no activity toward aspartyl and glutamyl peptides either. The intrinsic, very low level of trypsin activity toward hydrophobic residues was retained in the mutant enzyme with some modification. Computer graphic modeling provided a possible explanation of these results. Specifically, in one of the models, the $-NH_3^+$ group of Lys 189 is directed outside the pocket forming two or three hydrogen bonds with neighboring oxygen atoms of main chain carbonyl groups. In this orientation, the amino group of Lys 189 is not accessible to the carboxylate group of the substrate, and the bottom of the pocket becomes nonpolar.[127] This would make the binding and desolvation of any charged side chain thermodynamically unfavorable.

Although belonging to a different family of serine proteases, subtilisin has a similar specificity to chymotrypsin, but its binding pocket is more open. One side of this crevice is made up of Ser 125, Leu 126, and Gly 127, which corresponds to residues 214 to 216 in chymotrypsin. The other wall is a less regular surface than that found in chymotrypsin and is composed of Ala 152, Ala 153, and Gly 154.[14]

B. The Extended Binding Site: the S-P Interactions

In the case of polypeptide substrates, the binding site can be extended over several amino acid residues. This is of particular importance in the case of serine proteases having a small pocket and thus, weak binding interaction at the S_1 subsite, e.g., elastase, and α-lytic protease

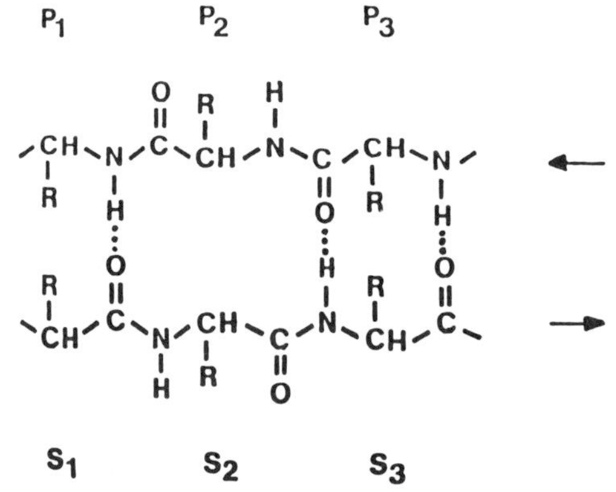

FIGURE 6. Schematic representation of the antiparallel β-pleated sheet in substrate binding to serine proteases. Adjacent strands run in opposite directions.

(Chapter 3, Section III. A). Because of the importance of secondary specificities in these instances, the assignment of primary specificity to the S_1 binding site may not be justified, contrary to the widely used practice (Chapter 2, Section V).

The structure of the extended binding site was first demonstrated in γ-chymotrypsin.[128] Polypeptides with an L-phenylalanine chloromethyl ketone at their carboxyl terminus were bound to the enzyme and located by difference Fourier technique. It was found that the backbone polypeptide segment of Ser 214, Trp 215, and Gly 216 forms an antiparallel β-structure with the polypeptide chain of the inhibitor. As it was mentioned before (Chapter 3, Section III.A), the same residues also constitute one wall of the binding pocket.

The most important hydrogen bonds to be formed between a serine protease and the acylating part of the substrate are depicted in Figure 6. These are found (1) between the backbone carbonyl oxygen of S_1 and the amido nitrogen of P_1, (2) between the backbone amido nitrogen of S_3 and the carbonyl oxygen of P_3, and (3) between the backbone carbonyl oxygen of S_3 and the amido nitrogen of P_3. This binding geometry appears to be common to most serine proteases studied so far. However, in trypsin an S_3–P_3 hydrogen bond cannot be formed because of a deletion in this region of the amino acid sequence which causes the carbonyl oxygen of residue 216 to turn inward.[120,129]

In addition to the antiparallel β-binding, hydrophobic interactions at subsites S_1, S_2, and S_4 were also observed, but the side chain of P_3 pointed away of the enzyme surface. These results were obtained from X-ray diffraction studies on the binding of specific tripeptide chloromethyl ketones to *Streptomyces griseus* protease B[130] and also from the binding of an aldehyde inhibitor, Ac-Pro-Ala-Pro-Phe-CHO, and two products, Ac-Pro-Ala-Pro-Phe-COO$^-$ and Ac-Pro-Ala-Pro-Tyr-COO$^-$ to *Streptomyces griseus* protease A.[131] The two products only bind at low pH, where an electrostatic interaction between the carboxylate ion of the product and the active site imidazolium ion is possible. The phenylalanine at position P_1 was used because the two microbial proteases, similar to chymotrypsin, preferentially bind hydrophobic side chains.[132,133] However, a substrate binding at S_4 subsite as observed in the microbial enzyme is not possible in chymotrypsin without disrupting the hydrogen bond from the peptide -NH- of P_3 Ala to the CO of Gly 216.[131] This is consistent with kinetic results showing that specific tetrapeptides bind to chymotrypsin with reduced affinity compared to the corresponding tripeptides.[132] Most recently the complex formed between *Streptomyces griseus* protease A and chymostatin has been described in detail.[125] Chymostatin is a small peptide inhibitor produced by different species of *Streptomyces*.[134] The unusual chemical structure of the inhibitor is illustrated in form [4]. The novel features include a cyclized arginine residue and a ureido group which reverses the direction of the polypeptide chain for the P_4 phenylalanine residue.

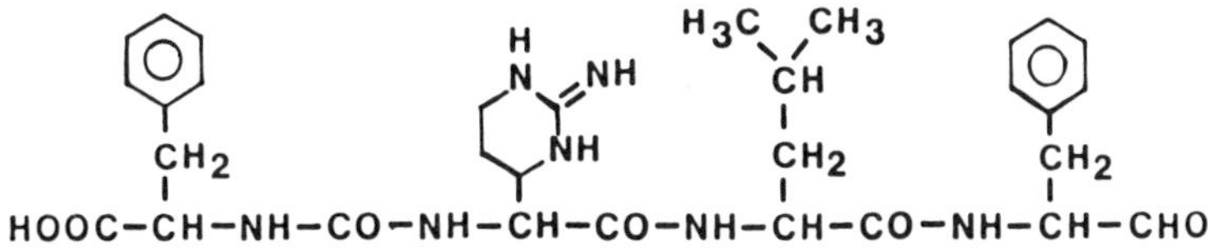

[4]

The X-ray diffraction analysis of the complex[125] has shown a more restrictive S_1 binding pocket for the P_1 phenylalanyl side chain than that found in chymotrypsin. The P_2 leucyl residue of the inhibitor is bound in a surface depression which is hydrophobic in nature. The leucyl side chain appears to fill completely the available volume of this binding subsite. The side chain of the cyclized arginine is projected away from the surface of the enzyme. An interesting feature of the inhibitor binding is that the positively charged side chain of the cyclized P_3 arginine and the terminal P_4 phenylalanine carboxylate ion appears to be stacked roughly coplanar.[125]

In the case of α-lytic protease, a crystallographic approach to studies of ligand binding was not possible due to packing interactions, which occlude the active site in the crystals.[124] However, kinetic data indicated that the active site of α-lytic protease extends over at least six subsites (S_4 to S_2').[135] The binding at this extended active site increases k_{cat}/K_m more than 10^6-fold on going from Ac-Ala-NH_2 to a hexapeptide amide. For the related pancreatic elastase, the increase is more than 10^7-fold,[136] whereas chymotrypsin exhibits an approximate 4000-fold increase for a similar series of substrate but with phenylalanine at position P_1.[133]

The differences in the specificity of the above three enzymes can be rationalized in terms of the different hydrophobic interactions at subsites S_1, S_2, and S_4. In chymotrypsin the S_1-P_1, in α-lytic protease the S_2-P_2, and in elastase the S_2-P_2 and in particular the S_4-P_4 interactions are important.[135,136] Kinetic investigations have also shown some contribution by individual subsites in the catalysis by trypsin.[137]

Serine proteases appear to share a specific way to interact with proline residues which can bind at subsites S_2 and S_4 but not at subsite S_3. This is evident from perusal of the antiparallel β-binding structure of enzyme-substrate complexes, which shows that the residues in position P_3 forms two hydrogen bonds with the main chain of the enzyme at residue 216 (Figure 6). However, an acylated proline at position P_3 would be unable to offer a free -NH- group to the main chain carbonyl oxygen.[138] A similar rationale pertains to the S_1-P_1 hydrogen bond which involves the -NH- group of the P_1 residue and the carbonyl oxygen of Ser 214 (chymotrypsin family) or Ser 125 (subtilisin family). This may be one of the reasons why the chymotrypsin- and subtilisin-type enyzmes do not hydrolyze substrates with proline at P_1 position. Another reason may arise from the steric hindrance caused by the rigid ring structure. There are special serine proteases with a "post-proline cleaving activity", but the structure of their active site is unknown.[138]

It was suggested that, when proline is at the P_2 position, chymotrypsin only cleaves the peptide containing the *trans*-isomer [5],[139] which in contrast to the *cis*-isomer [6], has its carbonyl oxygen in the proper position to take part in the antiparallel β-sheet formation. Later studies, however, indicated that chymotrypsin does hydrolyze peptides containing *cis*-proline at P_2 position, although at a lower catalytic efficiency by a factor of 20,000.[140] An even more expressed *cis-trans* isomeric specificity is encountered in the trypsin-catalyzed hydrolysis of substrates having proline at P_2' of the leaving group side.[140,141] Furthermore, proline-specific endopeptidase (post-proline cleaving enzyme)[142] and dipeptidyl peptidase

IV,[143] which hydrolyze peptides with proline in P_1 position, can also distinguish between the geometrical isomers.

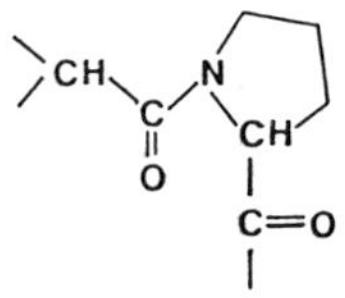

[5]

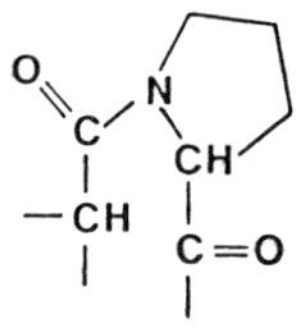

[6]

An extended substrate binding site similar to that observed in chymotrypsin was also found in subtilisin.[14] In the bacterial enzyme the β-binding segment involves Ser 125, Leu 126, and Gly 127,[144,145] which correspond to residues 214 to 216 in chymotrypsin. Moreover, the antiparallel β-segment was found to extend up to residue P_6 in the subtilisin-*Streptomyces* subtilisin inhibitor complex.[146]

A special feature of polypeptide binding to protease K has most recently been observed.[147] Protease K is produced by the fungus *Tritirachium album*. Its three-dimensional folding exhibits a high degree of homology with that of subtilisin. The catalytic triad is composed of Asp 38, His 68, and Ser 221, the corresponding residues being Asp 32, His 64, and Ser 221 in subtilisin. The binding mode of a dipeptide chloromethyl ketone was revealed by the difference Fourier technique. This showed a three-stranded antiparallel pleated sheet involving the central strand provided by the inhibitor and two other strands by the enzyme: (1) Ser 131, Leu 132, and Gly 133, and (2) Gly 99. The peptide carbonyl group of Gly 99 formed a hydrogen bond with the -NH- group of residue P_2. This residue is not involved in hydrogen bond formation in the case of the chymotrypsin family of enzymes (Figure 6).

A remarkable difference between protease K and the subtilisins is found in their cysteine content.[148] Whereas subtilisins do not contain cysteine residues, protease K has five cysteines, four of which are involved in disulfide bond formation and the remaining Cys 72 is free, and located in the vicinity of the catalytic groups. There are other serine proteases that contain a cysteine residue near the active site: thermitase,[149] thermomycolin,[150] and a protease from the *Bacillus thuringiensis*.[151] The importance of this free cysteine, if any, is not known.

Kinetic investigations on thermitase have shown that the affinity of peptide inhibitors to the enzyme increases with the elongation of the peptide chain up to four residues.[152] With elongation from the S_1 subsite toward the S_4 subsite, the enzyme exhibits diminished stereospecificity.[153]

The specificity characteristics of microbial proteases was reviewed earlier.[154]

C. Protein-Protease Inhibitors: the S′-P′ Interactions

An important approach aimed at studying the structure of extended binding sites utilizes X-ray diffraction measurements of complexes formed between proteases and protease inhibitors, such as discussed in Chapter 2, Section IV. These inhibitors possess a large

TG-PSTI

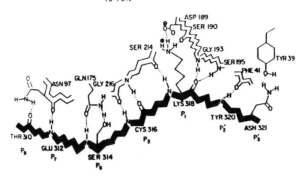

FIGURE 7. Schematic drawing of the primary contact region of trypsinogen-pancreatic secretory trypsin-inhibitor complex. (From Bolognesi, M., Gatti, G., Menegatti, E., Guarneri, M., Papamokos, E., and Huber, R., *J. Mol. Biol.*, 162, 839, 1982. With permission.)

polypeptide segment, which is complementary to the binding site of the target protease, on both the acyl (S subsites) and the leaving group (S' subsites) side. Actually, these complexes seem to provide the most reliable structural information about the leaving group side interactions. Most recently, peptide fluoromethyl ketone inhibitors have been designed to interact with chymotrypsin at both the S and S' subsites.[154a]

The three-dimensional structures of pancreatic trypsin inhibitor (Kunitz)[155] and of its various complexes with bovine β-tryspin, and trypsinogen have been determined at atomic resolution and reviewed.[15] The structure of soybean inhibitor (Kunitz) complexed with porcine trypsin has also been elucidated.[129] Later X-ray crystallographic investigations revealed the structures of a number of other protein-protease inhibitors, such as the *Streptomyces* subtilisin inhibitor[156] and its complex with subtilisin,[146,157] the complexes of the third domain of the turkey ovomucoid inhibitor with *Streptomyces griseus* protease B[158,159] and with human leukocyte elastase,[159a] the third domain of the Japanese quail ovomucoid,[160,161] the complex formed between the porcine Kazal inhibitor and bovine trypsinogen,[162,163] pancreatic kallikrein A complexed with bovine trypsin inhibitor (Kunitz)[122] and most recently the third domain of silver pheasant ovomucoid,[164] and two inhibitors of the potato inhibitor 1 family: chymotrypsin inhibitor 2 in complex with subtilisin Novo[165] and eglin from leeches in complex with subtilisin Carlsberg.[166-167a]

The basic pancreatic trypsin inhibitor (Kunitz) was the first protein-protease inhibitor analyzed by X-ray crystallography. The complementary binding site of the inhibitor was shown to extend from P_3 to P'_2 in the complex formed with trypsin.[15,120] Moreover, an S'_4–P'_4 hydrogen bond was also seen in the complex.[15] The porcine secretory trypsin inhibitor (Kazal) exhibited a different picture in the complex formed with bovine trypsinogen.[162,163] The contact loop of the inhibitor displayed a rather extended conformation involving residues P_9, P_7, P_6, P_5, P_3, P_1, P'_1, P'_2, and P'_3 (Figure 7). It should be noted that the binding site region of trypsinogen in the complex assumes a conformation similar to that of the active trypsin.

Other inhibitors of the Kazal family, the third domains of ovomucoid (Chapter 2, Section IV.B) from turkey,[158,159] Japanese quail,[160,161] and silver pheasant[164] have been analyzed crystallographically. They exhibit virtually the same structure as the Kazal inhibitor. Interestingly, some changes in the amino acid sequence occur near the reactive site (Figure 8). From its relatively high temperature factor, it appears that the reactive site loop is somewhat flexible, which may help the inhibitor in adapting itself to the binding site of slightly different serine proteases.

Inhibitor	P_3	P_2	P_1	P_1'	P_2'
Japanese quail	-Cys-Pro-Lys-Asp-Tyr-				
Turkey	-Cys-Thr-Leu-Glu-Tyr-				
Pheasant	-Cys-Thr-Met-Glu-Tyr-				

FIGURE 8. Amino acid sequences around the reactive site bond of ovomucoid inhibitors from Japanese quail, turkey, and silver pheasant. P_3 to P_2' stand for residues 16 to 20.

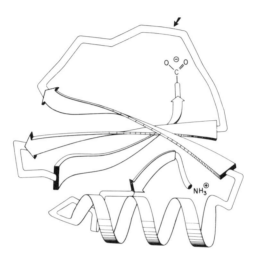

FIGURE 9. A representation of the secondary structural elements of chymotrypsin inhibitor 2. Small arrow at top indicates the reactive site. Large arrows indicate β-strands, wide ribbon is α-helix, and narrow ribbons are turns or unclassified structures. (From McPhalen, C. A., Svendsen, I., Jonassen, I., and James, M. N. G., *Proc. Natl. Acad. Sci. U.S.A.*, 82, 7242, 1985. With permission.)

Part of the contact loop (residues P_4 to P_9) of the *Streptomyces* subtilisin inhibitor is even more flexible than that of the Kazal inhibitors as indicated by the observation that the relevant electron densities are very weak (i.e., the thermal parameters are large) or almost nonexistent. In the complex with subtilisin BPN', however, the densities are clearly visible, indicating the rigidity of the loop. This complex represents again a new binding alternative, inasmuch as the antiparallel β-sheet extends over six residues.[146] It should be emphasized, however, that the steric structure near the scissile bond, as expected, is virtually the same for all complexes examined so far.

Chymotrypsin inhibitor 2, a protein from barley seeds, is a member of the potato inhibitor 1 family of the serine protease inhibitors (Chapter 2, Section IV.C). This family also includes eglin, an inhibitor from the leech *Hirudo medicinalis*. In contrast to most protease inhibitors, these two proteins lack stabilizing disulfide bonds. They have an almost identical three-dimensional structure which consists of a four-stranded mixed parallel and antiparallel β-sheet against an α-helix.[165-167] The interface between the helix and the sheet comprises the hydrophobic core. A scheme of the secondary structural elements is shown in Figure 9. This elaborate hydrogen-bonding architecture may partly compensate for the lack of disulfide bonds. A wide loop connecting two strands of the β-sheet contains the reactive site. The loop is held in its exposed conformation mainly by the side chains of two arginine residues protruding from the hydrophobic core. The arginines interact with the P_2 and P'_1 residues of the loop through hydrogen bond and electrostatic linkages.[165,167] The conformation of the four residues to either side of the reactive site bond is similar to that of the analogous residues of the Kazal-type inhibitors, although the overall polypeptide chain fold of the two types of inhibitors and the location of the reactive site in the respective polypeptide chains are basically different.

Residues P_1 to P_4 of the chymotrypsin inhibitor 2 and eglin form the central strand of a short three-stranded antiparallel β-sheet with Gly 100-Gly 102 and Ser 125-Gly 127 of subtilisin.[165,167] A similar sheet was also found in the complex formed between a dipeptide chloromethyl ketone and protease K[147] (Chapter 3, Section IV.B). On the other hand, in the *Streptomyces* subtilisin inhibitor the P_2 residue is proline which cannot form a hydrogen bond with the carbonyl oxygen of Gly 100. Instead, Gly 102-Tyr-104 form a β-sheet with the P_4 to P_6 residues.[146]

It is of interest that eglin in complex with subtilisin is shortened by seven residues at the amino terminus.[167] Residues 1 to 7 in intact eglin may be flexible and susceptible toward proteolysis. In this modified eglin the amino terminus (Lys 8) and the carboxy terminus (Gly 70) are linked through hydrogen bonds.

As it was mentioned at the beginning of this section, X-ray diffraction studies on protein-protease inhibitors provide the most valuable structural information about the interaction between the enzyme and the leaving group side (P' residues) of the substrate. Thus, the P'_1 side chain, which is alanine in the pancreatic trypsin inhibitor (Kunitz), and isoleucine in the soybean trypsin inhibitor (Kunitz), is in contact with a hydrophobic region of trypsin which includes the disulfide bridge formed between Cys 42 and Cys 58. The P'_2 residue is arginine in both inhibitors and forms a hydrogen bond with the backbone carbonyl oxygen of Phe 41.[15,120,129] The P'_2 residue is tyrosine in the Kazal-type inhibitors (Figure 7) and also in the chymotrypsin inhibitor 2 of the potato inhibitor 1 family of the serine proteases. In the complex formed between the latter inhibitor and subtilisin, the -NH- of P'_2, Tyr 61, forms a hydrogen bond to the carbonyl oxygen of Asn 218, and the side chain of Tyr 61 stacks with Phe 189 in a hydrophobic interaction.[165]

It appears that there are more important interactions between the enzyme and the substrate on the acyl group side (P residues) than on the leaving group side (P' residues). It is worth pointing out that this is conceivable in the light of the double displacement mechanism, i.e., catalysis involving a covalent acyl-enzyme. Notably, where the scissile bond is fixed predominantly by interactions at the leaving group side, then the specificity should be lost in the deacylation step, where the leaving group is absent.

IV. TRANSITION-STATE ANALOG INHIBITORS

In Chapter 2, Section VII, we discussed the importance of transition state theory in enzyme catalysis and pointed out the benefits of using transition-state analogs in mechanistic studies. As compared to substrates, the analogs bind much more tightly to the enzyme and do not transform into products. Transition-state analogs, thus, often serve as potent inhibitors. Their stable complexes with the target enzyme are suitable for X-ray crystallographic studies. It is worthy to emphasize that transition-state analogs in complexes with serine proteases resemble the metastable tetrahedral intermediate. Both the analogs and the intermediate are true covalent species. On the other hand, the transition state itself possesses *partially* broken and/or formed covalent bonds.

Some widely used serine protease inhibitors, such as diisopropyl phosphofluoridate and phenylmethanesulfonyl fluoride (Chapter 3, Section I.A), are oriented tetrahedrally at the active site. Also, phenylarsonic acids inhibit serine proteases probably by forming similar tetrahedral adducts.[168] However, because of the bulkiness of heteroatoms, phosphorus, sulfur, and arsenic, these inhibitors may not be considered as ideal models of the catalytic species. It appears that the tetrahedral adducts generated from boronic acid derivatives and peptide aldehydes bear a closer relationship to the structure of the true intermediate. There is, however, an important difference between these two analogs. Notably, the boronic acid derivative possesses a negative charge (Equation 3), whereas the hemiacetal adduct is a neutral species (Equation 4) as indicated by NMR studies.[169] Other NMR data suggest that a hemiketal adduct formed with trypsin might bear some negative charge at highly alkaline pH if it is stabilized by the protonated form of the active site imidazole.[170] It appears then that at acidic pH where most X-ray diffraction measurements were performed, the boronic acid adduct is a better transition-state analog than the hemiacetal, although the negative charge is on the boron atom, whereas in the true intermediate it is located on the carbonyl oxygen (Figure 2).

$$
\text{E-OH} + \begin{array}{c} \text{HO} \\ \diagdown \\ \text{B-R} \\ \diagup \\ \text{HO} \end{array} \rightleftharpoons \begin{array}{c} \text{OH} \\ | \\ \text{E-O-B}^- \text{-R} \\ | \\ \text{OH} \end{array} + \text{H}^+ \tag{3}
$$

$$
\text{E-OH} + \begin{array}{c} \text{O} \\ \diagdown\!\!\!\diagdown \\ \text{C-R} \\ \diagup \\ \text{H} \end{array} \rightleftharpoons \begin{array}{c} \text{OH} \\ | \\ \text{E-O-C-R} \\ | \\ \text{H} \end{array} \tag{4}
$$

Some boronic acid derivatives are very good inhibitors of serine proteases. For instance, *N*-dansyl-3-aminobenzene boronic acid exhibits a low K_i (~1 μM) and a significant increase in fluoroscence intensity upon binding to subtilisin.[171] The complex of boronic acid analog of *N*-acetyl-L-phenylalanine and chymotrypsin exhibits a similar K_i.[172] Peptide boronic acids are of particular interest because they are even more powerful inhibitors of several serine proteases, with K_i values in the nM range, and with the potential of their use as drugs.[173]

Kinetic investigations on the inhibition of chymotrypsin and subtilisin by boronic acid derivatives[174-178] suggested the formation of a covalent bond between the enzyme and inhibitor. It was not apparent from these studies whether the inhibitor is covalently linked either to the active site serine, or to the histidine, or to both. However, the X-ray crystallographic studies on the adducts formed between subtilisin BPN' and benzeneboronic acid or 2-phenylethane-boronic acid have shown that both complexes contain a covalent bond between OG of the catalytic Ser 221 and the inhibitor boron atom.[179] The boron atom was coordinated tetrahedrally with one of the two remaining oxygen atoms lying in the oxyanion hole and the other close, but not bound covalently to the histidine side chain.

Another important group of transition-state analogs, natural peptide aldehydes, was first discovered in the culture filtrate of various species of *Actinomycetes*.[180] These compounds inhibiting trypsin-like proteases are called leupeptins: acetyl-L-leucyl-L-leucyl-DL-argininal or propionyl-L-leucyl-L-leucyl-DL-argininal and their analogs, in which L-leucine is replaced by L-valine or L-isoleucine. Following leupeptins, several other potent aldehyde inhibitors have been isolated.[181] Their strong inhibitory effect is due to a covalent hemiacetal formation as indicated by NMR studies on the complex formed between the synthetic acetyl-L-phenylalaninal and chymotrypsin.[169,182]

The structural features of the hemiacetal were clearly shown by X-ray crystallographic studies performed with the complex formed between *Streptomyces griseus* protease A and a specific tetrapeptide aldehyde Ac-Pro-Ala-Pro-Phe-CHO.[131] According to these studies, the aldehyde forms a covalent hemiacetal with Ser 195. The carbonyl oxygen atom of the aldehyde is situated in the oxyanion hole. An unexpected observation that should be mentioned is that the side chain of His 57 undergoes a major conformational change upon inhibitor binding. This conformational change involves rotations about the CA–CB and CB–CG bonds of His 57, which expel the side chain into solvent.[131]

The conformational change observed with the above synthetic inhibitors does not occur with chymostatin (Chapter 3, Section III.B), a natural aldehyde inhibitor, when binding to *Streptomyces griseus* protease A.[125] First of all, the imidazole ring remains virtually in the same position as in the native enzyme. This is a consequence of the different locations of the hemiacetal group in the two complexes. In chymostatin complex, the hemiacetal is shifted away from His 57 toward the oxyanion hole, whereas in the complex formed with the synthetic peptide inhibitor, the hemiacetal group is prohibitorily close to the imidazole ring which is then forced into the surrounding solvent. It is of particular interest that not only the hemiacetal group, but the entire peptide chain of chymostatin is shifted away from His 57. The largest shift of inhibitor backbone occurs at the P_2 leucyl residue. The above results clearly point out the importance of binding effects remote from the catalytic site and their influence on the events that occur during enzymic reactions.

Another interesting result of high resolution X-ray analysis of the complex between chymostatin and *Streptomyces griseus* protease A is the finding of two different enantiomers (R, S) of the hemiacetal adduct.[125] Hemiacetal formation was shown to occur with either the aldehyde oxygen pointing into the oxyanion hole or alternatively oriented toward His 57. Since both resultant configurations are equally populated, there is an apparent lack of preference for a particular orientation of the incoming aldehyde with respect to the attacking OG of Ser 195. This observation argues against an assumed guiding by the oxyanion hole in substrate binding (see the Section V).

V. STEREOCHEMICAL ASPECTS OF THE CATALYSIS

The preceding chapters provided a large body of structural information about the catalytic site, zymogen activation, the binding modes of substrate analogs, and transition-state inhibitors. These data are extremely useful for delineation of the *approximate* stereochemistry of the catalytic process. So far, all of the crystallographic studies on serine proteases have dealt with enzyme complexes formed with inhibitors or substrate-like compounds which do not represent productive intermediates. These structures can only serve as starting points to speculate about the structures of the true intermediates. The difficulties of using the available X-ray structures are apparent from one of the best models of the tetrahedral intermediate, namely the peptide hemiacetal formed with *Streptomyces griseus* protease A,[131] where the side chain of His 57 moves too far from the tetrahedral adduct so that proton transfer between the two species seems to be impossible. In another hemiacetal complex formed between the same enzyme and chymostatin, the side chain of His 57 is found in a more favorable position.[125]

The basic characteristics of the catalysis by serine proteases, the bond formation and bond cleavage occur at the carbonyl carbon atom. Therefore, the elucidation of the geometry of the tetrahedral intermediate, i.e., the location of the carbonyl carbon atom of the substrate with respect to the catalytic groups, is of primary mechanistic importance. Several inhibitor structures were proposed to be similar to the true tetrahedral intermediate. Such structures are the trypsin-trypsin inhibitor complex,[120] boronic acid complexes of subtilisin,[179] peptide chloromethyl ketone,[128,145] and aldehyde derivatives.[125,131] Originally, the trypsin-trypsin inhibitor complex was thought to be a covalent tetrahedral adduct resembling the true catalytic intermediate.[120,129] However, the distance between the serine oxygen and Lys 15 carbonyl carbon of the inhibitor was later found to be 0.26 nm,[15] much greater than that required for a covalent bond. A similar distance (0.271 nm) was also found in the complex of *Streptomyces griseus* protease B with the third domain of the turkey ovomucoid inhibitor.[158] As the 0.26 nm distance is shorter by about 0.05 nm than a normal van der Waals contact, it was suggested that the trypsin-trypsin inhibitor complex represents an intermediate state of the nucleophilic addition reaction.[15] This suggestion was based on the considerable conversion of the carbonyl carbon of Lys 15 to a tetrahedron-like geometry.[120] However, in the complex formed with anhydrotrypsin, which lacks OG of Ser 195, the tetrahedron-like geometry of the Lys 15 carbonyl carbon is the same as in the native complex.[183] Conversion of the planar to a tetrahedron-like geometry, however, was not observed in the turkey ovomucoid inhibitor complexed with *Streptomyces griseus* protease B.[158,159] This difference between the two otherwise similar complexes shows the difficulty in drawing mechanistic conclusions from inhibitor structures. In fact, NMR studies on the same trypsin-trypsin inhibitor complex are inconsistent with the formation of even a partial covalent bond,[184,185] and thus, the short distance between the serine oxygen and the inhibitor carbonyl carbon atoms may rather be interpreted in terms of steric compression.[17]

Tetrahedron-like inhibitor structures have often been invoked to explain the stereochemistry of proton transfers to and from the NE2 of the active site histidine. For example in the trypsin-trypsin inhibitor complex the serine OG and the leaving group nitrogen are at a distance of 0.27 and 0.42 nm, respectively, from NE2,[15] whereas in a model derived from the complex formed between subtilisin and boronic acid derivatives the situation is just the opposite; the leaving atom is at a proper distance and the serine OG is too far apart.[179] A proper hydrogen bond distance (0.28 to 0.31 nm between the oxygen and nitrogen atoms) is of importance for proton transfer to occur.[186] None of the X-ray structures serving as model of the tetrahedral intermediate shows appropriate distances for the two subsequent proton transfers. To overcome this difficulty, it was suggested that a covalent bond is not formed between NE2 and the proton being transferred, rather the pathway traversed by the proton is defined by the very special electronic environment created by the proximity of the serine OG, the imidazole-aspartate couple, the oxyanion binding site, and the carbonyl group of the substrate.[131,187,188] It is difficult to understand in this mechanism how the proton would avoid being trapped by the free electron pair of NE2, i.e., covalent bond formation. Furthermore, proton in direct transit between the attacking and the leaving groups would be inconsistent with the well-established general base-acid catalysis since then the proton transfer would be a single step without the formation of a true tetrahedral intermediate. Thus, the following more simple explanations of the long distance proton transfers may be favored: (1) the inhibitor structures are different from the structure of the true catalytic intermediate and (2) the histidine side chain may move between the two proton transfer steps. A moving histidine mechanism is consistent with ^{1}H NMR studies indicating that the proton of ND1–H bond of the active site histidine, though hydrogen-bonded to the aspartate carboxyl of the catalytic triad, exchanges readily with the solvent water, at least in the case of subtilisin.[189] At present it is not possible to decide between the two alternatives.

Another stereochemical concept emphasizes the activation of both the acyl enzyme and the Michaelis complex. Accordingly, the acyl enzyme has a tetrahedron-like rather than a planar carbonyl carbon atom, with its carbonyl oxygen in the strongly polarizing electrostatic environment of the oxyanion binding site.[131] Furthermore, distortion from planarity was proposed for the peptide bond in the Michaelis complex,[125,131] such an activation of the peptide bond may arise from the oxyanion binding[15,131] and/or leaving group side interactions in the case of peptides possessing amino acid residue(s) also at the leaving group side. These possibilities, however, are not supported by any direct evidence. Moreover, if the major driving force for the hydrolysis of the peptide bond was the elimination of the resonance stabilization energy by the nonplanar distortion of the scissile bond, then this effect would be more important in amide than in ester hydrolysis. However, the ratio for the hydrolysis rates of an ester and the corresponding amide in chymotrypsin catalysis is similar to that found in the alkaline hydrolysis[2] where distortion of the peptide bond is not possible.

The substrate activation in the Michaelis complex, rather than the activation of the serine oxygen by general base catalysis, was considered to be important because crystallographic studies on most serine proteases demonstrated that the serine oxygen is too far apart and directionally in an unfavorable position to form a hydrogen bond with the histidine nitrogen.[14,96] However, it should be kept in mind that the position of the oxygen as found in the crystal is not necessarily relevant to the catalysis. Specifically, bound water molecules which affect the conformation of the serine side chain must be removed as the substrate approaches the active site. In fact, destabilization of the serine OH group by desolvation is essential for its reaction with the substrate. Notwithstanding the crystallographic data on free serine proteases, general base catalysis requires that the serine oxygen and the histidine nitrogen approach each other to a hydrogen bond distance.[186]

As binding and catalysis are closely related (Chapter 2, Section VII), the structural mobility and flexibility may be important in both processes. Interestingly, the static pictures provided by X-ray crystallography clearly demonstrate some flexibility. Thus, a tetrapeptide aldehyde, as compared to the corresponding carboxyl derivative, slightly moved within the antiparallel β-binding sheet toward the catalytic groups due to the formation of the covalent hemiacetal bond.[131] Moreover, two peptide aldehydes displayed somewhat different modes of binding within the same pleated sheet which is primarily caused by the different S_2-P_2 interactions.[125]

An S_1-P_1 hydrogen bond which is stronger in the covalent intermediates than in the Michaelis complex is probable for both the subtilisin and the chymotrypsin families of enzymes.[14] However, this is considered to be unlikely for chymotrypsin.[12] The refined structures of *Streptomyees griseus* protease A complexed with peptide inhibitors indicate that the S_1-P_1 hydrogen-bond formation is accompanied by a small conformational change in the main chain bearing the carbonyl oxygen of Ser 214, a part of the S_1 subsite.[131] The above contradictory results again show the difficulty of drawing conclusions from the inhibitor binding studies to the details of the catalytic mechanism. Certainly, a tighter binding in the transition state than in the Michaelis complex can be facilitated by the formation or strengthening of the S_1–P_1 hydrogen bond, in accord with the transition-state stabilization theory.[93]

The binding modes of the various inhibitors and substrate analogs are similar in many respects and offer a rough picture of how catalytic intermediates may be associated with the active site. However, even the high resolution X-ray crystallographic data of inhibitor complexes are insufficient to reveal the fine details of the stereochemistry of the true catalytic process.

VI. STABILIZATION OF THE TRANSITION STATE

On the reaction pathway in the serine protease catalysis from reactant to product, the tetrahedral intermediate is the highest energy species. Its formation involves the generation of an ion-pair: the tetrahedral adduct bearing the oxyanion and the imidazolium cation associated with the adduct (Figure 2 in Section I.C). The energy required by this charge separation is lowered by the sophisticated machinery of serine proteases, in particular by the aspartate ion of the catalytic triad and by the oxyanion binding site. This will be discussed in the following parts.

A. Charge Stabilization by the Aspartate of the Catalytic Triad

The catalytic triad, Ser . . . His . . . Asp (Chapter 3, Section II), has been found in all serine proteases examined so far. Only one report[190] claimed on the basis of amino acid sequence analysis that tonin, a kallikrein-related serine protease of the rat submaxillary gland, contained leucine in place of the catalytic aspartate residue. However, the nucleotide sequence coding for tonin recently showed that the aspartate residue of the catalytic triad is conserved also in this protease.[191] The recent X-ray crystallographic studies of tonin also indicate the presence of an aspartate residue in the catalytic triad.[191a]

The role of the catalytic triad in the catalysis has been a major issue since its discovery and proposal that it functions as a charge relay system transferring the negative charge from the aspartate through the imidazole to the serine oxygen.[86] The problems encountered with the catalytic triad have been reviewed with the conclusion that the catalytic triad is not a charge relay system.[17,18] The ^{13}C NMR study[192] on α-lytic protease that supported the charge relay idea was re-examined with different NMR techiques. The results have clearly shown that, contrary to the earlier experiment, it is the imidazole and not the aspartate that is protonated.[193,194] This was also found with very low field ^{1}H NMR studies on thiolsubtilisin.[189] This derivative of subtilisin (Chapter 4, Section V.B) contains an ion-pair at its active site which resembles the charge distribution of the transition state involving the negatively charged tetrahedral intermediate and the positive imidazolium ion. Hence, the relay of charge is not expected even in the transition state, but a hydrogen bond between the protonated histidine and the aspartate residue is more important here than at the other stages of the catalysis.[189] The proton was also located on the imidazole rather than on the aspartate by means of neutron diffraction technique, which is suitable for determining the position of the small hydrogen atom.[195]

As a proton relay from the highly basic serine OH group to the acidic aspartate is chemically unacceptable, it was proposed formerly that the role of the aspartate is to stabilize the developing charges both in the transition state and in the tetrahedral adduct.[196] It was pointed out that the negative tetrahedral intermediate, the positive imidazolium ion, and the negative aspartate generate a − + − symmetrical charge distribution, in which system the tetrahedral intermediate forming an ion-pair with the imidazolium ion can exist at a lower energy level than without the aspartate.[196,196a] In fact, such ionic triads can be stabilized by polarization effects.[197]

It should be noted that stabilization of the tetrahedral intermediate was suggested to occur by the α-helix dipole at the active site of subtilisin.[198] Indeed, an α-helix can stabilize a negatively charged group at its N-terminus. However, serine proteases of the chymotrypsin family do not contain any helix near their active site. This fact discounts a significant role of the α-helix even in subtilisin catalysis relative to the other means used by serine proteases for the stabilization of the tetrahedral adduct.

A recent protein engineering study on subtilisin has demonstrated a small but significant effect of a relatively remote (~1.5 nm) ionizing group (substitution of Ser for Asp 99) on the dissociation of the active site histidine residue.[199] Thus, the effect of aspartate interacting directly with the histidine can be substantial. Indeed, a most recent site-directed mutagenesis study on trypsin has shown that replacement of Asp 102 by Asn resulted in a mutant enzyme with catalytic activity approximately 10^4-fold lower at neutral pH than that of the native enzyme.[200] This observation confirms the crucial role of Asp 102 in the catalytic process.

B. Charge Stabilization by the Oxyanion Binding Site

Besides the aspartate of the catalytic triad, the oxyanion binding site can be another factor that stabilizes the charge which develops upon the formation of the tetrahedral adduct. Model building studies strongly suggested that in chymotrypsin the backbone -NH- groups of Gly 193 and Ser 195,[90] or in subtilisin the backbone -NH- of Ser 221 and the side chain amide group of Asn 155,[93] may form hydrogen bonds with the oxyanion. The oxyanion hole has been found in all serine proteases of known three-dimensional structure.[14,16]

The catalytic importance of the oxyanion binding site was tested by thionester substrates,[201] which contain a sulfur atom in place of the carbonyl oxygen (Chapter 1, Section X). The intrinsic chemical reactivities of the two types of esters are not very much different, inasmuch as their alkaline hydrolysis rates are similar[201] (Chapter 1, Section X). The delicate modification, i.e., replacement of oxygen by sulfur, could produce serious catalytic consequences in a stereochemically constraint position, such as the oxyanion binding site. This could be anticipated on the basis of studies on thiolsubtilisin,[8] which indicated that the ability of the protein to compensate for even relatively small distortions at certain locations can be rather limited. Specifically, conversion of subtilisin into thiolsubtilisin results in a virtually inactive enzyme toward peptide and protein substrates (Chapter 4, Section V). Similarly, substitution of sulfur for the carbonyl oxygen of the substrate inhibited the catalytic reaction. Notably, neither chymotrypsin nor subtilisin hydrolyzed the thionesters at a measurable rate. In the case of ethyl N-acetyl-phenylalaninate the rate constant for the thionester was smaller by a factor of more than 10^4 than that for the ester substrate.[201] This result substantiates the crucial role of the oxyanion binding site in a way independent of model building studies. Such an independent approach to the study of oxyanion stabilization is of importance because although an oxyanion hole is also apparent from X-ray analysis of papain, cysteine proteases readily hydrolyze thionesters (Chapter 4, Section V).

In connection with the oxyanion binding site, it seems to be appropriate to discuss the contribution of stereoelectronic control to the serine protease catalysis. This theory elaborated for small organic compounds (Chapter 1, Section XV) was applied to the mechanism of action of serine proteases.[202-205] Stereoelectronic control imposes restrictions on the orientation of the lone pair orbitals of the three heteroatoms of the tetrahedral intermediate: the leaving nitrogen of a peptide substrate, the oxyanion, and the serine OG (Figure 10). Specifically, cleavage of the C–O or C–N bond requires the assistance of antiperiplanar lone pair orbitals on each of the two remaining heteroatoms. However, the fulfillment of this requirement is complicated in the enzyme reaction due to hydrogen bond formation between the lone pairs of the oxyanion and the backbone -NH- groups. Since orbital overlapping is weakened with hydrogen-bonded lone pairs, these interactions may rule against the stereoelectronic effects[201] that prevail in simple organic reactions.

A further complication arises from the opposite requirements for the formation and decomposition of the tetrahedral intermediate. In accordance with the principle of microscopic reversibility, the developing lone pair orbitals on the heteroatoms must be antiperiplanar to the new bond. Hence, on the formation of the tetrahedral intermediate, the nonbonded pair of electrons on the leaving nitrogen atom should point toward the solvent and the N–H bond toward the active site imidazole. However, in the decomposition of the tetrahedral intermediate when the leaving nitrogen atom must accept a proton from the histidine, the nitrogen lone pair points to the wrong direction. To comply with the stereochemical requirements, one should invoke an inversion mechanism at the leaving nitrogen to interchange the orientations of the N–H bond and the lone pair orbital.[204,205] Such an inversion readily occurs in simple compounds. However, it is not clear how it would take place on the enzyme surface where binding interactions on the leaving group side could interfere. It is not impossible that energetically potent interactions overrule stereoelectronic control in enzyme catalysis. Indeed, the general validity of the stereoelectronic theory has been questioned in glycosidase catalysis.[206]

VII. THE NATURE OF THE TRANSITION STATE

A major goal of mechanistic investigations is to define the transition state of the reaction in question. Despite a number of attempts, this goal has not yet been reached, as will be apparent from the following discussion.

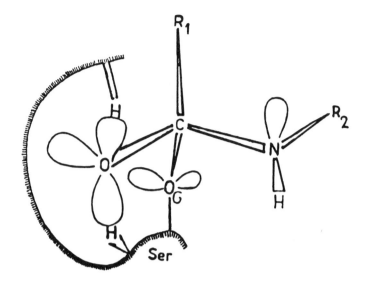

FIGURE 10.   Orientation of the lone pair orbitals of the tetrahedral intermediate in the active site of serine proteases. C is the tetrahedral carbon atom of the R_1 amino acid; OG represents the serine oxygen atom; N is the peptide nitrogen of the R_2 amino acid. The oxyanion is situated on the left side of the figure and forms two hydrogen bonds with the protein. The remaining free lone pair is antiperiplanar to the scissile C–N bond. (From Asbóth, B. and Polgár, L., *Biochemistry,* 22, 117, 1983. With permission.)

From the mechanistic and structural data treated in the preceding sections, in particular from the studies with transition-state analogs, we may envision how substrates bind to the serine proteases and how they are converted to products through elementary catalytic steps. The best picture is obtained with the tetrahedral intermediate, which is determined, apparently overdetermined, by interactions with the oxyanion binding site, the imidazolium ion, the serine OG, and the substrate specificity site. It appears from this picture that the active site of serine proteases is complementary to the tetrahedral rather than to a planar, trigonal carbon atom. Therefore, the tetrahedral intermediate is often referred to as a transition state-like species. However, as noted in Section IV, the tetrahedral intermediate is a true covalent adduct, whereas the transition-state complex contains partially formed and/or broken bonds. In addition, the transition-state complex, as the highest energy species, should be less stable compared to the tetrahedral intermediate. Nonetheless, the tetrahedral intermediate itself should be very labile. It is obviously the highest energy covalent structure both in acylation and deacylation. Thus, according to the Hammond postulate (Chapter 1, Section VIII.A) the transition state is more closely related to the tetrahedral adduct than to the acyl-enzyme or the Michaelis complex.

Since the tetrahedral intermediate is a high energy, labile species relative to the Michaelis complex, the equilibrium between these two species is expected to be shifted toward the Michaelis complex, preventing the accumulation of the tetrahedral adduct during catalysis, even if its breakdown is rate-limiting as in amide or anilide hydrolysis. Nevertheless, detection of the tetrahedral intermediate in the course of catalysis has been a controversial issue over the past decade. Several observations of some initial bursts occurring before the acyl-enzyme formation have been made in the reactions of *p*-nitroanilide substrates with chymotrypsin, trypsin, and elastase, which were interpreted in terms of accumulation of the tetrahedral adduct.[207-210] In other experiments, however, the intermediate could not be detected.[43,211-213] Moreover, it was claimed that the bursts observed previously were artifacts[212] like temperature/solvent induced isomerization of trypsin to a catalytically less efficient form.[213] In fact, there is no conclusive evidence in favor of the accumulation of the tetrahedral adduct. This is in accord with a recent approximate calculation, allowing very little chance of observing the accumulation of the tetrahedral adduct.[214]

Considerable efforts were devoted to study the nature of the transition state by measuring isotope effects (Chapter 1, Section IX) and the sensitivity of the reaction (Hammett ρ) to substituent effects (Chapter 1, Section VIII). However, the difficulty of interpretation of the data in complex enzymic reactions is clearly seen from the remarkably different results shown in the following. Notably, the Hammett ρ constants for the chymotrypsin-catalyzed acylations with phenyl ester substrates may be similar to or deviate substantially from that observed in the corresponding nonenzymic reactions.[215-218] Thus, the ρ value, calculated by using the second-order acylation rate constants for nonspecific substrates, is similar to that of the model reaction (~ 2). [215] However, the value is lower (~ 1) if the first-order acylation constants are considered, which indicates that the enzymic and nonenzymic transition states may be different even with nonspecific substrates.[218] Moreover, the chemical properties of the leaving group (e.g., protonated or not protonated imidazole) may affect the magnitude of the ρ value to a remarkable extent.[219]

In the case of specific substrates, the ρ values are lower (0.4 to 0.6)[217] compared with those for the reactions of nonspecific substrates, indicating less negative charge in the transition state. This is consistent with stabilization by the oxyanion binding site.[216] For substrates containing a poor leaving group, the reaction parameter may change more considerably. Specifically, anilide substrates exhibit a negative ρ[77,80] as opposed to the positive value encountered with ester substrates. The negative ρ, i.e., a rate decrease with increasing electron withdrawal, can be most straightforwardly rationalized in terms of proton transfer from the imidazolium ion to the anilide leaving group at the transition state. This implies different rate-determining steps for the two types of substrates: the formation of the tetrahedral intermediate being rate controlling with phenyl esters and its breakdown with anilides. It should be noted that p-nitroanilide substrates, which possess a very strong electron withdrawing group, deviate considerably from the correlation.[220,221] The correlation for a linear relationship is often very poor because the effects of structural changes in a set of congeners is a multivariate problem. Besides electronic effects, other factors, such as hydrophobicity, molar refractivity, and steric properties of the substituent, should be taken into consideration.[222]

The Brönsted coefficient was also determined in order to delineate the charge distribution of the transition state (Chapter 1, Section VIII.A) in aminolysis of the nonspecific furoyl-chymotrypsin[223] and the highly specific acetyltyrosyl-chymotrypsin.[224] These aminolysis reactions are of mechanistic importance because the principle of microscopic reversibility requires an identical transition state for amide bond cleavage. It was found that the reaction of nonaromatic amines is virtually independent of amine basicity (Brönsted slope is near zero). This is consistent with there being little or no positive charge development on the amine in the transition state. If there were significant charge developments, electron-donating groups would stabilize this charge and the reaction rate would correlate with amine basicity. On the other hand, aniline nucleophiles show a significant dependence on amine pK_a (Brönsted slope is equal to 0.52). Unfortunately, these data cannot be interpreted in a straightforward manner.[224] Two differences, however, may be emphasized: (1) some interaction between the aromatic amines and the enzyme, which does not exist with the nonaromatic amines; and (2) the generally lower pK_a values for the aromatic amines.

The hydrolysis of amide and peptide substrates appears to have some special features. It was suggested that in amide hydrolysis a normal tetrahedral intermediate may not be formed[225] because of the high pK_a estimated for the leaving group (8 to 11) relative to that of the imidazole (~ 7). It was argued that the more basic leaving group will accept the proton in the general acid catalyzed step prior to the formation of the transition state.[225] Although this is a reasonable assumption for simple systems, in serine proteases the hydrogen bonds from the oxyanion binding site reduces the proton affinity of the leaving group, and the negative aspartate interacting with the catalytic histidine enhances the proton affinity of the imidazole, i.e., both factors facilitate the residence of proton on the imidazole rather than on the leaving group. In fact, in thiolsubtilisin where a thiolate-imidazolium ion-pair mimicks the ion-pair formed between the negative tetrahedral intermediate and the imidazolium ion, the apparent pK_a of the imidazole is 10.1.[226] ^{1}H NMR studies on transition-state analogs in chymotrypsin,[227] trypsin,[228] and subtilisin[189] have arrived at a similar conclusion. Accordingly, the proton affinity of the imidazole should be raised during the formation of the tetrahedral intermediate, whereby the reaction pathway through a real tetrahedral adduct becomes possible.

Secondary kinetic isotope effects and proton inventory studies (Chapter 1, Section IX) have also provided some insight into the nature of the transition state of serine proteases although the interpretation of the data is rather difficult in the complex enzymic reactions. Secondary isotope effects are assumed to signal the "tetrahedrality" of the transition state. Such studies were performed by using substrates in which deuterium was substituted for the hydrogen atom of a formyl group (α-deuterium isotope effect) or for the hydrogens linked to the β-carbon of acetyl and glycyl groups β-deuterium isotope effect).[229-231] The "tetrahedrality" of the transition state varied with the substrate. However, a transition state closer to the tetrahedron than to a planar structure, was generally observed.

The results obtained with proton inventory studies[232] also depended on the substrate used.[231,233] In the reactions of the less specific substrates only one moving proton was found, in accordance with the action of a single general base. On the other hand, with the more specific substrates two or more protonic sites contributed to the catalysis. The results were interpreted to mean that remote-subsite interactions with the more specific substrates cause a structural compression in the transition state.[231] This is consistent with the other observations that part of the binding energy may be utilized for the rate enhancement,[234,235] primarily in the reactions of the more specific substrates.

As it may appear from the foregoing discussion, a quantitative description of the transition state of serine protease catalysis has not yet been achieved. A major problem in this respect is that a unique mechanism, as far as the fine details are concerned, may not exist. This is obvious for substrates possessing leaving groups of different basicity, i.e., phenyl esters, alkyl esters, and amines. However, several lines of evidence show that interactions between the enzyme and the P_2, P_3, . . . P_n residues of the substrate can also significantly affect the nature of the transition state. Indeed, we have already discussed (Chapter 3, Section VI) that two peptide aldehydes differing at position P_2 gives rise to different hemiacetal positions with respect to the catalytic groups. Also linear free-energy relationships and kinetic isotope effects may signal different transition states for different substrates. Different transition states are most conceivable to occur with enantiomeric substrates.[60] Thus, there should be a multitude of transition states and tetrahedral intermediates differing in steric and electronic properties. These structures are restricted to a limited space within which the serine protease mechanism is valid. This is an average, approximate mechanism that can be proposed on the basis of model building studies. Each individual substrate has its own individual transition state structure. For the different congeners, the difference is probably a few hundredths of a nanometer at most, representing details too fine to be readily accessible to our current methods.

We have already mentioned that the central carbon atom of the tetrahedral adduct is very precisely guided by its four covalently bonded atoms, each imposing a directional constraint: (1) the serine OG, (2) the oxyanion in the oxyanion hole, (3) the leaving group in a position where it can accept the proton from the imidazolium ion, and (4) the C_α whose position is dictated by the S-P interactions, particularly by those occurring in the binding pocket, and the β-antiparallel pleated sheet. Even three atoms could be sufficient for defining the position of the tetrahedral intermediate. As for the rough stereochemistry, this is indeed the case. However, a minute play is allowed for the tetrahedral adduct, inasmuch as hydrogen bonds are flexible and rotations by the enzymic side chains can compensate for small changes.[60] The four interactions determining the actual location of the tetrahedral intermediate are in a kind of balance which is primarily influenced by the interaction between the acyl group and the binding site. Thus, within the framework of the general serine protease mechanism, slightly different kinds of stereochemistry, as well as transition states, should be distinguished for the individual substrates. The former has now been fairly well characterized; the details of the mechanism remain to be established.

REFERENCES

1. **Cunningham, L.,** The structure and mechanism of action of proteolytic enzymes, in *Comprehensive Biochemistry,* Vol. 16, Florkin, M. and Stotz, E. H., Eds., Elsevier, Amsterdam, 1965, 85.
2. **Bender, M. L. and Kézdy, F. J.,** Mechanism of action of proteolytic enzymes, *Annu. Rev. Biochem.,* 34, 49, 1965.
3. **Bender, M. L. and Killheffer, J. V.,** Chymotrypsins, *Crit. Rev. Biochem.,* 1, 149, 1973.
4. **Hess, G. P.,** Chymotrypsin. Chemical properties and catalysis, in *The Enzymes,* Vol. 3, 3rd ed., Boyer, P. D., Ed., Academic Press, New York, 1971, 213.
5. **Keil, B.,** Trypsin, in *The Enzymes,* Vol. 3, 3rd ed., Boyer, P. D., Ed., Academic Press, New York, 1971, 249.
6. **Hartley, B. S. and Shotton, D. M.,** Pancreatic elastase, in *The Enzymes,* Vol. 3, 3rd ed., Boyer, P. D., Ed., Academic Press, New York, 1971, 323.
7. **Markland, F. S., Jr. and Smith, E. L.,** Subtilisins: primary structure, chemical and physical properties, in *The Enzymes,* Vol. 3, 3rd ed., Boyer, P. D., Ed., Academic Press, New York, 1971, 561.
8. **Polgár, L. and Bender, M. L.,** Simulated mutation at the active site of biologically active proteins, *Adv. Enzymol. Relat. Areas Mol. Biol.,* 33, 381, 1970.
9. **Philipp, M. and Bender, M. L.,** Kinetics of subtilisin and thiolsubtilisin, *Mol. Cell. Biochem.,* 51, 5, 1983.
10. **Blow, D. M. and Steitz, T. A.,** X-ray diffraction studies of enzymes, *Annu. Rev. Biochem.,* 39, 63, 1970.
11. **Blow, D. M.,** The structure of chymotrypsin, in *The Enzymes,* Vol. 3, 3rd ed., Boyer, P. D., Ed., Academic Press, New York, 1971, 185.
12. **Blow, D. M.,** Structure and mechanism of chymotrypsin, *Acc. Chem. Res.,* 9, 145, 1976.
13. **Kraut, J.,** Subtilisin: X-ray structure, in *The Enzymes,* Vol. 3, 3rd ed., Boyer, P. D., Ed., Academic Press, New York, 1971, 547.
14. **Kraut, J.,** Serine proteases: structure and mechanism of catalysis, *Annu. Rev. Biochem.,* 46, 331, 1977.
15. **Huber, R. and Bode, W.,** Structural basis of the activation and action of trypsin, *Acc. Chem. Res.,* 11, 114, 1978.
16. **James, M. N. G.,** An X-ray crystallographic approach to enzyme structure and function, *Can. J. Biochem.,* 58, 251, 1980.
17. **Polgár, L. and Halász, P.,** Review article. Current problems in mechanistic studies of serine and cysteine proteinases, *Biochem. J.,* 207, 1, 1982.
18. **Steitz, T. A. and Shulman, R. G.,** Crystallographic and NMR studies of serine proteases, *Annu. Rev. Biophys. Bioeng.,* 11, 419, 1982.
19. **Polgár, L.,** Structure and function of serine proteases, in *New Comprehensive Biochemistry,* Neuberger, A. and Brocklehurst, K., Eds., Elsevier, Amsterdam, 1987, chap. 3.
20. **Jansen, E. F., Nutting, M.-D. F., and Balls, A. K.,** Mode of inhibition of chymotrypsin by diisopropyl fluorophosphate. I. Introduction of phosphorus, *J. Biol. Chem.,* 179, 201, 1949.
21. **Schaffer, N. K., May, S. C., Jr., and Summerson, W. H.,** Serine phosphoric acid from diisopropyl-phosphoryl chymotrypsin, *J. Biol. Chem.,* 202, 67, 1953.
22. **Oosterbaan, R. A., Kunst, P., van Rotterdam, J., and Cohen, J. A.,** The reaction of chymotrypsin and diisopropylphosphorofluoridate. II. The structure of two DP-substituted peptides from chymotrypsin-DP, *Biochim. Biophys. Acta,* 27, 556, 1958.
23. **Fahrney, D. E. and Gold, A. M.,** Sulfonyl fluorides as inhibitors of esterases. I. Rates of reaction with acetylcholinesterase, α-chymotrypsin, and trypsin, *J. Am. Chem. Soc.,* 85, 997, 1963.

24. **Laura, R., Robinson, D. J., and Bing, D. H.,** (p-Amidinophenyl) methanesulfonyl fluoride, an irreversible inhibitor of serine proteases, *Biochemistry,* 19, 4859, 1980.

25. **Aldridge, W. N.,** Acetylcholinesterase and other esterase inhibitors, in *Enzyme Inhibitors as Drugs,* Sandler, M., Ed., Macmillan Press, New York, 1980, 115.

26. **Gorenstein, G. D. and Findlay, J. B.,** ^{31}P NMR of diisopropyl phosphoryl α-chymotrypsin and catechol cyclic phosphate α-chymotrypsin. Direct observation of two conformational isomers, *Biochem. Biophys. Res. Commun.,* 72, 640, 1976.

27. **Reeck, G. R., Nelson, T. B., Paukstelis, J. V., and Mueller, D. D.,** Comparisons of diisopropyl fluorophosphate derivatives of chymotrypsin and chymotrypsinogen by phosphorus-31 nuclear magnetic resonance, *Biochem. Biophys. Res. Commun.,* 74, 643, 1977.

28. **Porubcan, M. A., Westler, W. M., Ibanez, I. B., and Markley, J. L.,** (Diisopropylphosphoryl) serine proteinases. Proton and phosphorus-31 nuclear magnetic resonance-pH titration studies, *Biochemistry,* 18, 4108, 1979.

29. **Van der Drift, A. C. M.,** Physico-Chemical Characterization of Atropin Esterase from *Pseudomonas putida.* A Comparison with Other Serine Hydrolyses, *Doctoral thesis,* State University of Utrecht, The Netherlands, 1983.

30. **Jordan, F., Polgár, L., and Tous, G.,** Proton magnetic resonance studies of the states of ionization of histidines in native and modified subtilisins, *Biochemistry,* 24, 7711, 1985.

31. **Hartley, B. S. and Kilby, B. A.,** The reaction of p-nitrophenyl esters with chymotrypsin and insulin, *Biochem. J.,* 56, 288, 1954.

32. **Oosterbaan, R. A., van Adrichem, M., and Cohen, J. A.,** An acetyl-peptide from acetylchymotrypsin, *Biochim. Biophys. Acta,* 63, 204, 1962.

33. **Bender, M. L., Schonbaum, G. R., and Zerner, B.,** Spectrophotometric investigations of the mechanism of α-chymotrypsin-catalyzed hydrolyses. Detection of the acyl-enzyme intermediate, *J. Am. Chem. Soc.,* 84, 2540, 1962.

34. **Bernhard, S. A. and Tashjian, Z. H.,** Acyl intermediates in the α-chymotrypsin-catalyzed hydrolysis of indoleacryloylimidazole, *J. Am. Chem. Soc.,* 87, 1806, 1965.

35. **Bender, M. L. and Zerner, B.,** The formation of acyl-enzyme intermediate in the α-chymotrypsin-catalyzed hydrolyses of non-labile trans-cinnamic acid esters, *J. Am. Chem. Soc.,* 84, 2550, 1962.

36. **Zerner, B., Bond, R. P. M., and Bender, M. L.,** Kinetic evidence for the formation of acyl-enzyme intermediates in the α-chymotrypsin-catalyzed hydrolyses of specific substrates, *J. Am. Chem. Soc.,* 86, 3674, 1964.

37. **Zerner, B. and Bender, M. L.,** The kinetic consequences of the acyl-enzyme mechanism for the reactions of specific substrates with chymotrypsin, *J. Am. Chem. Soc.,* 86, 3669, 1964.

38. **Bender, M. L., Clement, G. E., Gunter, C. R., and Kézdy, F. J.,** The kinetics of α-chymotrypsin reactions in the presence of added nucleophiles, *J. Am. Chem. Soc.,* 86, 3697, 1964.

39. **Berezin, I. V., Kazanskaya, N. F., and Klyosov, A. A.,** Determination of the individual rate constants of α-chymotrypsin-catalyzed hydrolysis with the added nucleophilic agent, 1,4-butanediol, *FEBS Lett.,* 15, 121, 1971.

40. **Bernhard, S. A., Lau, S. J., and Noller, H.,** Spectrophotometric identification of acyl enzyme intermediates, *Biochemistry,* 4, 1108, 1965.

41. **Charney, E. and Bernhard, S. A.,** Optical properties and the chemical nature of acyl-chymotrypsin linkages, *J. Am. Chem. Soc.,* 89, 2726, 1967.

42. **Bernhard, S. A. and Gutfreund, H.,** The optical detection of transients in trypsin- and chymotrypsin-catalyzed reactions, *Proc. Natl. Acad. Sci. U.S.A.,* 53, 1238, 1965.

43. **Glazer, A. N.,** Spectral studies of the interaction, α-chymotrypsin and trypsin with proflavine, *Proc. Natl. Acad. Sci. U.S.A.,* 54, 171, 1965.

44. **Brandt, K. G., Himoe, A., and Hess, G. P.,** Investigations of the chymotrypsin-catalyzed hydrolysis of specific substrates. III. Determination of individual rate constants and enzyme-substrate binding constants for specific amide and ester substrates, *J. Biol. Chem.,* 242, 3973, 1967.

45. **Hirohara, H., Bender, M. L., and Stark, R. S.,** Acylation of α-chymotrypsin by oxygen and sulfur esters of specific substrates: kinetic evidence for a tetrahedral intermediate, *Proc. Natl. Acad. Sci. U.S.A.,* 71, 1643, 1974.

46. **Himoe, A., Brandt, K. G., and Hess, G. P.,** Investigations of the chymotrypsin-catalyzed hydrolysis of specific substrates. II. Characterization of the spectral changes of the enzyme at 290 mμ and determination of over-all enzyme-substrate dissociation constants, *J. Biol. Chem.,* 242, 3963, 1967.

47. **Christensen, U. and Ipsen, H. H.,** Steady-state kinetics of plasmin- and trypsin-catalyzed hydrolysis of a number of tripeptide-p-nitroanilides, *Biochim. Biophys. Acta,* 569, 177, 1979.

48. **Stein, R. L., Viscarello, B. R., and Wildonger, R. A.,** Catalysis by human leucocyte elastase. II. Rate-limiting deacylation for specific p-nitroanilides and amides, *J. Am. Chem. Soc.,* 106, 796, 1984,.

49. **Fastrez, J. and Fersht, A. R.,** Demonstration of the acyl-enzyme mechanism for the hydrolysis of peptides and anilides by chymotrypsin, *Biochemistry,* 12, 2025, 1973.

50. **Epand, R. M.,** Evidence against the obligatory formation of an acyl enzyme intermediate in the α-chymotrypsin catalyzed reactions of amides, *Biochem. Biophys. Res. Commun., 37,* 313, 1969.

51. **Antonov, V. K., Ginodman, L. M., Rumsh, L. D., Kapitannikov, Y. V., Barshevskaya, T. N., Yavashev, L. P., Gurova, A. G., and Volkova, L. I.,** Studies on the mechanism of action of proteolytic enzymes using heavy oxygen exchange, *Eur. J. Biochem., 117,* 195, 1981.

52. **Hammond, B. R. and Gutfreund, H.,** Two steps in the reaction of chymotrypsin with acetyl-L-phenylalanine ethyl ester, *Biochem. J., 61,* 187, 1955.

53. **Cunningham, L. W. and Brown, C. S.,** The influence of pH on the kinetic constants of α-chymotrypsin-catalyzed esterolysis, *J. Biol. Chem., 221,* 287, 1956.

54. **Bender, M. L., Clement, G. E., Kézdy, F. J., and Zerner, B.,** Sigmoid and bell-shaped pH-rate profiles in α-chymotrypsin-catalyzed hydrolyses. A mechanistic correlation, *J. Am. Chem. Soc., 85,* 358, 1963.

55. **Bender, M. L., Clement, G. E., Kézdy, F. J., and Heck, H.d'A.,** The correlation of the pH (pD) dependence and the stepwise mechanism of α-chymotrypsin-catalyzed reactions, *J. Am. Chem. Soc., 86,* 3680, 1964.

56. **Himoe, A., Parks, P. C., and Hess, G. P.,** Investigations of the chymotrypsin-catalyzed hydrolysis of specific substrates. I. The pH dependence of the catalytic hydrolysis of N-acetyl-L-tryptophanamide by three forms of the enzyme at alkaline pH, *J. Biol. Chem., 242,* 919, 1967.

57. **Hess, G. P., McConn, J., Ku, E., and McConkey, G.,** Studies of the activity of chymotrypsin, *Philos. Trans. R. Soc. London Ser. B, 257,* 89, 1970.

58. **Kraut, J.,** Chymotrypsinogen: X-ray structure, in *The Enzymes,* Vol. 3, 3rd. ed., Boyer, P. D., Ed., 1971, 165.

59. **Polgár, L.,** Symmetry and asymmetry in the mechanisms of hydrolysis by serine proteases and their thiol analogues, *Acta Biochim. Biophys. Acad. Sci. Hung., 7,* 319, 1972.

60. **Polgár, L. and Fejes, J.,** Mechanism-controlled stereospecificity. Acylation of subtilisin with enantiomeric alkyl and nitrophenyl ester substrates, *Eur. J. Biochem., 102,* 531, 1979.

61. **Weil, L., James, S., and Buchert, A. R.,** Photooxidation of crystalline chymotrypsin in the presence of methylene blue, *Arch. Biochem. Biophys., 46,* 266, 1953.

62. **Koshland, D. E., Jr., Strumeyer, D. H., and Ray, W. J., Jr.,** Amino acids involved in the action of chymotrypsin, *Brookhaven Symp. Biol., 15,* 101, 1962.

63. **Schoellman, G. and Shaw, E.,** Direct evidence for the presence of histidine in the active center of chymotrypsin, *Biochemistry, 2,* 252, 1963.

64. **Shaw, E.,** Chemical modification by active-site-directed reagents, in *The Enzyme,* Vol. 1, 3rd ed., Boyer, P. D., Ed., Academic Press, New York, 1970, 91.

65. **Powers, J. C.,** Reaction of serine proteases with halomethyl ketones, *Methods Enzymol., 46,* 197, 1977.

66. **Nakagawa, Y. and Bender, M. L.,** Methylation of histidine-57 in α-chymotrypsin by methyl p-nitrobenzenesulfonate. A new approach to enzyme modification, *Biochemistry, 9,* 259, 1970.

67. **Henderson, R.,** Catalytic activity of α-chymotrypsin in which histidine-57 has been methylated, *Biochem. J., 124,* 13, 1971.

68. **Ryan, D. S. and Feeney, R. E.,** The interaction of inhibitors of proteolytic enzymes with 3-methylhistidine-57-chymotrypsin, *J. Biol. Chem., 250,* 843, 1975.

69. **Dixon, G. H., Neurath, H., and Pechére, J.-F.,** Proteolytic enzymes, *Annu. Rev. Biochem., 27,* 489, 1958.

70. **Wootton, J. F. and Hess, G. P.,** Spectroscopic studies of α-chymotrypsin catalyzed reactions. I. Spectral changes at 245 mμ, *J. Am. Chem. Soc., 83,* 4234, 1962.

71. **Bender, M. L. and Hamilton, G. A.,** Kinetic isotope effects of deuterium oxide on several α-chymotrypsin-catalyzed reactions, *J. Am. Chem. Soc., 84,* 2570, 1962.

72. **Hubbard, C. D. and Kirsch, J. F.,** Acylation of chymotrypsin by active esters of nonspecific substrates. Evidence for a transient acylimidazole intermediate, *Biochemistry, 11,* 2483, 1972.

73. **Hubbard, C. D. and Shoupe, T. S.,** Mechanisms of acylation of chymotrypsin by phenyl esters of benzoic acid and acetic acid, *J. Biol. Chem., 252,* 1633, 1977.

74. **Polgár, L.,** Deuterium isotope effects on acylation of subtilisin, *Biochim. Biophys. Acta, 321,* 639, 1973.

75. **Bender, M. L.,** The mechanism of α-chymotrypsin-catalyzed hydrolyses, *J. Am. Chem. Soc., 84,* 2582, 1962.

76. **Bender, M. L. and Kézdy, F. J.,** The current status of the α-chymotrypsin mechanism, *J. Am. Chem. Soc., 86,* 3704, 1964.

77. **Bender, M. L. and Glasson, W. A.,** The kinetics of the α-chymotrypsin-catalyzed hydrolysis and methanolysis of acetyl- L-phenylalanine methyl ester. Evidence for the specific binding of water on the enzyme surface, *J. Am. Chem. Soc., 82,* 3336, 1960.

78. **Bruice, T. C.,** The mechanisms for chymotrypsin, *Proc. Natl. Acad. Sci. U.S.A., 47,* 1924, 1961.

79. **Inagami, T., York, S. S., and Patchornik, A.,** An electrophilic mechanism in the chymotrypsin catalyzed hydrolysis of anilide substrates, *J. Am. Chem. Soc., 87,* 126, 1965.

80. **Parker, L. and Wang, J. H.,** On the mechanism of action at the acylation step of the α-chymotrypsin-catalyzed hydrolysis of anilides, *J. Biol. Chem.,* 243, 3729, 1968.

81. **Green, A. L. and Nicholls, J. D.,** The reactivation of phosphorylated chymotrypsin, *Biochem. J.,* 72, 70, 1959.

82. **Cohen, W. and Erlanger, B. F.,** Studies on the reactivation of diethylphosphorylchymotrypsin, *J. Am. Chem. Soc.,* 82, 3928, 1960.

83. **Wedler, F. C., Killian, F. L., and Bender, M. L.,** Reaction of acetyl-α-chymotrypsin and other esters with an ionizable nucleophile, monoisonitrosoacetone, *Proc. Natl. Acad. Sci. U.S.A.,* 65, 1120, 1970.

84. **Polgár, L.,** On the mechanism of proton transfer in the catalysis by serine proteases, *J. Theor. Biol.,* 31, 165, 1971.

85. **Matthews, B. W., Sigler, P. B., Henderson, R., and Blow, D. M.,** Three-dimensional structure of tosyl-α-chymotrypsin, *Nature (London),* 214, 652, 1967.

86. **Blow, D. M., Birktoft, J. J., and Hartley, B. S.,** Role of a buried acid group in the mechanism of action of chymotrypsin, *Nature (London),* 221, 337, 1969.

87. **Birktoft, J. J. and Blow, D. M.,** Structure of crystalline α-chymotrypsin. The atomic structure of tosyl-α-chymotrypsin at 2 Å resolution, *J. Mol. Biol.,* 68, 187, 1972.

88. **Sigler, P. B., Blow, D. M., Matthews, B. W., and Henderson, R.,** Structure of crystalline α-chymotrypsin. II. A preliminary report including a hypothesis for the activation mechanism, *J. Mol. Biol.,* 35, 143, 1968.

89. **Steitz, T. A., Henderson, R., and Blow, D. M.,** Structure of crystalline α-chymotrypsin. III. Crystallographic studies of substrates and inhibitors bound to the active site of α-chymotrypsin, *J. Mol. Biol.,* 46, 337, 1969.

90. **Henderson, R.,** Structure of crystalline α-chymotrypsin. IV. The structure of indoleacryloyl-α-chymotrypsin and its relevance to the hydrolytic mechanism of the enzyme, *J. Mol. Biol.,* 54, 341, 1970.

91. **Ingles, D. W. and Knowles, J. R.,** The stereospecificity of α-chymotrypsin, *Biochem. J.,* 108, 561, 1968.

92. **Matta, M. S. and Staley, D. D.,** Specificities of α-chymotrypsin and subtilisin Carlsberg. The α-acylamido effect in β-phenylpropionates and their rigid analogs, *J. Biol. Chem.,* 249, 732, 1974.

93. **Robertus, J. D., Kraut, J., Alden, R. A., and Birktoft, J. J.,** Subtilisin; a stereochemical mechanism involving transition-state stabilization, *Biochemistry,* 11, 4293, 1972.

94. **Tsukada, H. and Blow, D. M.,** Structure of α-chymotrypsin refined at 1.68 Å resolution, *J. Mol. Biol.,* 184, 703, 1985.

95. **Blevins, R. A. and Tulinsky, A.,** The refinement and structure of the dimer of α-chymotrypsin at 1.67 Å resolution, *J. Biol. Chem.,* 260, 4264, 1985.

96. **Cohen, G. H., Silverton, E. W., and Davies, D. R.,** Refined crystal structure of γ-chymotrypsin at 1.9 Å resolution. Comparison with other pancreatic serine proteases, *J. Mol. Biol.,* 148, 449, 1981.

97. **Chambers, J. L. and Stroud, R. M.,** The accuracy of refined protein structures: comparison of two independently refined models of bovine trypsin, *Acta Crystallogr. Sect. B,* 35, 1861, 1979.

98. **Bode, W. and Schwager, P.,** The refined crystal structure of bovine β-trypsin at 1.8 Å resolution. II. Crystallographic refinement, calcium binding site, benzamidine binding site and active site at pH 7.0, *J. Mol. Biol.,* 98, 693, 1975.

99. **Fujinaga, M., Delbaere, L. T. J., Brayer, G. D., and James, M. N. G.,** Refined structure of α-lytic protease at 1.7 Å resolution. Analysis of hydrogen bonding and solvent structure, *J. Mol. Biol.,* 184, 479, 1985.

100. **Read, R. J., Brayer, G. D., Jurasek L., and James, M. N. G.,** Critical evaluation of comparative model building of *Streptomyces griseus* trypsin, *Biochemistry,* 23, 6570, 1984.

101. **Sielecki, A. R., Hendrickson, W. A., Broughton, C. G., Delbaere, L. T. J., Brayer, G. D., and James, M. N. G.,** Protein structure refinement: *Streptomyces griseus* serine protease A at 1.8 Å resolution, *J. Mol. Biol.,* 134, 781, 1979.

102. **Bode, W., Chen, Z., Bartels, K., Kutzbach, C., Schmidt-Kastner, G., and Bartunik, H.,** Refined 2 Å X-ray crystal structure of porcine pancreatic kallikrein A, a specific trypsin-like serine proteinase. Crystallization, structure determination, crystallographic refinement structure, and its comparison with bovine trypsin, *J. Mol. Biol.,* 164, 237, 1983.

103. **Sawyer, L., Shotton, D. M., Campbell, J. W., Wendell, P. L., Muirhead, H., Watson, H. C., Diamond, R., and Ladner, R. C.,** The atomic structure of crystalline porcine pancreatic elastase at 2.5 Å resolution: comparison with the structure of α-chymotrypsin, *J. Mol. Biol.,* 118, 137, 1978.

104. **Matthews, D. A., Alden, R. A., Birktoft, J. J., Freer, S. T., and Kraut, J.,** Re-examination of the charge relay system in subtilisin and comparison with other serine proteases, *J. Biol. Chem.,* 252, 8875, 1977.

105. **Wright, C. S., Alden, R. A., and Kraut, J.,** Structure of subtilisin BPN' at 2.5 Å resolution, *Nature (London),* 221, 235, 1969.

106. **Drenth, J., Hol, W. G. J., Jansonius, J. N., and Koekoek, R.,** Subtilisin novo. The three-dimensional structure and its comparison with subtilisin BPN', *Eur. J. Biochem.,* 26, 177, 1972.

107. **Delbaere, L. T. J., Brayer, G. D., and James, M. N. G.,** The 2.8 Å resolution structure of *Streptomyces griseus* protease B and its homology with α-chymotrypsin and *Streptomyces griseus* protease A, *Can. J. Biochem.,* 57, 135, 1979.

108. **Kay, J. and Kassel, B.,** The autoactivation of trypsinogen, *J. Biol. Chem.,* 246, 6661, 1971.

109. **Morgan, P. H., Robinson, N. C., Walsh, K. A., and Neurath, H.,** Inactivation of bovine trypsinogen and chymotrypsinogen by diisopropylphosphorofluoridate, *Proc. Natl. Acad. Sci. U.S.A.,* 69, 3312, 1972.

110. **Gertler, A., Walsh, K. A., and Neurath, H.,** Catalysis by chymotrypsinogen. Demonstration of an acyl-zymogen intermediate, *Biochemistry,* 13, 1302, 1974.

111. **Kerr, M. A., Walsh, K. A., and Neurath, H.,** A proposal for the mechanism of chymotrypsinogen activation, *Biochemistry,* 15, 5566, 1976.

112. **Lonsdale-Eccles, J. D., Neurath, H., and Walsh, K. A.,** Probes of the mechanism of zymogen catalysis, *Biochemistry,* 17, 2805, 1978.

113. **Freer, S. T., Kraut, J., Robertus, J. D., Wright, H. T., and Xuong, Ng. H.,** Chymotrypsinogen: 2.5 Å crystal structure, comparison with α-chymotrypsin, and implications for zymogen activation, *Biochemistry,* 9, 1997, 1970.

114. **Wang, D., Bode, W., and Huber, R.,** Bovine chymotrypsinogen A. X-ray crystal structure analysis and refinement of a new crystal form at 1.8 Å resolution, *J. Mol. Biol.,* 185, 595, 1985.

115. **Kossiakoff, A. A., Chambers, J. L., Kay, L. M., and Stroud, R. M.,** Structure of bovine trypsinogen at 1.9 Å resolution, *Biochemistry,* 16, 654, 1977.

116. **Fehlhammer, H., Bode, W., and Huber, R.,** Crystal structure of bovine trypsinogen at 1.8 Å resolution. II. Crystallographic refinement, refined crystal structure and comparison with bovine trypsin *J. Mol. Biol.,* 111, 415, 1977.

117. **Walter, J., Steigemann, W., Singh, T. P., Bartunik, H., Bode, W., and Huber, R.,** On the disordered activation domain in trypsinogen: chemical labeling and low temperature crystallography, *Acta Crystallogr. Sect. B,* 1462, 1982.

118. **Bode, W. and Huber, R.,** Induction of the bovine trypsinogen-trypsin transition by peptides sequentially similar to the N-terminus of trypsin, *FEBS Lett.,* 68, 231, 1976.

119. **Bode, W.,** The transition of bovine trypsinogen to trypsin-like state upon strong ligand binding. The binding of pancreatic trypsin inhibitor and of isoleucine-valine and of sequentially related peptides to trypsinogen and p-gaunidinobenzoate-trypsinogen, *J. Mol. Biol.,* 127, 357, 1979.

120. **Rühlmann, A., Kukla, D., Schwager, P., Bartels, K., and Huber, R.,** Structure of the complex formed by bovine trypsin and bovine pancreatic trypsin inhibitor. Crystal structure determination and stereochemistry of the contact region, *J. Mol. Biol.,* 77, 417, 1973.

121. **Krieger, M., Kay, L. M., and Stroud, R. M.,** Structure and specific binding of trypsin: comparison of inhibited derivatives and a model for substrate binding, *J. Mol. Biol.,* 83, 209, 1974.

122. **Chen, Z. and Bode, W.,** Refined 2.5 Å X-ray crystal structure of the complex formed by porcine kallikrein A and the bovine pancreatic trypsin inhibitor. Crystallization, Patterson search, structure determination, refinement, structure and comparison with its components and with the bovine trypsin-pancreatic trypsin inhibitor complex, *J. Mol. Biol.,* 164, 283, 1983.

123. **Shotton, D. M. and Watson, H. C.,** Three-dimensional structure of tosyl-elastase, *Nature (London),* 225, 811, 1970.

124. **Brayer, G. D., Delbaere, L. T. J., and James, M. N. G.,** Molecular structure of the α-lytic protease from *Myxobacter 495* at 2.8 Å resolution, *J. Mol. Biol.,* 131, 743, 1979.

125. **Delbaere, L. T. J. and Brayer, G. D.,** The 1.8 Å structure of the complex between chymostatin and *Streptomyces griseus* protease A. A model for serine protease catalytic tetrahedral intermediates, *J. Mol. Biol.,* 183, 89, 1985.

126. **Craik, C. S., Largman, C., Fletcher, T., Roczniak, S., Barr, P. J., Fletterick, R., and Rutter, W. J.,** Redesigning trypsin: alteration of substrate specificity, *Science,* 228, 291, 1985.

127. **Graf, L., Craik, C. S., Patthy, A., Roczniak, S., Fletterick, R., and Rutter, W. J.,** Selective alteration of substrate specificity by replacement of Asp 189 with Lys in the binding pocket of trypsin, *Biochemistry,* 26, 2616, 1987.

128. **Segal, D. M., Powers, J. C., Cohen, G. H., Davies, D. R., and Wilcox, P. E.,** Substrate binding site in bovine chymotrypsin A$_\gamma$. A crystallographic study using peptide chloromethyl ketones as site-specific inhibitors, *Biochemistry,* 10, 3728, 1971.

129. **Sweet, R. M., Wright, H. T., Janin, J., Clothia, C. H., and Blow, D. M.,** Crystal structure of the complex of porcine trypsin with soybean trypsin inhibitor (Kunitz) at 2.6 Å resolution, *Biochemistry,* 13, 4212, 1974.

130. **James, M. N. G., Brayer, G. D., Delbaere, L. T. J., Sielecki, A. R., and Gertler, A.,** Crystal structure studies and inhibition kinetics of tripeptide chloromethyl ketone inhibitors with *Streptomyces griseus* protease B, *J. Mol. Biol.,* 139, 423, 1980.

131. **James, M. N. G., Sielecki, A. R., Brayer, G. D., Delbaere, L. T. J., and Bauer, C.-A.,** Structures of product and inhibitor complexes of *Streptomyces griseus* protease A at 1.8 Å resolution. A model for serine protease catalysis, *J. Mol. Biol.,* 144, 43, 1980.

132. **Bauer, C.-A., Thompson, R. C., and Blout, E. R.,** The active centers of *Streptomyces griseus* protease 3 and α-chymotrypsin: enzyme-substrate interactions remote from the scissile bond, *Biochemistry,* 15, 1291, 1976.

133. **Bauer, C.-A.,** Active centers of Streptomyces griseus protease 1, *Streptomyces griseus* protease 3, and α-chymotrypsin: enzyme-substrate interaction, *Biochemistry,* 17, 375, 1978.

134. **Umezawa, H.,** Structures and activities of protease inhibitors of microbial origin, *Methods Enzymol.,* 45, 678, 1976.

135. **Bauer, C.-A., Brayer, G. D., Sielecki, A. R., and James, M. N. G.,** Active site of α-lytic protease. Enzyme-substrate interactions, *Eur. J. Biochem.,* 120, 289, 1981.

136. **Thompson, R. C. and Blout, E. R.,** Dependence of the kinetic parameters for elastase-catalyzed amide hydrolysis on the length of peptide substrates, *Biochemistry,* 12, 57, 1973.

137. **Pozsgay, M., Szabó, G. Cs., Bajusz, S., Simonsson, R., Gáspár, R., and Elődi, P.,** Investigation of the substrate-binding site of trypsin by the aid of tripeptidyl-p-nitroanilide substrates, *Eur. J. Biochem.,* 115, 497, 1981.

138. **Walter, R., Simmons, W. H., and Yoshimoto, T.,** Proline specific endo- and exopeptidases, *Mol. Cell. Biochem.,* 30, 111, 1980.

139. **Fisher, G., Bang, H., Berger, E., and Schellenberger, A.,** Conformational specificity of chymotrypsin toward proline-containing substrates, *Biochim. Biophys. Acta,* 791, 87, 1984.

140. **Lin, L.-N. and Brandts, J. F.,** Isomer- specific proteolysis of model substrates: influence that the location of the proline residue exerts on cis/trans specificity, *Biochemistry,* 24, 6533, 1985.

141. **Lin, L.-N. and Brandts, J. F.,** Determination of cis-trans proline isomerization by trypsin proteolysis. Application to a model pentapeptide and to oxidized ribonuclease A, *Biochemistry,* 22, 553, 1983.

142. **Lin, L.-N. and Brandts, J. F.,** Evidence showing that a proline-specific endopeptidase has an absolute requirement for a trans peptide bond immediately preceding the active bond, *Biochemistry,* 22, 4480, 1983.

143. **Fischer, G., Heins, J., and Barth, A.,** The conformation around the peptide bond between the P_1- and P_2-positions is important for catalytic activity of some proline-specific proteases, *Biochim. Biophys. Acta,* 742, 452, 1983.

144. **Robertus, J. D., Alden, R. A., Birktoft, J. J., Kraut, J., Powers, J. C., and Wilcox, P. E.,** An X-ray crystallographic study of the binding of peptide chloromethyl ketone inhibitors to subtilisin BPN', *Biochemistry,* 11, 2439, 1972.

145. **Poulos, T. L., Alden, R. A., Freer, S. T., Birktoft, J. J., and Kraut, J.,** Polypeptide halomethyl ketones bind to serine proteases as analogs of the tetrahedral intermediate: X-ray crystallographic comparison of lysine- and phenylalanine-polypeptide chloromethyl ketone-inhibited subtilisin, *J. Biol. Chem.,* 251, 1097, 1976.

146. **Hirono, S., Akagawa, H., Mitsui, Y., and Iitaka, Y.,** Crystal structure at 2.6 Å resolution of the complex of subtilisin BPN' with *Streptomyces* subtilisin inhibitor, *J. Mol. Biol.,* 178, 389, 1984.

147. **Betzel, C., Pal, G. P., Struck, M., Jany, K.-D., and Saenger, W.,** Active-site geometry of proteinase K. Crystallographic study of its complex with a dipeptide chloromethyl ketone inhibitor, *FEBS Lett.,* 197, 105, 1986.

148. **Jany, K.-D., Lederer, G., and Mayer, B.,** Amino acid sequence of proteinase K from the mold *Tritirachium album* limber. Proteinase K — a subtilisin-related enzyme with disulfide bonds, *FEBS Lett.,* 199, 139, 1986.

149. **Meloun, B., Baudys, M., Kostka, V., Hausdorf, G., Frömmel, C., and Höhne, W. E.,** Complete primary structure of thermitase from *Thermoactinomyces vulgaris* and its structural features related to the subtilisin-type proteinases, *FEBS Lett.,* 183, 195, 1985.

150. **Gaucher, G. M. and Stevenson, K. J.,** Thermomycolin, *Methods Enzymol.,* 45, 415, 1976.

151. **Stepanov, V. M., Chestukhina, G. G., Rudenskaya, G. N., Epremyan, A. S., Osterman, A. L., Khodova, O. M., and Belyanova, L. P.,** A new subfamily of microbial serine proteinases? Structural similarities of *Bacillus thuringiensis* and *Thermoactinomyces vulgaris* extracellular serine proteinases, *Biochem. Biophys. Res. Commun.,* 100, 1680, 1981.

152. **Fittkau, S., Smalla, K., and Pauli, D.,** Thermitase — eine thermostabile Serinprotease. IV. Kinetische Untersuchungen der Bindung von N-Azylpeptidketonen als Substratanaloge Inhibitoren, *Biomed. Biochim. Acta,* 43, 887, 1984.

153. **Fittkau, S. and Brutsceck, M.,** Thermitase — eine thermostabile Serinprotease. V. Zur Stereospezifität der Bindung von Peptidliganden, *Biomed. Biochim. Acta,* 43, 901, 1984.

154. **Morihara, K.,** Comparative specificity of microbial proteinases, *Adv. Enzymol. Relat. Areas Mol. Biol.,* 41, 179, 1974.

154a. **Imperiali, B. and Abeles, R. H.,** Extended binding inhibitors of chymotrypsin that interact with leaving group subsites S_1'–S_3', *Biochemistry,* 26, 4474, 1987.

155. **Deisenhofer, J. and Steigemann, W.,** Crystallographic refinement of the structure of bovine pancreatic trypsin inhibitor at 1.5 Å resolution, *Acta Crystallogr. Sect. B,* 31, 238, 1975.

156. **Mitsui, Y., Satow, Y., Watanabe, Y., and Iitaka, Y.,** Crystal structure of a bacterial protein proteinase inhibitor (*Streptomyces* subtilisin inhibitor) at 2.6 Å resolution, *J. Mol. Biol.,* 131, 697, 1979.

157. **Hirono, S., Nakamura, K. L., Iitaka, Y., and Mitsui, Y.,** Crystal structure of the complex of subtilisin BPN' with its protein inhibitor *Streptomyces* subtilisin inhibitor. The structure at 4.3 Å resolution, *J. Mol. Biol.,* 131, 855, 1979.

158. **Fujinaga, M., Read, R. J., Sielecki, A., Ardelt, W., Laskowski, M., Jr, and James, M. N. G.,** Refined crystal structure of the molecular complex of *Streptomyces griseus* protease B, a serine protease, with the third domain of the ovomucoid inhibitor from turkey, *Proc. Natl. Acad. Sci. U.S.A.,* 79, 4868, 1982.

159. **Read, R. J., Fujinaga, M., Sielecki, A. R., and James, M. N. G.,** Structure of the complex of *Streptomyces griseus* protease B and the third domain of the turkey ovomucoid inhibitor at 1.8 Å resolution, *Biochemistry,* 22, 4420, 1983.

159a. **Bode, W., Wei, A.-Z., Huber, R., Meyer, E., Travis, J., and Neumann, S.,** X-ray crystal structure of the complex of human leukocyte elastase (PMN elastase) and the third domain of the turkey ovomucoid inhibitor, *EMBO J.,* 5, 2453, 1986.

160. **Weber, E., Papamokos, E., Bode, W., Huber, R., Kato, I., and Laskowski, M., Jr.,** Crystallization, crystal structure analysis and molecular model of the third domain of Japanese quail ovomucoid, a Kazal type inhibitor, *J. Mol. Biol.,* 149, 109, 1981.

161. **Papamokos, E., Weber, E., Bode, W., Huber, R., Empie, M. W., Kato, I., and Laskowski, M., Jr.,** Crystallographic refinement of Japanese quail ovomucoid, a Kazal-type inhibitor, and model building studies of complexes with serine proteases, *J. Mol. Biol.,* 158, 515, 1982.

162. **Bolognesi, M., Gatti, G., Menegatti, E., Guarneri, M., Papamokos, E., and Huber, R.,** Three-dimensional structure of the complex between pancreatic secretory trypsin inhibitor (Kazal type) and trypsinogen at 1.8 Å resolution. Structure solution, crystallographic refinement and preliminary structural interpretation, *J. Mol. Biol.,* 162, 839, 1982.

163. **Antonini, E., Ascenzi, P., Bolognesi, M., Gatti, G., Guarneri, M., and Menegatti, E.,** Interaction between serine (pro)enzymes, and Kazal and Kunitz inhibitors, *J. Mol. Biol.,* 165, 543, 1983.

164. **Bode, W., Epp, O., Huber, R., Laskowski, M., Jr., and Ardelt, W.,** The crystal and molecular structure of the third domain of silver pheasant ovomucoid (OMSVP₃), *Eur. J. Biochem.,* 147, 387, 1985.

165. **McPhalen, C. A., Svendsen, I., Jonassen, I., and James, M. N. G.,** Crystal and molecular structure of chymotrypsin inhibitor 2 from barley seeds in complex with subtilisin Novo, *Proc. Natl. Acad. Sci. U.S.A.,* 82, 7242, 1985.

166. **McPhalen, C. A., Schnebli, H. P., and James, M. N. G.,** Crystal and molecular structure of the inhibitor eglin from leeches in complex with subtilisin Carlsberg, *FEBS Lett.,* 188, 55, 1985.

176. **Bode, W., Papamokos, E., Musil, D., Seemueller, U., and Fritz, H.,** Refined 1.2 Å crystal structure of the complex formed between subtilisin Carlsberg and the inhibitor eglin c. Molecular structure of eglin and its detailed interaction with subtilisin, *EMBO J.,* 5, 813, 1986.

167a. **Bode, W., Papamokos, E., and Musil, D.,** The high-resolution X-ray crystal structure of the complex formed between subtilisin Carlsberg and eglin c, an elastase inhibitor from leech *Hirudo medicinalis.* Structural analysis, subtilisin structure and interface geometry, *Eur. J. Biochem.,* 166, 673, 1987.

168. **Glazer, A. N.,** Inhibition of "serine" esterases by phenylarsonic acids. α-Chymotrypsin and the subtilisins, *J. Biol. Chem.,* 243, 3693, 1968.

169. **Shah, D. O., Lai, K., and Gorenstein, D. G.,** ¹³C NMR Spectroscopy of "transition-state-analogue" complexes of N-acetyl-L-phenylalanie and α-chymotrypsin, *J. Am. Chem. Soc.,* 106, 4272, 1984.

170. **Malthouse, J. P. G., Primrose, W. U., Mackenzie, N. E., and Scott, A. I.,** ¹³C NMR Study of the ionization within a trypsin-chloromethyl ketone inhibitor complex, *Biochemistry,* 24, 3478, 1985.

171. **Philipp, M. and Maripuri, S.,** Inhibition of subtilisin by substituted arylboronic acids, *FEBS Lett.,* 133, 36, 1980.

172. **Matteson, D. S., Sadhu, K. M., and Lienhard, G. E.,** (R)-1-Acetamido-2-phenylethaneboronic acid. A specific transition-state analogue for chymotrypsin, *J. Am. Chem. Soc.,* 103, 5241, 1981.

173. **Kettner, C. A. and Shenvi, A. B.,** Inhibition of the serine proteases leukocyte elastase, pancreatic elastase, cathepsin G, and chymotrypsin by peptide boronic acids, *J. Biol. Chem.,* 259, 15106, 1984.

174. **Antonov, V. K., Ivanina, T. V., Berezin, I. V., and Martinek, K.,** N-alkylboronic acids as bifunctional reversible inhibitors of α-chymotrypsin, *FEBS Lett.,* 7, 23, 1970.

175. **Koehler, K. A. and Lienhard, G. E.,** 2-Phenylethaneboronic acid, a possible transition-state analog for chymotrypsin, *Biochemistry,* 10, 2477, 1971.

176. **Rawn, J. D. and Lienhard, G. E.,** The binding of boronic acids to chymotrypsin, *Biochemistry,* 13, 3124, 1974.

177. **Philipp, M. and Bender, M. L.,** Inhibition of serine proteases by arylboronic acids, *Proc. Natl. Acad. Sci. U.S.A.,* 68, 478, 1971.

178. **Lindquist, R. N. and Terry, C.,** Inhibition of subtilisin by boronic acids, potential analogs of tetrahedral reaction intermediates, *Arch. Biochem. Biophys.,* 160, 135, 1974.

179. **Matthews, D. A., Alden, R. A., Birktoft, J. J., Freer, S. T., and Kraut, J.,** X-ray crystallographic study of boronic acid adducts with subtilisin BPN'(Novo). A model for the catalytic transition state, *J. Biol. Chem.,* 250, 7120, 1975.

180. **Aoyagi, T., Miyata, S., Nanbo, M., Kojima, F., Matsuzaki, M., Ishizuka, M., Takeuchi, T., and Umezawa, H.,** Biological activities of leupeptins, *J. Antibiot.,* 22, 558, 1969.

181. **Umezawa, H. and Aoyagi, T.,** Trends in research of low molecular weight protease inhibitors of microbial origin, in *Proteinase Inhibitors, Medical and Biological Aspects,* Katunuma, N., Umezawa, H., and Holzer, H., Eds., Japan Scientific Societies Press, Tokyo, 1983, 3.

182. **Chen, R., Gorenstein, D. G., Kennedy, W. P., Lowe, G., Nurse, D., and Schultz, R. M.,** Evidence for hemiacetal formation between N-acyl-L- phenylalaninals and α-chymotrypsin by cross-saturation nuclear magnetic resonance spectroscopy, *Biochemistry,* 18, 921, 1979.

183. **Huber, R., Bode, W., Kukla, K., Kohl, U., and Ryan, C. A.,** The structure of the complex formed by bovine trypsin and bovine pancreatic trypsin inhibitor. III. Structure of the anhydro-trypsin-inhibitor complex, *Biophys. Struct. Mech.,* 1, 189, 1975.

184. **Baillargeon, M. W., Laskowsky, M., Jr., Neves, D. E., Porubcan, M. A., Santini, R. E., and Markley, J. L.,** Soybean trypsin inhibitor (Kunitz) and its complex with trypsin. Carbon-13 nuclear magnetic resonance studies of the reactive site arginine, *Biochemistry,* 19, 5703, 1980.

185. **Richarz, R., Tschesche, H., and Wüthrich, K.,** Carbon-13 nuclear magnetic resonance studies of the selectively isotope-labeled reactive site peptide bond of the basic pancreatic trypsin inhibitor in the complexes with trypsin, trypsinogen, and anhydrotrypsin, *Biochemistry,* 19, 5711, 1980.

186. **Hine, J.,** Hydrogen-bonded intermediates and stepwise mechanism for proton-exchange reactions between oxygen atoms in hydroxylic solvents, *J. Am. Chem. Soc.,* 94, 5766, 1972.

187. **van Duijnen, P.Th.,** Quantum chemistry and enzymes, *J. Mol. Catal.,* 11, 263, 1981.

188. **van Duijnen, P.Th.,** On the inactivity of thiol-subtilisin. The role of the intramolecular electric field, *Biophys. Chem.,* 13, 133, 1981.

189. **Jordan, F. and Polgár, L.,** Proton nuclear magnetic resonance evidence for the absence of a stable hydrogen bond between the active site aspartate and histidine residues of native subtilisins and for its presence in thiolsubtilisins, *Biochemistry,* 20, 6366, 1981.

190. **Lazure, C., Leduc, R., Seidah, N. G., Thibault, G., Genest, J., and Chrétien, M.,** Amino acid sequence of rat submaxillary tonin reveals similarities to serine proteases, *Nature (London),* 307, 555, 1984.

191. **Ashley, P. L. and McDonald, R. J.,** Kallikrein-related mRNAs of the rat submaxillary gland: nucleotide sequences of four distinct types including tonin, *Biochemistry,* 24, 4512, 1985.

191a. **Fujinaga, M. and James, M. N. G.,** Rat submaxillary gland serine protease, tonin. Structure solution and refinement at 1.8 Å resolution, *J. Mol. Biol.,* 195, 373, 1987.

192. **Hunkapiller, M. W., Smallcombe, S. H., Whitaker, D. R., and Richards, J. H.,** Carbon nuclear magnetic resonance studies of the histidine residue in α-lytic protease. Implications for the catalytic mechanism of serine proteases, *Biochemistry,* 12, 4732, 1973.

193. **Bachovchin, W. W. and Roberts, J. D.,** Nitrogen-15 nuclear magnetic resonance spectroscopy. The state of histidine in the catalytic triad of α-lytic protease. Implications for the charge-relay mechanism of peptide-bond cleavage by serine proteases, *J. Am. Chem. Soc.,* 100, 8041, 1978.

194. **Bachovchin, W. W., Kaiser, R., Richards, J. H., and Roberts, J. D.,** Catalytic mechanism of serine proteases: reexamination of the pH dependence of the histidyl $^{1}J_{13C2-H}$ coupling constant in the catalytic triad of α-lytic protease, *Proc. Natl. Acad. Sci. U.S.A.,* 78, 7323, 1981.

195. **Kossiakoff, A. A. and Spencer, S. A.,** Direct determination of the protonation states of aspartic acid-102 and histidine-57 in the tetrahedral intermediate of the serine proteases: neutron structure of trypsin, *Biochemistry,* 20, 6462, 1981.

196. **Polgár, L.,** On the role of hydrogen-bonding system in the catalysis by serine proteases, *Acta Biochim. Biophys. Acad. Sci. Hung.,* 7, 29, 1972.

196a. **Náray-Szabó, G., Kapur, A., Mezey, P. G., and Polgár, L.,** Molecular orbital analysis of the catalytic process of serine proteinases: effect of environment on protonation of the histidine-aspartate diad of subtilisin, *J. Mol. Struct.,* 90, 137, 1982.

197. **Warshel, A.,** Energetics of enzyme catalysis, *Proc. Natl. Acad. Sci. U.S.A.,* 75, 5250, 1978.

198. **Hol, W. G. J., van Duijnen, P.Th., and Berendsen, H. J. C.,** The α-helix dipole and the properties of proteins, *Nature (London),* 273, 443, 1978.

199. **Thomas, P. G., Russel, A. J., and Fersht, A. R.,** Tailoring the pH-dependence of enzyme catalysis using protein engineering, *Nature (London),* 318, 375, 1985.

200. **Craik, C. S., Roczniak, S., Largman, C., and Rutter, W. J.,** The catalytic role of the active site aspartic acid in serine proteases, *Science,* 237, 909, 1987.

201. **Asbóth, B. and Polgár, L.,** Transition-state stabilization at the oxyanion binding site of serine and thiol proteinases: hydrolyses of thiono and oxygen esters, *Biochemistry,* 22, 117, 1983.

202. **Bizzozero, S. A. and Zweifel, B. O.,** The importance of the conformation of the tetrahedral intermediate for the α-chymotrypsin-catalyzed hydrolysis of peptide substrates, *FEBS Lett.,* 59, 105, 1975.

203. **Petkov, D. D., Christova, E., and Stoineva, I.,** Catalysis and leaving groups binding in anilide hydrolysis by chymotrypsin, *Biochim. Biophys. Acta,* 527, 131, 1978.

204. **Dugas, H. and Penney, C.,** *Bioorganic Chemistry,* Springer-Verlag, Berlin, 1981, 238.

205. **Bizzozero, S. A. and Dutler, H.,** Stereochemical aspects of peptide hydrolysis catalyzed by serine proteases of the chymotrypsin type, *Bioorg. Chem.,* 10, 46, 1981.

206. **Hosie, L. and Sinnott, M. L.,** Effects of deuterium substitution α and β to the reaction centre, ^{18}O substitution in the leaving group, and aglycone acidity on hydrolyses of aryl glucosides and glucosyl pyridinium ions by yeast α glucosidase. A probable failure of the antiperiplanar-lone-pair hypothesis in glycosidase catalysis, *Biochem. J.,* 226, 437, 1985.

207. **Hunkapiller, M. V., Forgac, M. D., and Richards, J. H.,** Mechanism of action of serine proteases: tetrahedral intermediate and concerted proton transfer, *Biochemistry,* 15, 5581, 1976.

208. **Petkov, D. D.,** Detection of a tetrahedral intermediate in the trypsin-catalysed hydrolysis of specific ring-activated anilides, *Biochim. Biophys. Acta,* 523, 538, 1978.

209. **Fink, A. L. and Meehan, P.,** Detection and accumulation of tetrahedral intermediates in elastase catalysis, *Proc. Natl. Acad. Sci. U.S.A.,* 76, 1566, 1979.

210. **Compton, P. and Fink, A. L.,** The detection, accumulation and stabilization of a tetrahedral intermediate in trypsin catalysis, *Biochem. Biophys. Res. Commun.,* 93, 427, 1980.

211. **Fink, A. L.,** Cryoenzymology of chymotrypsin: the detection of intermediates in the catalysis of a specific anilide substrate, *Biochemistry,* 15, 1580, 1976.

212. **Markley, J. L., Travers, F., and Balny, C.,** Lack of evidence for a tetrahedral intermediate in the hydrolysis of nitroanilide substrates by serine proteinases, *Eur. J. Biochem.,* 120, 477, 1981.

213. **Compton, P. D. and Fink, A. L.,** Low-temperature reactions of trypsin with p-nitroanilide substrates: tetrahedral intermediate formation or enzyme isomerization, *Biochemistry,* 23, 2989, 1984.

214. **Fastrez, J.,** On the stability of tetrahedral intermediates within the active sites of serine and cysteine proteases, *Eur. J. Biochem.,* 135, 339, 1983.

215. **Bender, M. L. and Nakamura, K.,** The effect of structure on the rates of some α-chymotrypsin-catalyzed reactions, *J. Am. Chem. Soc.,* 84, 2577, 1962.

216. **Williams, A.,** Chymotrypsin-catalyzed phenyl ester hydrolysis. Evidence for electrophilic assistance on carbonyl oxygen, *Biochemistry,* 9, 3383, 1970.

217. **Williams, R. E. and Bender, M. L.,** Substituent effects on the chymotrypsin-catalyzed hydrolysis of specific ester substrates, *Can. J. Biochem.,* 49, 210, 1971.

218. **Ikeda, K., Kunugi, Sh., and Ise, N.,** Study of the substituent effect on α-chymotrypsin-catalyzed hydrolysis of phenyl acetates by using a stopped-flow titration method, *Arch. Biochem. Biophys.,* 217, 37, 1982.

219. **Kogan, R. L. and Fife, T. H.,** Electronic effects in the acylation of α-chymotrypsin by substituted N-benzoyl-imidazoles, *Biochemistry,* 23, 2983, 1984.

220. **Bundy, H. F. and Moore, C. L.,** Chymotrypsin-catalyzed hydrolysis of m- , p- , and o-nitroanilides of N-benzoyl-L-tyrosine, *Biochemistry,* 5, 808, 1966.

221. **Inagami, T., Patchornik, A., and York, S. S.,** Participation of an acidic group in the chymotrypsin catalysis, *J. Biochem. (Tokyo),* 65, 809, 1969.

222. **Hansch, C. and Leo, A.,** *Substituent Constants for Correlation Analysis in Chemistry and Biology,* Wiley-Interscience, New York, 1979, chap. 1, 2, 3, and 5.

223. **Inward, P. W. and Jencks, W. P.,** The reactivity of nucleophilic reagents with furoyl-chymotrypsin, *J. Biol. Chem.,* 240, 1986, 1965.

224. **Zeeberg, B. and Caplow, M.,** Transition state charge distribution in reactions of an acetyltyrosylchymotrypsin intermediate, *J. Biol. Chem.,* 248, 5887, 1973.

225. **Komiyama, M. and Bender, M. L.,** Do cleavages of amides by serine proteases occur through a stepwise pathway involving tetrahedral intermediates?, *Proc. Natl. Acad. Sci. U.S.A.,* 76, 557, 1979.

226. **Polgár, L., Halász, P., and Moravcsik, E.,** On the reactivity of the thiol group of thiolsubtilisin, *Eur. J. Biochem.,* 39, 421, 1973.

227. **Robillard, G. and Shulman, R. G.,** High resolution nuclear magnetic resonance studies of the active site of chymotrypsin. II. Polarization of histidine 57 by substrate analogues and competitive inhibitors, *J. Mol. Biol.,* 86, 541, 1974.

228. **Malthouse, J. P. G., Primrose, W. U., Mackenzie, N. E., and Scott, I.,** ^{13}C NMR study of the ionization within a trypsin-chloromethyl ketone inhibitor complex, *Biochemistry* , 24, 3478, 1985.

229. **Kovach, I. M., Hogg, J. L., Raben, T., Halbert, K., Rodgers, J., and Schowen, R. L.,** The β-hydrogen secondary isotope effect in acyl transfer reactions. Origins, temperature dependence, and utility as a probe of transition-state structure, *J. Am. Chem. Soc.,* 102, 1991, 1980.

230. **Lehmann, G., Quinn, D., and Cordes, E. H.,** Kinetic α-deuterium isotope effects for acylation of chymotrypsin by 4-methoxyphenyl formate and for deacylation of formylchymotrypsin, *J. Am. Chem. Soc.,* 102, 2491, 1980.

231. **Stein, R. L., Elrod, J. P., and Schowen, R. L.,** Correlative variations in enzyme-derived and substrate-derived structures of catalytic transition states. Implications for the catalytic strategy of acyl-transfer enzymes, *J. Am. Chem. Soc.,* 105, 2446, 1983.
232. **Venkatasubban, K. S. and Schowen, R. L.,** The proton inventory technique, *CRC Crit. Rev. Biochem.,* 17, 1, 1985.
233. **Elrod, J. P., Hogg, J. L., Quinn, D. M., Venkatasubban, K. S., and Schowen, R. L.,** Protonic reorganization and substrate structure in catalysis by serine proteases, *J. Am. Chem. Soc.,* 102, 3917, 1980.
234. **Jencks, W. P.,** Binding energy, specificity, and enzymic catalysis: the Circe effect, *Adv. Enzymol. Relat. Areas Mol. Biol.,* 43, 219, 1975.
235. **Fersht, A. R., Leatherbarrow, R. J., and Wells, T. N. C.,** Binding energy and catalysis: a lesson from protein engineering of the tyrosyl-tRNA synthetase, *Trends Biochem. Sci.,* 11, 321, 1986.

Chapter 4

CYSTEINE PROTEASES

I. GENERAL PROPERTIES OF ENZYMES AND THEIR INHIBITORS

Cysteine proteases are widely distributed in nature. They are encountered, for instance, in plants, animals, and microorganisms. Papain, a plant enzyme, is the protagonist of cysteine proteases. It had been studied most extensively and almost exclusively until the last 10 years when mammalian cathepsins and calpains raised considerable interest. Most recently, viral proteases have become the subject of a number of investigations.

A. Plant Proteases

Several reviews have been devoted to plant proteases, the emphasis being on papain in general,[1,2,2a] on its tertiary structure,[3,4] or on the mechanism of its action.[5-7a] Besides papain, two other components of the papaya latex, chymopapain and papaya peptidase,[1,2] bromelain of the pineapple,[1] and ficin[1] of different species of the genus *Ficus* have also been studied in some detail. Actinidin from the ripe fruit of the Chinese gooseberry has been completely sequenced.[8] A substantial part of the amino acid sequence of stem bromelain has also been determined.[9,10] Numerous plant proteases, such as asclepains from the milkweed,[11] have been shown to occur in multiple forms.

Following papain, actinidin is the second cysteine protease whose three-dimensional structure has been solved.[12,13] The two structures are very closely related. X-ray crystallographic studies at low resolution have also shown that the overall molecular architecture of calotropin, a cysteine protease from the latex of the madar plant, *Calotropis gigantea,* is similar to that of papain and actinidin.[14] The high resolution structure of calotropin has now been completed at the Freie Universität, Berlin.[228] It may be surprising that the physiological role of the most thoroughly studied papain and that of the related plant cysteine proteases is uncertain. It is possible that these enzymes protect the plant against invasion by fungi and insects.[15]

There has been some confusion concerning the term papain. Originally, it was introduced to describe the proteolytic principle of the latex of the melon-like green fruit of the small tropical tree *Carica papaya.* The latex contains a number of protein components, including several proteases, lysozyme, glutamine cyclotransferase, and endo-1,3-β-glucanase.[2] The latex or its partially purified form is used extensively in the industry under the name of papain, for example, as a chill-proofing agent to prevent the formation of a chill-haze in beer on cold storage, or as a meat tenderizer. However, in the biochemical literature, the term papain is applied to the individual enzyme first crystallized from the latex.[16] This single component of the latex is meant by papain throughout this book.

The papain molecule is composed of a single polypeptide chain of 212 amino acid residues. Its amino acid sequence has been determined,[17,18] yielding a relative molecular mass of 23,350. The enzyme contains three disulfide bonds (Cys 22-Cys 63, Cys 56-Cys 95, and Cys 153-Cys 200) and a free cysteine (Cys 25) which takes part in the catalysis . Commercial papain, as prepared by the method of Kimmel and Smith,[19,20] contains variable amounts of a free thiol group between 0.1 and 0.5 mole -SH per mole of enzyme. The catalytic activity is proportional to the thiol content of the enzyme. A part of commercial papain can be activated by thiol compounds or other reducing agents,[1,2] but an enzyme that contains 1.0 free thiol group per mole of protein cannot be obtained through activation. Actually, the commercial preparation has three forms of papain: active, activatable, and inactive which cannot be activated.[21] The activatable enzyme contains a mixed disulfide derived from the thiol group of the enzyme and a free cysteine.[22-24] The thiol group of the inactive enzyme exists in a higher oxidation state, probably in a sulfonic acid form.

To separate inactive papain from active papain, three procedures have generally been used. (1) One method employs affinity chromatography on a column consisting of the inhibitor Gly-Gly-Tyr-(*O*-benzyl)-Arg covalently linked to Sepharose.[25-27] The same method can also be used for the purification of papaya peptidase A, but binding of this enzyme requires higher ionic strength than that of papain.[28] (2) Another method uses *p*-aminophenylmercuric acetate attached to Sepharose.[29] This binds only the active enzyme, which can be eluted with a solution of mercuric chloride or thiol compounds, like cysteine or mercaptoethanol. (3) A further method employs covalent chromatography by thiol-disulfide interchange.[30,31]

With the use of the above three methods, pure papain containing 1.0 mol of thiol group per mole of enzyme can be obtained. The thiol content of papain and other related cysteine proteases can be determined by active site titration (Chapter 2, Section VIII.D), preferably with 2,2′-dipyridyl disulfide.

The specificity of papain is substantially different from that of most serine proteases (Chapter 3, Section IV) in that the discriminating primary binding site of papain is associated with the S_2 rather than the S_1 subsite.[2] This explains that studies using synthetic derivatives of single amino acids as substrates could not find a specific substrate for papain. Nonetheless, it is widely assumed that papain is specific for arginine. In fact, benzoylarginine ethyl ester (Bz-Arg-OEt or BAEE) is generally used as a reference substrate for papain. However, the specificity constant (k_{cat}/K_m) is not too much different for Bz-Arg-OMe and the corresponding citrulline derivative.[32] This finding also indicates that the positive charge on the arginine side chain does not make an important contribution in the binding to papain.[32,33] The importance of a large hydrophobic residue in binding to the S_2 subsite is apparent from a comparison of the specificity constants obtained with the corresponding derivatives of Phe-Gly and Arg, which shows that the dipeptide substrates react with papain at a rate more than two orders of magnitude higher than the corresponding Arg derivative.[33,34] The *N*-benzoyl and *N*-benzyloxycarbonyl derivatives of single amino acid substrates are fairly good substrates of papain, which is presumably due to the binding of the *N*-acyl groups to the S_2 subsite. Indeed, the benzyloxycarbonyl group which, as compared with the benzoyl group, resembles more the phenylalanyl side chain, yields a better substrate when combined with a single amino acid.[32,33]

Besides the S_2 primary binding site, other subsites are also important in the papain catalysis. This enzyme has an extended binding site that is capable of binding about seven amino acid residues of a peptide substrate.[35,36] The subsites are located on both sides of the catalytic site, probably four on the N-terminal side (S_1 to S_4) and three on the C-terminal side (S_1' to S_3')[35] The interactions between papain and four N-terminal residues of the substrate have been demonstrated by detailed kinetic studies.[37] As for the C-terminal side, the S_1' subsite has been shown to exhibit a strong preference for hydrophobic residues.[38]

Fewer studies were carried out with chymopapain and papaya peptidase A, two other proteases of papaya latex. Chymopapain (pI $\sim$ 10.1 to 10.6) and papaya peptidase A (pI $>$ 11) are more basic proteins than papain (pI $=$ 9.55).[2] Indeed, recent determination of the pI of 11.7 for papaya peptidase A establishes this protease as an enzyme of extreme basicity.[39] Due to their considerably different basicities, the three enzymes are readily resolved by ion-exchange chromatography.[40,41] However, chymopapain, which gives rise to the largest peak, is not homogeneous. Roughly, it contains two components, chymopapain A (the less basic component) and chymopapain B (the more basic component), but these chymopapains can be further fractionated, especially by fast protein liquid chromatography (FPLC).[42,43] At present, we have not enough information to appropriately classify the different enzyme forms of the complex elution band containing chymopapain.[44-46] It is very probable that most of the different forms are artifacts developed during latex processing. Others may be individual enzymes highly homologous with the known plant proteases, as indicated by the isolation of cDNA clones for two new cysteine enzymes from the leaf tissue cDNA library of *Carica papaya*.[47]

In contrast to chymopapain, papaya peptidase A can be readily purified as a fully active enzyme although a substantial amount of this enzyme is inactive in the latex.[48] For papaya peptidase A, the following names have recently been suggested: papaya proteinase A,[44] papaya proteinase Ω,[45] and papaya proteinase III.[46]

In contrast to papain and papaya peptidase A which are monothiol proteases, chymopapain is classified as a dithiol protease.[2,49] However, chymopapain may contain more than two free thiol groups.[43]

Compared with papain, little is known about the specificity characteristics of chymopapain and papaya peptidase A.[2] It appears, however, that the S_2 subsite of cysteine proteases of the papain family has a preference for hydrophobic residues.[50]

Chymopapain preparations are used in the treatment of prolapsed intervertebral disks. The procedure, termed "chemonucleosis" or "discolysis", involves injection of the enzyme directly into the nucleus pulposus of the affected disk.[51,52]

It is interesting to note that the above plant proteases usually occur together with their inhibitors. However, very limited data are available concerning these inhibitors. The primary structure of one of the bromelain inhibitors has been reported.[53] The inhibitor was shown to contain a relatively large number of disulfide bridges. Furthermore, it was suggested that inhibitors may interfere with the purification of papain from crude latex.[19,26]

B. Cathepsins

Besides plant cysteine proteases, lysosomal cysteine proteases have recently raised considerable interest. Specifically, these enzymes play a major role in intracellular breakdown of proteins and may also be involved in controlling cellular functions by limited proteolysis.[54,55] Furthermore, they may be implicated in tumor invasion and metastasis.[56] Three of the lysosomal cysteine proteases are fairly well characterized. These are cathepsins B, H, and L. The amino acid sequences of cathepsins B and H are highly homologous to that of papain, indicating that the plant and mammalian cysteine proteases have evolved from a common ancestral protein.[57]

Cathepsin B is the most extensively studied lysosomal cysteine protease. It can be isolated from several mammalian species and various tissues including spleen,[58-62] liver,[55,63,64] parathyroid gland,[66] lymph nodes,[66] breast tumor,[67] and brain.[68] It is probably present in all mammalian cells. Cathepsin B is a glycoprotein[64] with very low[61] or low[69] mannose content.

Cathepsin H and cathepsin L were discovered much later[70-72] than cathepsin B. All three enzymes are purified together through several steps until they are separated by ion exchange chromatography. A new efficient method, covalent affinity chromatography, has recently been introduced for the purification of cathepsin B.[72,74] All three cathepsins are very unstable above pH 7. Some of their characteristics are compared in Table 1.

The three cathepsins can be distinguished by using appropriate substrates shown in Table 1. It is seen that the recommended substrate for cathepsin B contains a pair of arginine.[59] Earlier, cathepsin B was generally assayed with benzoyl-DL-Arg-*p*-nitroanilide and benzoyl-Arg-2-naphthylamide. However, these substrates are much less sensitive and less specific as they are also susceptible to cathepsin H. Z-Phe-Arg-4-methyl-7-coumarylamide is also an excellent substrate of cathepsin B, but it is even better for cathepsin L. As with papain, the S_2 subsite appears to be important also with cathepsin B. Both Phe and Arg as the P_2 residue greatly increase the k_{cat}/K_m value.[55]

A remarkable observation concerning the activity of cathepsin B is its peptidyldipeptidase activity. It has been shown that it cleaves the C-terminal dipeptides sequentially from glucagon[75] and muscle aldolase[76] with broad specificity, and it is this activity of cathepsin B that inactivates aldolase (see Table 1).

Table 1
A COMPARISON OF THE PROPERTIES OF LYSOSOMAL
CYSTEINE PROTEINASES CATHEPSINS B, H, AND L

	Cathepsin B	Cathepsin H	Cathepsin L
Recommended test substrate	Z-Arg-Arg-NMec	Arg-NMec	Z-Phe-Arg-NMec
(pH optimum)	(pH 6.0)	(pH 6.8)	(pH 5.5)
Aminopeptidase activity	No	Yes	No
Endopeptidase activity	Moderate	Variable	High
Peptidyldipeptidase activity	Yes	No	No
Aldolase inactivation	Yes	No	Yes
Leupeptin sensitivity	High	Low	High
Z-Phe-Phe-CHN$_2$ sensitivity	Low	Low	High
pI	4.5—5.5	6.0—7.1	5.5—6.1
Concanavalin A-Sepharose	Not bound	Bound	Bound

From Barrett, A. J. and Kirschke, H., *Methods Enzymol.*, 80, 535, 1981. With permission.

Cathepsin H is unique among endopeptidases characterized so far, inasmuch as it exhibits an aminopeptidase activity,[71,77] at least with synthetic substrates. For example, Arg-2-naphthylamide and Arg-4-methyl-7-coumarylamide are the best substrates yet discovered for cathepsin H; Leu-2-naphthylamide and benzoyl-Arg-2-naphthylamide are somewhat less susceptible.[78]

It has recently been found[79] that cathepsin L occurs in human liver tissue homogenate mainly as a "latent" enzyme. Therefore, an acid treatment at pH 4.2, which converts the "latent" form into the active enzyme, is of critical importance to obtain the active protease in high yield.[79] As for its specificity, cathepsin L is of particular interest because it has the highest capacity among lysosomal proteases to degrade proteins, including insoluble collagen[80] and elastin.[81] Synthetic substrates of cathepsin L, e.g., Z-Phe-Arg-4-methyl-7-coumarylamide, succinyl-Tyr-Met-2-naphthylamide, and others[54,55] are also hydrolyzed by cathepsin B. Nevertheless, there are considerable differences in the kinetic specificities, cathepsin L being more active. For a good substrate of cathepsin L, a hydrophobic residue at the P$_2$ position seems to be essential, as also found in the reactions of papain and cathepsin B.

The complete amino acid sequences of rat liver lysosomal cathepsins B and H have recently been determined.[57] Cathepsin B consists of 252 amino acid residues. Its molecular mass without the carbohydrate group is 27,411, as calculated from the amino acid sequence. Cathepsin H is significantly smaller possessing 220 amino acid residues, which gives a molecular mass of 24,000, omitting the carbohydrate. Most recently, the amino acid sequence of cathepsin B isolated from human liver[82] and beef spleen[83] has also been determined. No major alteration in the primary structure was found when compared with that of the rat liver enzyme. Recent identification of cDNA clones coding for cathepsin B has revealed that the

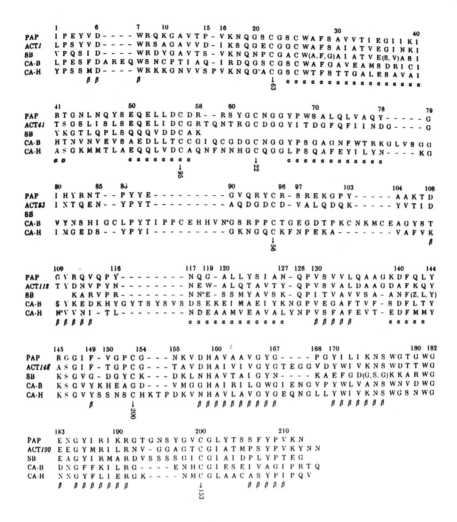

FIGURE 1.   Amino acid sequences of papain (PAP), actinidin (ACT), stem bromelain (SB), cathepsin B (CA-B), and cathepsin H (CA-H). Symbols: α, α-helix; β, β-sheet; ↓, S-S bond; c, carbohydrate attachment at the preceding residue. The actinidin residue numbers are marked at the beginning of each line. Residues 86 to 87 of papain should be Y-P instead of P-Y. Slight corrections for the amino acid sequence of papain have recently been suggested.[87d] (From Kamphuis, I. G., Drenth, J., and Baker, E. N., *J. Mol. Biol.*, 182, 317, 1985. With permission.)

enzyme is synthesized as a larger precursor that should be proteolytically processed to its mature form.[84] Larger forms of cathepsin B may be involved in the processing of proinsulin to insulin.[85]

In purified cathepsins, the polypeptide chain is often cleaved as a consequence of limited proteolysis.[86] The sites of the limited proteolysis, where cathepsins B and H can be split into light and heavy chains, are different in the two proteases. However, both sites involve an asparaginyl bond, which is near the amino terminus in cathepsin B and the carboxyl terminus in cathepsin H. It is not known whether the limited proteolysis has any effect on the catalytic properties of the enzymes. It is likely that the entire carbohydrate moiety of each enzyme is linked to a single asparaginyl residue: residue 111 in cathepsin B and residue 115 in cathepsin H (numberings relate to the individual enzymes).[57]

The amino acid sequences of cathepsin B and H are highly homologous to that of papain and other plant proteases, as seen from Figure 1.[57,87] It is evident that the proteins have a relatively high degree of identity in the amino and carboxyl terminal regions, but much less in the central region. Cathepsin B has substantial insertions in this region, which mainly accounts for its larger molecular mass. A more recent study of the cDNA encoding a precursor of cathepsin B has revealed two minor errors in the original sequence: Trp instead of Gly 78, and insertion of Gly-Arg between the light and heavy chains.[84]

Cloning of the cDNA for cathepsin B has revealed the structure of the precursor protein.[87a] This primary translation product contains a signal peptide (17 residues), a propeptide (62 residues), the mature enzyme (254 residues) and a C-terminal extension peptide (6 residues). Similar studies involving cloning of the cDNA for rat cathepsin H,[87b] rat cathepsin L,[87c] and papain[87d] have also been established the structure of the N-terminal extension peptides, i.e., the signal and the propeptides. The complete cDNA sequences of three other cysteine proteases have also been reported. Two of the sequences are for developmentally regulated proteins of the cellular slime mold *Dictyostelium discoideum,* namely CP1[88] and CP2.[89] The third sequence is for aleurin, a gibberelic acid-regulated protein of barley aleurone.[90] The propeptide regions of the cysteine proteases show significant homology which is, however, lower than the homology observed in the mature protease regions.[87b,91]

Cathepsins, as cysteine proteases, react readily with many thiol-blocking reagents. There are two, relatively new classes of inhibitors which are reasonably selective to the active site thiol of cathepsins and related cysteine proteases at acidic pH (pH 4 to 6). These two classes of compounds, peptidyl-diazomethanes and epoxysuccinyl-peptides, do not possess highly reactive functional groups, but their preliminary binding to specificity subsites facilitates reactions. The peptidyl-diazomethanes (or peptidyl-diazomethyl ketones) appear to be specific for cysteine proteases.[92,93] As seen in Table 1, Z-Phe-Phe-CHN$_2$ is a poor inhibitor of cathepsin B (k_{2nd} = 185 M^{-1} sec^{-1}), whereas it is an excellent inhibitor of cathepsin L (k_{2nd} = 116,000 M^{-1} sec^{-1}).[79] This renders it possible to inhibit cathepsin L selectively in the presence of cathepsin B when a low concentration (0.5 μM) of inhibitor is used. On the other hand, cathepsin B is most sensitive to Z-Phe-Thr(*O*-benzyl)-CHN$_2$ (k_{2nd} = 30,000 M^{-1} sec^{-1}).[94] For cathepsin H, a good diazomethane inhibitor has not yet been described. The epoxide inhibitors, such as E-64, are discussed in Chapter 2, Section VIII.D, concerning the active site titration of cysteine proteases.

The protein inhibitors of cathepsins have virtually not been investigated until recently, which may be surprising in the light of the extensive studies to serine protease inhibitors (Chapter 2, Section IV and Chapter 3, Section III.C). This imbalance is now being adjusted by a rapid progress in this new field. Most interestingly, some of the protein inhibitors have actually been known for a long time, but the inhibitory function of the known protein has been recognized only lately. As a review of these important results is not available at present, we shall consider cysteine protease inhibitors in some detail.

Three groups of related cysteine protease inhibitors, stefins, cystatins, and kininogens, have been discovered so far.[95] Stefins and cystatins are low molecular mass inhibitors (11,000 to 15,000) relative to kininogens (56,000 to 160,000). The evolutionary relationship among the three groups is apparent from their amino acid sequence homology. Stefin is the simplest inhibitor that contains no disulfide bridges. Cystatin is similar to stefin, but possesses two disulfide bridges. Gene triplication of an acestral inhibitor has generated the kininogen heavy chain which contains three cystatin-like domains.

In contrast to the serine protease inhibitors, the protein inhibitors of cysteine proteases form complexes also with inhibited target enzymes. This can be exploited in the purification of inhibitors by using carboxymethylated papain as a ligand for affinity chromatography.[96,97]

Stefins constitute the latest group of cysteine protease inhibitors that have been discovered and characterized.[95] Four members of the stefin group have been sequenced and proven to be highly homologous. These are human stefin A,[98] human stefin B,[99] rat stefin α,[100] and rat stefin β.[101]

Cystatin was first isolated from egg white.[102,103] Its amino acid sequence has recently been determined[104,105] together with the sequences of three other members of this group: human cystatin C,[106] human cystatin S,[107] and bovine colostrum cystatin.[108] Cystatin C has been proved to be identical with human urinary γ-trace.[106] Furthermore, it has been found that hereditary hemorrhage with amyloidosis occurring in Icelanders is caused by selective deposition of fragments of cystatin C in the walls of the brain arteries, which leads to fatal cerebral hemorrhages in most cases.[109] The deposited fragment is devoid of the N-terminal decapeptide of cystatin C, and exhibits one amino acid difference (Gln for Leu) probably at the potential reactive site of the inhibitor.[109]

The inhibitory action of kininogens on cysteine proteases has not been recognized until most recently. Kininogens are the large precursor molecules of the vasoactive kinins involved in the regulation of blood pressure.[110] For example, the nonapeptide bradykinin is released from kininogen on the action of the serine protease, kallikrein. As a consequence, the single chain glycoprotein gives rise to one heavy chain and one light chain which are held together by a disulfide bond. Two types of kininogens, a high- and a low-M_r form, are present in plasma and secretions. They share the heavy chains preformed at the N-termini and the consecutive kinin segments, but they differ in their light chain portions which are located at the C-termini.[110,111]

It is interesting to note that cysteine protease inhibitors, called α-cysteine protease inhibitors, have already been isolated from plasma.[112-115] Two major forms have been identified, namely α_1 (high M_r) and α_2 (low M_r) cysteine protease inhibitor.[114] Enzymological, immunological, and amino acid sequence data have provided compelling evidence for the identity of α-cysteine protease inhibitors and kininogens.[116-122]

The primary structure derived from the cDNA sequence for human α_2-cysteine protease inhibitor, which has been found to be homologous to that of bovine low-M_r kininogen,[116] contains the nonapeptide sequence of bradykinin and two internally repeated sequences which are homologous to stefins and cystatins. A comparison of these homologous regions has shown the presence of a highly conserved sequence, namely Gln-Val-Val-Ala-Gly, which could represent the reactive site of the inhibitors.[116] A third repeated sequence, which is located at the N-terminal portion of the heavy chain, has also been recognized.[117,123] This sequence, however, exhibits less pronounced homology to cystatins, in particular at the potential reactive site.

The three individual domains of the heavy chain of low-M_r kininogens could be separated by limited proteolysis.[123] Each domain was assayed for inhibition of papain, calpain, and cathepsin L. It was found that the N-terminal domain did not inhibit these enzymes. The middle domain inhibited calpain and, to a lesser extent, papain and cathepsin L. The C-terminal domain contained the tighter binding site for the latter two enzymes.[123]

No information is available about the tertiary structure of either cysteine protease protein inhibitor. However, suitable crystals for X-ray diffraction studies have recently been obtained from chicken egg white cystatin.[124]

C. Calpains

Compared to the plant proteases and cathepsins, calpains are high molecular mass, intracellular, nonlysosomal cysteine proteases that have recently been discovered.[125,125a] They are dimers composed of one heavy subunit (approximately 80 kdaltons) and one light (approximately 30 kdaltons) subunits. As for the name calpain, *cal* stands for calcium since the enzyme activity depends on Ca^{2+} concentration, and *pain* alludes to the fact that this enzyme is a cysteine protease of the papain family. There are two types of calpain: one with a high Ca^{2+} sensitivity (in the micromolar range) called calpain I, and the other with low Ca^{2+} sensitivy (in the millimolar range) called calpain II. An alternative name of calpains, namely Ca^{2+}-activated neutral proteases[126] (μCANP and mCANP), has also been used, although the calpain term is recommended by the Nomenclature Committee.[127]

Calpains are widely distributed in mammalian and avian cells. Their role in the metabolism has not yet been elucidated. Located principally in the cytosol, calpains are presumably regulatory proteases which cause irreversible effects on the target proteins by limited proteolysis. For example, it has been shown by immunostaining that calpain I is located almost exclusively in the B cells of islets of human endocrine pancreas, which suggests that calpain I may be involved in the secretion of insulin.[128] In fact, Ca^{2+} also plays an important role in this process.[129] Furthermore, calpain may also be involved in the activation of protein kinase C.[125a]

Calpain hydrolyzes a variety of peptides with preference for Tyr, Met, or Arg residue as the P_1 and Leu or Val as the P_2 residue.[130] No great difference in specificities has been found between calpain I and calpain II. It is worthy of note that the k_{cat}/K_m values for the calpains are considerably smaller than those for papain. On the other hand, when casein is used as substrate, calpain becomes a somewhat more efficient catalyst than papain. It appears then that in addition to the P_1 and P_2 residues, other structural features of the substrate should also control the specificity of calpain. This specificity characteristic remains to be established.

In common with other cysteine proteases of the papain family, calpain is sensitive to leupeptin, antipain, and E-64, but its inactivation only occurs in the presence of calcium ions.[131,132] The simple alkylating agent, iodoacetate also requires Ca^{2+} for inactivating calpain.[131] The rate of inactivation is enhanced in the presence of Ca^{2+} more than 1000 times. It is a remarkable finding that in the presence of Ca^{2+}, calpain is not only activated but also inactivated during incubation at 30°C, probably as a consequence of autodigestion.[131]

Calpains are encountered in tissue homogenates with their endogenous inhibitors, calpastatins (M_r = 270,000 to 300,000).[133,134] The complexes formed between them do not exhibit any Ca^{2+}-dependent proteolytic activity. Calpastatin is extremely heat stable. The cDNA sequence encoding the inhibitor shows four consecutive internal repeats corresponding to about 140 amino acid residues.[134a,134b] This suggests that the inhibitor is a multi-domain protein. Calpastatin is a highly specific inhibitor of calpain; it does not even inhibit papain. Resolution of the complex into calpain and calpastatin can be achieved by ion exchange or gel chromatography in the presence of EDTA. Calpain I, calpain II, and calpastatin can also be separated on a phenyl-Sepharose column if the tissue used for isolation has been perfused with EDTA before homogenization.[135] These observations suggest that a Ca^{2+}-induced conformational change in calpain molecule is necessary not only for the proteolytic activity, but also for binding the calpastatin molecule. The enzyme is also inhibited by other cysteine protease inhibitors of protein nature, such as kininogen.[116]

The complete amino acid sequence of the heavy subunit of chicken calpain (705 amino acid residues),[136] as well as that of the very similar human enzyme,[137] has been deduced from the nucleotide sequence of cloned complementary DNA. As illustrated in Figure 2, the protein contains four distinct domains (I to IV from the N-terminus). Domain II involving the residues 81 to 320 (M_r is about 28,000) displays significant structural homology with plant proteases and cathepsins B and H. This is particularly clear around the active site residues: Cys-108 and His-265 (Figure 3a).

Domain IV, involving the residues 561 to 705 (M_r is about 17,000), has four consecutive E-F hand structures typical for Ca^{2+}-binding proteins,[138,139] such as calmodulin, troponin C, and myosin light chain (Figures 2 and 3b). However, there is an important difference between the regulation of enzyme activity by domain IV and calmodulin. Specifically, the former is covalently linked to the target protease (domain II) through domain III, whereas the latter is not bound covalently to the target protein. Although the function of domains I and III has not yet been elucidated, the established sequences of domains II and IV indicate that calpain has evolved by gene fusion of proteins with different functions and evolutionary origins.[136]

The gene structure of the heavy chain of chicken calpain is of particular interest. It is

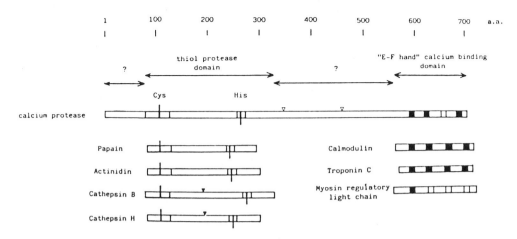

FIGURE 2. Schematic representation of the structure of the heavy chain of calcium protease (calpain) and comparison with cysteine proteases and calcium-binding proteins. The four double arrows show domains I to IV from left to right. Upward and downward projecting lines show the active site Cys and His residues, respectively. The highly homologous amino acid sequences around the active site Cys and His residues are indicated by boxes. Predicted calcium-binding sites composed of 12 amino acid residues are represented as solid boxes, and potential sequences for calcium-binding sites are shown as open boxes. ▼, Attachment site of carbohydrate in cathepsins; ▽, potential attachment sites of in carbonydrate in calcium protease. Numbers indicate the positions of amino acid residues in calcium protease. (From Ohno, S., Emori, Y., Imajoh, S., Kawasaki, H., Kisaragi, M., and Suzuki, K., *Nature (London)*, 312, 566, 1984. With permission.)

about 10 kilobases (kb) long and is composed of 21 exons.[140] Introns are encountered at all domain junctions. In domain II, the highly homologous sequence around Cys-108 is split by one intron, and three introns exist between Cys-108 and His-265. In contrast, each of the four E-F hand structures of domain IV is encoded by one exon.

The complete primary structure of the light subunit of calpain has also been derived from the nucleotide sequence of the complementary DNA.[141,142] This protein, which is composed of 266 amino acid residues, contains two distinct domains. The N-terminal portion has an unusually elongated polyglycyl sequence, the functional significance of which is unknown. The C-terminal portion contains four E-F hand structures. Considering the presence of similar structures in the heavy subunit, it appears that the proteolytic activity of calpain is controlled through the concerted action of multiple Ca^{2+}-binding sites present in two different subunit molecules.[141,142]

D. Microbial Proteases

Relatively few cysteine proteases are found in bacteria. One of their best characterized representatives, streptococcal protease, has been known for a long time.[143] The enzyme is produced by many strains of *Streptococci* in zymogen form, which is secreted into the culture fluids and is converted to active enzyme by an autocatalytic reaction.[144] This is initiated by reduction of the zymogen with the thiol groups of the bacterial cell wall, and it is followed by proteolytic cleavages carried out by the reduced zymogen which possesses a slight proteolytic activity.[145] The activation of the zymogen can also be affected by the action of trypsin or subtilisin. Upon activation of the zymogen to the final form of the active protease, a significant portion of the polypeptide chain is removed resulting in a decrease of the molecular mass from 34,000 to 28,000. Both the zymogen and the active protease are readily crystallized.[146]

There is an unusual mixed disulfide in the zymogen: protein–S–S–CH₃. The biosynthesis of the mixed disulfide appears to involve the formation of a protein–S–SH followed by transmethylation rather than a direct transfer of an intact methanethiol to the cysteinyl residue of the zymogen.[147]

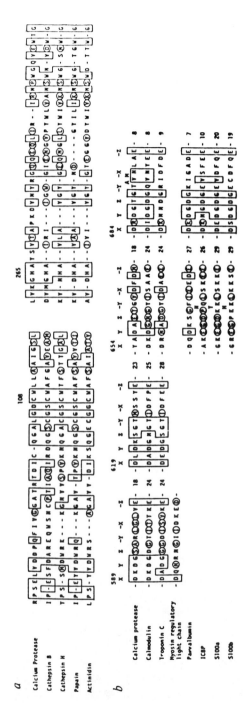

FIGURE 3.   (a), Amino acid sequences around the active site Cys and His residues in calcium protease (calpain). Corresponding sequences of cysteine proteases are aligned for comparison. The identical residues in the sequences are boxed. Circled residues represent favored amino acid substitutions. Numbers above the sequence refer to amino acid positions in the sequence of calpain. (b), Amino acid sequences of four predicted calcium-binding sites. Corresponding sequences of other calcium-binding proteins are aligned for comparison. Residues corresponding to the calcium-binding loop region are shown, where X, Y, Z, -Y, -X, and -Z represent the calcium coordinating positions. Numbers shown at the top of the sequences designate the residues of calpain at position X. Numbers shown between amino acid sequences refer to the numbers of amino acid residues present between calcium-binding loops. Numbers shown in the right-hand row indicate residue numbers encountered up to the C-termini following the calcium-binding loops. ICBP stands for bovine intestinal calcium-binding protein, and S100a and S100b are α and β subunits, respectively, of bovine brain S100 protein. (From Ohno, S., Emori, Y., Imajoh, S., Kawasaki, H., Kisaragi, M., and Suzuki, K., *Nature (London)*, 312, 566, 1984. With permission.)

The enzyme favors the cleavage at hydrophobic residues in the peptide chain.[148] Thus, in contrast to papain, not only the P_2 and P_1', but also the P_1 residue should have a nonpolar large side chain for the most specific action.

The amino acid sequence of streptococcal protease has been determined.[149] The enzyme has a polypeptide chain larger (253 residues) than that of papain (212 residues). Most of the additional amino acid residues are found at the N-terminus. The active site cysteine is the residue 47 rather than the residue 25 in papain. The sequence is homologous to that of papain near the active site, but for the other regions, the homology is very limited so that a meaningful evolutionary relationship cannot be established.

Another well-characterized cysteine protease of bacterial origin is clostripain which is derived from the culture filtrate of *Clostridium histolyticum*.[150] This filtrate is a rich source of proteolytic enzymes, of which collagenases have been studied most extensively. Clostripain generated interest when its high specificity toward arginine side chain became apparent, and this property has been exploited in amino acid sequence studies.

Commercial collagenase preparations serve as convenient starting materials for the preparation of clostripain.[151,152] It is interesting to note that calcium ions are important for the enzyme action. Therefore, EDTA, which is generally used to protect cysteine proteases in solution, completely inhibits the action of clostripain.[150]

Clostripain is composed of two polypeptide chains with relative molecular masses of approximately 45,000 and 12,500,[153] giving a large molecular mass compared with that of most extracellular proteases. The primary structure of the light chain has recently been determined.[154] It contains 133 amino acid residues ($M_r = 15,400$). No homology can be found between clostripain and enzymes of the papain family. Cysteine residues are located in the positions 12 and 112, and they may form a disulfide bridge in the nonactivated enzyme.[154]

The remarkable specificity of clostripain for arginine residues is obvious from the cleavage of parvalbumin, which contains 12 lysines and 1 arginine, and only the arginine bond cleaved by the enzyme.[153] An attractive hypothesis has been put forward to explain the narrow specificity of clostripain. This assumes that two carboxyl groups are located at the binding site, and both interact with the delocalized positive charge of the guanidine group.[150]

An apparently unique family of cysteine proteases includes viral proteases, in particular picornavirus proteases (Chapter 2, Section III). These viruses, such as poliovirus, encephalomyocarditis virus, rhinovirus, and foot and mouth disease virus, are small RNA viruses of plus strand polarity. The amino acid sequences of several picornavirus proteases have been deduced from the corresponding nucleotide sequences. A comparative study, which involved the proteases derived from three picornavirus subgroups and cowpea mosaic virus, has revealed only a single region of significant amino acid homology.[155] Within this homolgous region, one cysteine and one histidine have been strongly conserved, and it is assumed that these residues constitute a part of the active site. The so-called 3C protease of the poliovirus contains Cys 147 and His 161 as the active site residues. This has been supported by site-specific mutagenesis studies which showed that mutation of Cys 147 or His 161 produced an inactive enzyme, while the mutation of the nonconserved Cys 153 had a negligible effect on the proteolytic activity.[156] The peculiar structure of this protease, which consists of 183 amino acid residues, is apparent from the fact that both catalytically competent residues are located in the C-terminal region. In the enzymes of the papain family, the two active site residues are far away from each other in the primary structure. The specificity of protease 3C is also different from that of papain and the related cysteine proteases, inasmuch as the viral enzyme does not require a large hydrophobic P_2 residue, but it specifically hydrolyzes the Gln-Gly bond.[157]

II. MINIMAL MECHANISM: ACYL-ENZYME FORMATION

The formation of an acyl-thiolenzyme intermediate in the papain catalysis has long been known from chemical and kinetic studies.[1] In this respect, the mechanism of action of papain is similar to that of chymotrypsin, and the minimal kinetic scheme for the hydrolysis reactions is the same for papain and serine proteases (Chapter 2, Section VIII.A, Equation 10). Kinetic evidence for an acyl-enzyme intermediate was obtained from closely similar k_{cat} values for substrates of inherently different reactivities, such as analogous alkyl and aryl esters.[158-160] The simplest interpretation of this observation is that the rate-determining step does not involve the departure of the leaving group, and so reflects deacylation (see also serine proteases, Chapter 3, Section I.A., Table 1).

In contrast to serine proteases, where k_{cat} for ester hydrolysis is much higher than k_{cat} for the corresponding amide substrate, papain hydrolyzes benzoyl-Arg-OEt (BAEE) and benzoyl-Arg-NH$_2$ (BAA) at similar rates within a factor of 3.[161] This had been interpreted in terms of the formation of a common acyl-enzyme intermediate.[162] However, the similar k_{cat} values are not necessarily the consequence of rate-limiting deacylation. In fact, the acylation (k_2) and deacylation (k_3) rate constants are not too much different, and thus, both contribute to the value of k_{cat} for both BAEE and BAA hydrolyses.[163] As for the k_2/k_3 ratio, conflicting data have been reported[163,164] which may be reconciled by invoking nonproductive binding[165,166] or conformational changes.[167]

The most compelling evidence for the existence of a thiolester intermediate has been provided by its direct spectroscopic observation in the hydrolysis of the chromophoric synthetic substrates, methyl thionhippurate[168] and *N-trans*-cinnamoylimidazole.[169] Under suitable conditions, *N-trans*-cinnamoyl papain is sufficiently stable to permit an examination of its properties. It has been shown that methylamine is 670-fold more reactive toward *N-trans*-cinnamoyl papain than toward *N-trans*-cinnamoyl chymotrypsin.[169] This is consistent with the greater propensity of thiolesters than oxygen esters to react with nitrogen nucleophiles (see Chapter 1, Section X).

The catalytic groups possibly involved in the formation and decomposition of the acyl-thiolenzyme have been studied by means of pH-rate profile measurements. Thus, analysis of the pH dependence of the individual rate constants k_2 and k_3 has shown that acylation exhibits a bell-shaped pH-rate profile, whereas deacylation follows a sigmoid pH-dependence curve.[160,163] The second-order acylation rate constant, $k_{cat}/K_m = k_2/K_s$, also conforms to a bell-shaped curve, suggesting that the binding constant, K_s, is independent of pH. The bell-shaped curve is characterized by two ionizing groups with pK_a values near 4 and 8.5. The deacylation step for a number of substrates has been demonstrated to be dependent on a base of pK_a of about 4. On the basis of this value and of the lack of dependence of the pK_a on temperature, it was supposed in earlier studies that the catalytic group was a carboxylate ion.[163,170] However, X-ray crystallographic studies (see the next section) have demonstrated that the imidazole group of His 159 is in the immediate vicinity of the sulfur atom of Cys 25. The closest carboxyl group, i.e., that of Asp 158, has been found significantly farther and in an unfavorable position to interact with Cys 25. Nonetheless, the idea of the involvement of Asp 158 in the catalysis was maintained in a few studies. Based on the magnitude of the Hammett ρ value for the rates of deacylation of substituted benzoyl papains (for this parameter see Chapter 1, Section VIII.B), it was suggested that deacylation of nonspecific acyl-enzymes are catalyzed by a carboxylate ion as a general base.[171] This mechanism was extended to the deacylation of specific substrates as well.[172] It has been pointed out, however, that because of the difficulties associated with the interpretation of the ρ values, the direct involvement of the imidazole group in the deacylation step cannot be excluded.[173] Another argument in favor of carboxylate participation,[170,171] the near-zero heat of ionization has also been shown to be consistent with the action of the imidazole as a general base.[173] Whereas the catalytic role of imidazole is consistent with all experimental data, the direct involvement of the carboxyl group in the catalysis is ruled out by the fact that cathepsins B and H, which are homologous to papain (see the preceding section) do not contain the corresponding aspartate residue in their amino acid sequence. Consequently, the pK_a of 4 can unambiguously be assigned to His 159. This value is lower than the usual pK_a of an imidazole group. However, the side chain of His 159 is covered to a great extent by the indolyl group of Trp 177 (see the next section), ensuring a more hydrophobic environment around the imidazole group, and this can account for the low pK_a.

The two pK_a values, about 4 and 8.5, observed in the pH-rate profile of papain acylation have often been attributed to the His 159 and Cys 25. In the light of the formation of a thiolate-imidazolium ion-pair between these two residues (Section IV), however, it is more appropriate to assign these pK_a values to the formation and the decomposition of the ion-pair.[174] Notably, these pK_a values do not reflect true ionizations because the proton of the thiol group is donated to the imidazole rather than to the solvent water.

It is interesting to note that the *acidic limb* of the bell-shaped curve does not fit to the dissociation of a single ionization group,[175-178] but rather it is modulated apparently by two ionizing groups according to Equation 1.[176,178]

$$\log k = \log \frac{k(\text{lim})}{1 + [H^+]/K_b + [H^+]^2/K_aK_b} \tag{1}$$

where $k(\text{lim})$ is the pH-independent maximal rate constant (k_{cat}/K_m), and K_a and K_b are macroscopic dissociation constants.

Figure 4a shows that the deviation from single group ionization is not great but significant. On the other hand, papaya peptidase A (Figure 4b) exhibits a simple dissociation (Chapter 2, Section VIII.C, Equation 21). The two pK_a values modulating papain catalysis are fairly close, existing in the pH range between 3.5 to 4.0 for most substrates.[176,178] It is usually assumed that the two groups modulating the curve are His 159 and Asp 158.[176,178] Fluorescence and ^{19}F NMR studies of alkyl and alkylthio derivatives of Cys 25 also indicate the existence of two ionizable groups (presumably His 159 and Asp 158) in the vicinity of Cys 25.[179,180] The effect of Asp 158 is also apparent in the reaction of 2,2'-dipyridyl disulfide with papain.[181] As we have pointed out above, the catalytic competence of Asp 158 is discounted by its absence in cathepsins B and H.

III. THE TERTIARY STRUCTURE OF PAPAIN

A clear description of the three-dimensional structure of papain as determined from X-ray crystallography at 0.28 nm resolution was given earlier.[3,4] Refinement at 0.165 nm of the structure has recently been accomplished.[182] A characteristic feature of the papain molecule is its binuclear nature which implies that the molecule is constructed around two hydrophobic cores. This is apparent from Figure 5 showing the folding of the main chain into two virtually independent lobes. Residues 11 to 111 and 112 to 208 give rise to lobe L and lobe R, respectively. The chain crosses three times the boundary between the two lobes: at residues 11, 111, and 208. The two ends of the chain, residues 1 to 10 and 209 to 212, assist the connection of the two lobes by attaching themselves to the R and L half-molecules, respectively.

Between lobes L and R, there is a large groove, approximately 2 nm long, which can accommodate polypeptide substrates consisting of several amino acid residues. The side chain of the essential Cys 25 is found in the groove. The sulfur atom of Cys 25 is in the plane of the imidazole ring of His 159, which is situated on the opposite wall of the groove. These and some other residues of the active site region are depicted in Figure 6. It is seen

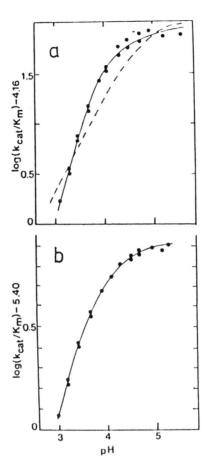

FIGURE 4.   The pH rate profiles of acylation of papain (a) and papaya peptidase A (b) with benzyloxycarbonylglycine *p*-nitrophenyl ester. For the papain reaction, the solid line was calculated according to Equation 1 involving two pK_a values, the dashed line is a normal curve for one ionizing group. The papaya peptidase A reaction depends on a single ionizing group. (From Schack, P. and Kaarsholm, N. C., *Biochemistry*, 23, 631, 1984. With permission.)

that the imidazole ring is hydrogen bonded to the side chain of Asn 175. This hydrogen bond is shielded from the solvent by Trp 177. It is important to note that in serine proteases the analog residue of Asn 175 is the negatively charged aspartate of the catalytic triad (Chapter 2, Sections II and VIII.A). It is also seen from Figure 6 that the side chain of Gln 19 points towards the sulfur atom of Cys 25. Residue 23 can only be Gly because the accessibility of the sulfur atom would be seriously obstructed by any other side chain. At a distance of 0.67 nm from the imidazole ring, the side chain carboxyl group of Asp 158 is encountered. The left-hand side of the active site groove (Figure 5) includes four aromatic residues: Tyr 61, Tyr 67, Trp 69, and Phe 207. The right-hand side wall contains Val 133 and Val 157. Thus, hydrophobic regions are encountered on both sides of the cleft.

Structural information about the binding of substrate to papain has been obtained from difference Fourier maps with specific chloromethyl ketone inhibitors: benzyloxycarbonyl-

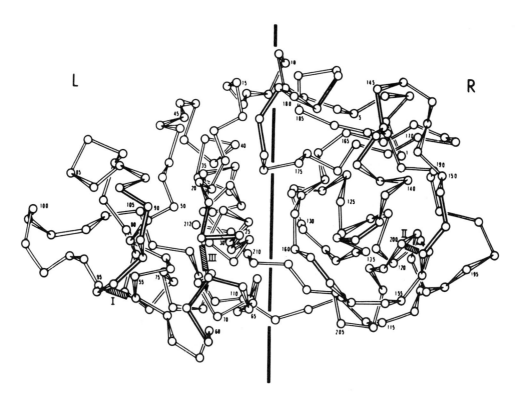

FIGURE 5. The main folding of papain. The characters L and R represent the left and right lobes, respectively. The Roman numerals designate disulfide bridges. (From Drenth, J., Jansonius, J. N., Koekoek, R., and Wolthers, B. G., *Adv. Protein Chem.*, 25, 79, 1971. With permission.)

Phe-Ala-CH$_2$Cl, benzyloxycarbonyl-Gly-Phe-Gly-CH$_2$Cl, and acetyl-Ala-Ala-Phe-Ala-CH$_2$Cl, all containing Phe at the P$_2$ position.[183] The specific P$_2$ phenylalanine residue has been shown to be involved in hydrophobic contact with the side chains of Val 133 and Val 157. On the opposite (left) side of the groove, the P$_2$ carbonyl oxygen forms a hydrogen bond to the backbone -NH- of Gly 66. A second hydrogen bond is also generated between the -NH- of P$_2$ and the carbonyl oxygen of Gly 66. The P$_1$ -NH- group points towards the main chain carbonyl oxygen of Asp 158 at the right side. Thus, the peptide bond between the P$_1$ and P$_2$ residues appears to cross-link the active site groove, although the distance for this S$_1$-P$_1$ hydrogen bond may be too long in the inhibited papain. The importance of the cross-link, i.e., the existence of the S$_1$-P$_1$ and S$_2$-P$_2$ hydrogen bonds, is supported by kinetic studies using substrates not capable of forming the appropriate hydrogen bonds.[184] Specifically, the -NH- or the - CONH- groups have been replaced by -CH$_2$- and -CH$_2$CH$_2$- groups, respectively, which has resulted in considerable decreases in the acylation rate constants. In addition to this possible hydrogen bond linkage, the hydrophobic side chain of the P$_2$ residue is bound to the right-hand wall of the cleft, and the side chain of the P$_3$ residue appears to interact with the left-hand side of the cleft. These interactions result in a further cross-link, which is hydrophobic in nature.

The most interesting impact on the structure of papain, which is caused by the inhibitor binding, is that the active site groove widened by about 0.1 nm. This effect was predicted from model building studies for the binding of substrates.[184]

The structure of actinidin has also been solved at high resolution (0.17 nm).[13] A detailed comparison of the papain and actinidin structures[87] has shown that having 48% identity in their amino acid sequence, the two structures, especially the positions of the main chain

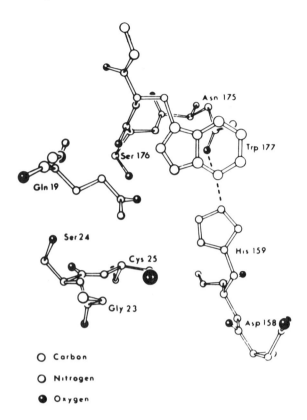

FIGURE 6.   Some amino acid residues at the active site region
of papain. (From Drenth, J., Jansonius, J. N., Koekoek, R., and
Wolthers, B. G., *Adv. Protein. Chem.*, 25, 79, 1971. With
permission.)

atoms, are very similar. Insertions and deletions alter the conformation over a limited range,
involving two to three residues only. Great structural similarities are found for the active
site residues. By combining the three-dimensional structural information with sequence
homologies of stem bromelain, cathepsin B, and cathepsin H, it has been concluded that
the overall folding pattern of the polypeptide chain is grossly the same in all five proteases,
and that they utilize the same catalytic mechanism.[87] Recent determination of the tertiary
structure of calotropin D1 has indicated that the folding of the main chain is very similar
to that of both papain and actinidin. Most interestingly, the L domain shows higher structural
homology with that of papain, whereas the R domain does so with that of actinidin.[228]

It is of mechanistic importance that Asp 158 in papain and actinidin is substituted by Asn
in bromelain and cathepsin H, and by Gly in cathepsin B. Thus, the implication of Asp 158
in a serine protease-like mechanism[171,172] (see also Section V.A of this chapter) is considered
very unlikely. Furthermore, Asn 175, which is assumed to orient the imidazole ring of His
159, is retained in all sequences except for bromelain. The corresponding serine OG of
bromelain could also perform the orienting role.[87] Another peculiarity of bromelain is that
Trp 177 which shields the His 159-Asp 175 hydrogen bond from the solvent in the other
four enzymes is replaced by a Lys residue. However, an error in the incomplete sequence
may also explain the unique substitutions for residues 175 and 177.

As mentioned before, Val 133 and Val 157 are responsible for the S_2 substrate specificity
of papain. Val 133 is replaced by Ala in all other sequences. Val 157 is substituted by Leu
and Gly in bromelain and cathepsin B, respectively (Figure 1). The change of Ala 162 of

papain to an Arg residue in cathepsin B, has been suggested to explain the capability of this enzyme to act as a peptidyl dipeptidase.[57] Such a binding mode, however, would demand that either the S_2-S_3 binding region of the groove be overcrowded by accepting both the P and P' residues, or the substrate be bound with its N- and C-termini reversed. The catalytic consequences of either binding mode are unclear. The aminopeptidase activity of cathepsin H was interpreted in terms of an electrostatic interaction between the positively charged amino terminus of the substrate and the glutamic acid of cathepsin H in place of Gln 135 of papain.[185] However, perusal of the active site model of papain shows that the γ- carboxylate and the α-ammonium ions are too far from one another to interact directly during the catalysis.[229]

IV. THIOLATE-IMIDAZOLIUM ION-PAIR, THE NUCLEOPHILE IN THE CATALYSIS

The X-ray crystallographic studies discussed in the preceding chapter have shown that the SG of Cys 25 and the ND1 of His 159 are at a distance of 0.34 nm, corresponding to a van der Waals contact. A number of authors,[1,5,160,186,187] apparently by analogy with the serine protease mechanism, suggested that the nucleophilic attack by the sulfur atom on the substrate carbonyl carbon is promoted by the imidazole group as a general base. This implies that all three reacting species, i.e., the sulfur atom, the carbonyl carbon, and the imidazole base, must be included in the same transition state (Chapter 1, Section V). The first evidence that this is not the case has come from the studies on the alkylation of Cys 25 with neutral haloacetamides.[188] The uncharged character of the alkylating agent is essential to avoid complications due to electrostatic effects. The pH-alkylation rate profile for the papain-chloroacetamide reaction (Figure 7) does not conform to a simple dissociation curve found with ordinary thiol groups, but rather gives a doubly sigmoid curve with two pK_a values, 4.0 and 8.4. This is consistent with the reaction with the free thiolate ion at high pH and by facilitation by the imidazole group at low pH.[188] Clearly, three protonation states can be distinguished. (1) Below pH 4 both catalytic groups (Cys 25 and His 159) are protonated. (2) Between the pH values 4 and 8.4 a mono-protonated thiol-imidazole couple exists, the thiol component of which exhibits a considerably enhanced reaction rate relative to a simple thiol group at the same pH. (3) Above pH 8.4 neither the thiol nor the imidazole is protonated. It is of importance that during the catalytic reaction, the formation of the acyl-enzyme also depends on two pK_a values, which are similar to those observed in alkylation, but the pH-acylation rate profile is bell-shaped rather than doubly sigmoid, and exhibits maximal catalytic activity around pH 6 (Section II). This finding suggests that only the mono-protonated form of the thiol-imidazole couple is catalytically competent.

As for the nature of the assistance of the thiol reaction by the imidazole, kinetic isotope studies have provided useful information.[188] In the case of general base catalysis, the alkylation in the acidic pH-range is expected to proceed in water about three times as fast as in deuterium oxide (Chapter 1, Section IX) as actually found in the serine protease catalysis (Chapter 3, Section I.B). However, the lack of deuterium isotope effect in alkylation,[188] as well as in acylation by various substrates,[189] is inconsistent with general base catalysis. Hence, the results have been interpreted in terms of a thiolate-imidazolium ion-pair, implying that the proton is already on the imidazole when the thiolate ion attacks the substrate.

The presence of the thiolate ion as part of the ion-pair has been demonstrated by a spectrophotometric method which is based on measuring the absorption band of the thiolate ion during alkylation.[190,191] It is known that the dissociated, but not the nondissociated or alkylated form of the thiol group displays a characteristic absorbence in the ultraviolet (UV) spectrum. A considerable degree of ionization of Cys 25 has been detected as seen in Figure 8, which shows the pH dependence of the absorbence change on alkylation of papain. The

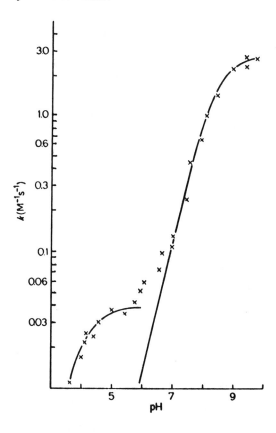

FIGURE 7.   pH-Dependence of the reaction of papain with chloroacetamide. The solid lines represent theoretical curves with parameters $pK_1 = 4.00$, k(limit)$_1$ = 0.039 M^{-1} sec^{-1}, and $pK_2 = 8.40$, k(limit)$_2$ = 2.95 M^{-1} sec^{-1}. (From Polgár, L., *Eur. J. Biochem.*, 33, 104, 1973. With permission.)

solid line is a theoretical curve calculated from Equation 2 which describes a doubly sigmoid event.

$$\epsilon^{obs} = \epsilon_{IP}\left[\frac{1}{1 + [H^+]/K_1}\right]\left[\frac{1}{1 + K_2/[H^+]}\right]$$
$$+ \epsilon_{S^-}\left[\frac{1}{1 + [H^+]/K_2}\right] \tag{2}$$

where K_1 and K_2 stand for the apparent ionization constants on which the absorbence change depends, ϵ_{IP} and ϵ_{S^-} are the molar absorption coefficients for the thiolate ion of the ion-pair and for the free thiolate ion, respectively. The first term of Equation 2 gives the contribution of the ion-pair, represented by a bell-shaped curve (dashed line in Figure 8), while the second term accounts for the contribution of the free thiolate ion, represented by a simple dissociation curve (dotted line in Figure 8). The two pK_a values (3.7 and 8.4) are similar to those determined from the pH-rate profiles of alkylation and acylation. The lower absorbence at 250 nm of the thiolate in the ion-pair ($\epsilon_{IP} = 1150$) relative to that of the free thiolate ion ($\epsilon_{S^-} = 2300$) is consistent with the perturbation of the negative charge on the sulfur by the positive charge of the imidazolium ion. Similar ϵ_{IP} values were found also for thiolsubtilisin [190] and D-glyceraldehyde-3-phosphate dehydrogenase,[192] indicating similar

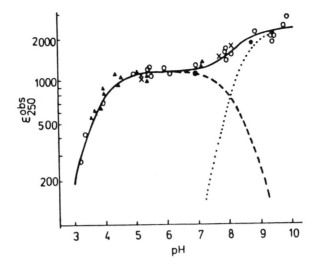

FIGURE 8. The pH-dependence of the molar absorbence of the essential thiol group of papain. Determined from alkylations with chloroacetamide (○), iodoacetamide (△), and chloroacetate (x) at ionic strength of 0.07 or 0.38 (full symbols: ●, ▲). For calculation of the theoretical lines see text. (From Polgár, L., *FEBS Lett.*, 47, 15, 1974. With permission.)

thiolate-imidazolium interactions in all these enzymes. Thus, it appears that ion-pair formation may be a general mode of activation of the sulfhydryl group of various thiolenzymes.

Further support for ion-pair formation in papain has been obtained from fluorescence studies which indicate that His 159 exists in the protonated form up to pH 8.6.[193-195] Specifically, it is the charged His 159 that quenches the fluorescence of Trp 177, which partly covers the imidazole ring. This quenching has been explained in terms of ion-pair formation between His 159 and Cys 25.[188,194,195] The protonation of the imidazole in papain[196] and in thiolsubtilisin[197] has also been confirmed by proton NMR studies. Potentiometric difference titration between native and S-methyl-thiolated papain has also supported the formation of thiolate-imidazolium ion-pair in the pH range of the catalytic activity.[195,198]

The role of a positively charged imidazolium ion is also apparent in the reactions of papain with negatively charged alkylating agents. These reactions involving chloroacetate or iodoacetate, exhibit bell-shaped rather than doubly sigmoid pH-rate profiles.[199-202] At the maximum of the curve, the rate constant is substantially higher than that found with the corresponding neutral haloacetamide. Also, the reaction is faster at low pH (at the maximum) than at high pH where the more nucleophilic free thiolate ion reacts. The rate enhancement is apparently a consequence of a favorable interaction between the positive imidazolium ion of papain and the negatively charged carboxylate of the alkylating agent. The pH dependence of the alkylation with negatively charged reagents and that of the acylation with substrates display similar bell-shaped curves, in accordance with the dependence of both reactions on the same ion-pair.

The reaction between the ion-pair and haloacetates requires a precise stereochemistry, owing to the assistance by the carboxylate ion. This can be exploited in the study of the ion-pair geometry of the various cysteine proteases.[202] Indeed, as seen from Table 2 there is a difference of more than 1000-fold in the rate constants for the reaction of iodoacetate with the ion-pair of papain and thiolsubtilisin, whereas the neutral iodoacetamide reacts at similar rates (with only a twofold difference) with the two enzymes. As for the ion-pairs, the similar chloroacetate reactions indicate that ficin[203] and even the evolutionarily distant

Table 2
SECOND-ORDER RATE CONSTANTS
FOR ALKYLATIONS OF THE ION-PAIR
OF PAPAIN AND THIOLSUBTILISIN[202]

	$k_{iodoacetate}$ $(M^{-1} sec^{-1})$	$k_{iodoacetamide}$ $(M^{-1} sec^{-1})$
Papain	1100	14.5
Thiolsubtilisin	0.84	6.3

streptococcal protease[204] are similar to papain. On the other hand, cathepsins B and H display different pH-rate profiles in their reactions with iodoacetate.[62] In contrast to papain, where the pH optimum for acylation with substrate and alkylation with iodoacetate is not very different, cathepsins are alkylated in a much wider pH range than acylated. In both cathepsins, the ion-pair may assume at least two forms with changing pH: one with a high and one with a low reactivity toward iodoacetate. The geometry and the reactivity of the ion-pair of cathepsin B are probably modulated primarily by the ionization of the nearby Glu 205.[62] It is of particular note that the active site form of cathepsin B, which exists at low pH and is alkylated more readily than the high pH form, accounts for the peptidyl dipeptidase activity, whereas the high pH form is mainly responsible for the endopeptidase activity.[62] The negatively charged alkylating agent thus appears to be of great value in revealing slight differences in the active sites of the different cysteine proteases.

An unexpected alkylation reaction has been observed with a cysteine protease isolated from germinating bean cotyledons.[205] The amino acid sequence of this protease and its relationship to papain are not known. The enzyme appears to be highly specific for Asn residues at the P_1 position of the substrate polypeptide chain. The rate constant for the reaction of iodoacetate with the bean protease is very low (0.11 M^{-1} sec^{-1}), lower than that observed in the thiolsubtilisin reaction (Table 2). Thus, in this respect, the bean protease resembles thiolsubtilisin, an artificial enzyme lacking true protease activity, rather than papain or the related cysteine proteases.

From the magnitude of solvent deuterium isotope effects on the binding of inhibitory aldehydes to papain, it has recently been concluded that instead of a thiolate-imidazolium, a thiol-imidazole pair exists in free papain.[206] Apart from the several pieces of evidence listed above in favor of ion-pair formation, in particular, the detection of the thiolate ion on the basis of its UV absorbence, we have already pointed out the difficulty of interpreting the isotope effects on enzyme reactions (Chapter 1, Section IX). Specifically, the neutral thiohemiacetals used as reference compounds are substantially different from the negatively charged tetrahedral intermediate that should be formed in the papain catalysis. Furthermore, the critical assumption that the fractionation factor of the free imidazolium ion and that of the imidazolium ion of the ion-pair would be the same, cannot be accepted in the absence of any supporting evidence.

An important mechanistic consequence of the ion-pair formation in cysteine proteases of the papain family is that their mechanism is different from that of serine proteases. Originally, it was suggested that the catalysis by cysteine proteases proceeds by general base and subsequent general acid catalysis both in acylation and deacylation, just as found with serine proteases (Chapter 3, Section I.C). However, ion-pair formation and the lack of deuterium isotope effect in papain acylation require the catalysis to proceed according to Equation 3.

$$E + (S \text{ or } P) \underset{\text{no catalysis}}{\overset{\text{no catalysis}}{\rightleftarrows}} THI \underset{\text{general base}}{\overset{\text{general acid}}{\rightleftarrows}} ES' \qquad (3)$$

Equation 3 implies that in acylation (forward reaction) formation of the tetrahedral inter-mediate (THI) is not promoted by general base catalysis, in accordance with the residence of proton on the imidazole group of the ion-pair. General acid catalysis in the breakdown of the tetrahedral intermediate is indicated by the negative Hammett ρ constant found in the papain-catalyzed hydrolysis of anilide substrates.[207] In deacylation (reverse reaction), His 159 operates as a general base catalyst, showing the expected deuterium isotope effect. The principle of microscopic reversibility (Chapter 1, Section XIV and Chapter 3, Section I.C) demands that in the deacylatin step the expulsion of the thiolate ion from the tetrahedral intermediate should proceed without general acid catalysis, as shown by Equation 3. Thus, in contrast to the serine protease catalysis, papain catalysis involves nonrepetitive mecha-nisms for acylation and deacylation.[188]

The existence of a THI on the reaction path (Equation 3) also follows from nitrogen isotope effects on the papain-catalyzed hydrolysis of benzoyl-Arg-NH$_2$.[208] Furthermore, the magnitude of the isotope effects indicates that the breakdown of the THI to acyl-enzyme, rather than its formation from the enzyme-substrate complex, is the rate-determining step.

V. STEREOCHEMICAL PROBLEMS IN THE CATALYSIS

A. Possible Stereochemistry of the Catalysis

In the preceding chapter, we have seen that there is a fairly good picture of the serine protease catalysis, which involves a reliable stereochemistry of the hydrolytic reaction. As for the cysteine protease, our knowledge is less complete. The most trustworthy enzyme-substrate interactions may be inferred from the X-ray studies of some S-alkylated derivatives of papain (Section III). These studies have yielded information about the S-P interactions between the enzyme and the substrates examined. There are no experimental data about the nature of the S'-P' interactions, as well as about the stereochemistry of the catalytic process. The available data are inappropriate with respect to the latter problem because the papain derivatives obtained by alkylations with peptide chloromethyl ketones have an extra meth-ylene group between the carbonyl carbon and the sulfur atoms. We may recall that in the serine protease reactions the chloromethyl ketones alkylate the imidazole group (Chapter 3, Section IV), whereas in the cysteine protease reactions they alkylate the highly nucleophilic sulfur atom. Therefore, the models of acyl-papain and the THI could only be constructed by removing the extra methylene group.[183] By analogy with the mechanism established for serine proteases, the THI was poised so that the oxyanion could be stabilized by two hydrogen bonds, one from the backbone -NH- group of Cys 25, and the other from the -CONH$_2$ group of Gln 19. An alternative stereochemical mechanism was also proposed[184,209] in which the oxyanion existed in a sterically unrestricted environment, i.e., without hydrogen bonding to an oxyanion binding site.

The importance of the oxyanion binding in cysteine protease catalysis has been tested by thionester substrates.[50] It has been found that the substitution of sulfur for the P$_1$ carbonyl oxygen of benzyloxycarbonyl-Phe-Gly-OEt affects the second-order rate constants of acyl-ation only slightly and diminishes the catalytic activity by about one order of magnitude in deacylation. These results contrast with those obtained with serine proteases (Chapter 3, Section VI. B), where the hydrolysis of thionesters is not detectable.[210] Like papain, the related enzymes chymopapain, papaya peptidase A, and ficin can also hydrolyze thiones-ters.[50] These experiments do not support that an oxyanion binding site would be as important in the cysteine protease catalysis as in the serine protease catalysis. However, it has been argued that thionesters probably can be accommodated in the oxyanion hole if some ad-justment of the enzymic groups is permitted.[211] Concerning this possibility it should be noted that some flexibility of the oxyanion binding site arising from the side chain of Gln 19 may not account for the activity of papain towards thionester substrates. Specifically, the serine protease, subtilisin also has a flexible side chain, Asn 155, in the corresponding position, but it cannot catalyze the hydrolysis of thionesters.[50] Consequently, although the oxyanion may or may not be accepted by the oxyanion hole, the catalytic contribution of the oxyanion binding site is apparently much less with the cysteine than with the serine proteases.

As a model of the tetrahedral intermediate, the thiohemiacetal formed between papain and *N*-acetylphenylalanyl-[1-^{13}C]- glycinal has been examined.[212] This complex has been studied with ^{13}C NMR (nuclear magnetic resonance) and two closely spaced signals have been observed. They have been attributed to the formation of two diastereomeric tetrahedral thiohemiacetals, one of which interacts with the oxyanion binding site. The generation of two different tetrahedral conformations at the active site indicates that the binding of the aldehyde is not as specific as expected for a proper model of the catalytic intermediate. Indeed, the hemiacetal complex is a protonated species, whereas the THI formed in the enzymic reaction presumably bears a negative charge.

The strongest evidence in favor of the oxyanion binding site may be that Gln 19 has been conserved during evolution. However, one may conclude from the above apparently contradictory data that the role of Gln 19 in the catalysis and that the exact position of the THI with respect to the catalytic residues remain at issue.

A stereochemical mechanism analogous to that performed by the catalytic triad of serine proteases was earlier suggested for the papain action.[172] The key feature of this mechanism involved two conformational states of the enzyme, one with the imidazole of His 159 hydrogen bonded to Asn 175 (catalytically inactive "up" conformation), as found by X-ray diffraction measurements, and the other with the imidazole protonated and electrostatically interacting with the carboxylate of Asp 158 and the thiolate of Cys 25 (catalytically active "down" position). As proposed, the substrate would only bind to the "up" conformation, and during the catalytic action, it would move between the "up" and "down" positions. There are, however, several difficulties in this two-state mechanism. (1) There is no indication that the "up" position would be inactive. In fact, the imidazole is located in an ideal position to activate the thiol group of Cys 25. (2) According to the proposed mechanism, the "down" position should be the predominant form of papain and the "up" position could only be found in the inactive crystals grown at high pH (9.3). However, the crystallographic studies of actinidin at pH 6,[13] where the analogous "down" conformation, if it indeed exists, is expected to be seen, have shown that the active site structure is closely similar to the "up" conformation of papain. (3) Cathepsins B and H do not contain an asparatate, but rather a neutral residue in place of Asp 158 (Section III). This rules out a catalytically essential role of Asp 158 in the cysteine protease catalysis.

It is clear from the above discussion that the stereochemical mechanism of cysteine protease catalysis is less well-established than that of the serine protease catalysis. The underlying differences in the mechanisms of the two protease families will be considered later in this chapter (Section VI).

B. Thiolsubtilisin, a Cysteine Enzyme with Distorted Active Site

The similarity of the catalytic groups involved in the reactions of serine and cysteine proteases has raised the possibility of transforming one type of enzymes into the other type, which should allow a better insight into their mechanism of action. Subtilisin BPN was the first serine enzyme that had been converted into a cysteine enzyme.[213,214]

The transformation of subtilisin into thiolsubtilisin was carried out in three steps according to Equation 4. The first step involved an almost instantaneous and stoichiometric reaction of subtilisin with the inhibitor, phenylmethanesulfonyl fluoride (PMFS), at pH 7.0. The phenylmethanesulfonyl group was then displaced from the enzyme with a high concentration of thiolacetate (0.7 M) at pH 5.25. The resulting acetyl-thiolsubtilisin spontaneously deacetylated to give thiolsubtilisin. The last step in the synthesis was facilitated by the enzymic activity inherent in thiolsubtilisin, since this step was identical to the deacylation step in hydrolysis of an acetyl-enzyme intermediate.

$$\text{subtilisin–OH} \xrightarrow[{-\,HF}]{+\,C_6H_5CH_2SO_2F}$$

$$\text{subtilisin–OSO}_2CH_2C_6H_5 \xrightarrow[{-\,C_6H_5CH_2SO_3^{-}}]{+\,CH_3C(=O)S^{-}}$$

$$\text{subtilisin–SC(=O)CH}_3 \xrightarrow[{-\,CH_3CO_2H}]{+\,H_2O} \text{subtilisin–SH} \qquad (4)$$

Pure enzyme, free of contaminating native subtilisin, was obtained by agarose mercurial chromatography.[215] Amino acid sequence analysis of the active site peptide confirmed that the catalytically competent serine residue had been replaced by a cysteine residue.[216] This conversion of the active site has proved to be the first site-specific mutagenesis,[217] and it has provided important mechanistic information about enzyme catalysis,[217,218] as it is discussed below.

Detailed kinetic analyses[219-221] have indicated that thiolsubtilisin catalyzes the hydrolysis of activated ester substrates, such as p-nitrophenyl acetate, but does not cleave the alkyl ester and amide bonds. Comparison of p-nitrophenyl acetate hydrolyses catalyzed by subtilisin and thiolsubtilisin shows different pH-rate profiles obtained for acylation (k_{cat}/K_m), and this can be explained in terms of the thiolate-imidazolium ion-pair formation in thiolsubtilisin.[190] This implies an acylation mechanism similar to that found with papain rather than subtilisin. The similar values for the acylation rate constants obtained with subtilisin and thiolsubtilisin show that the thiolenzyme is as reactive as the native subtilisin. However, the good substrates of subtilisin, e.g., benzyloxycarbonylglycine p-nitrophenyl ester, are poor substrates of thiolsubtilisin, even if they possess good leaving groups.

The deacylation mechanism of thiolsubtilisin seems to be identical to that of subtilisin and papain, inasmuch as all processes are catalyzed by a general base. The pK_a values for the base are similar with subtilisin and thiolsubtilisin, i.e., higher (about 7) than with papain (about 4). Here again, the specific substrates of subtilisin are poor substrates of thiolsubtilisin, whereas the nonspecific substrates are relatively good substrates of the modified enzyme.

Other serine proteases, like subtilisin Carlsberg,[222] a protease from *Aspergillus oryzae*,[223] and *Streptomyces griseus* trypsin,[224] have also been converted into the corresponding thiolenzyme with similar alteration in the catalytic activity.

Replacement of an oxygen atom by the larger sulfur atom is a very delicate modification. Yet the 0.04 nm difference in the radii of the two atoms is sufficient to cause a distortion in the active site, leading to the destruction of the catalytic capability of the enzyme. The activated ester substrates have a good leaving group and, therefore, they do not require general acid catalysis for their reactions to form an acyl-enzyme. In other words, they are hydrolyzed through a simpler mechanism than alkyl esters and amide substrates.[225] Since a distorted active site is expected to catalyze easier a simple reaction than a complex one, it is conceivable that thiolsubtilisin is only active toward activated substrates. Furthermore, a coupling between the binding and catalytic sites makes additional restrictions for the specific substrates, whereas the nonspecific substrates, which do not use the entire binding site, can be more readily accommodated to the distorted catalytic site. The inadequate geometries of the ground and transition states can account for the failure of thiolsubtilisin to operate as a cysteine protease. Accordingly, the stereochemistry of the catalysis, at least in serine proteases, must be very precise. The high sensitivity of the catalysis to the oxygen-sulfur exchange has already been noted in connection with the thionester hydrolysis catalyzed by serine proteases (Chapter 3, Section VI.B).

VI. DIFFERENCES IN THE MECHANISMS OF SERINE AND CYSTEINE PROTEASES

Besides an underlying relationship between the mechanisms of serine and cysteine proteases, i.e., acyl-enzyme formation and catalytic assistance by an imidazole group, there are several important differences which have already been discussed in this chapter. Notably, (1) acylation is a general base-catalyzed nucleophilic attack with serine and a simple nucleophilic attack with cysteine proteases. (2) Stabilization of the THI by an oxyanion binding site is well-established in serine protease catalysis. On the other hand, studies with thionester substrates have shown that even if it exists, an oxyanion binding site contributes much less to the cysteine protease catalysis. (3) A charge stabilizing system involving a negatively charged aspartate residue is essential for the serine protease catalysis, whereas an analogous system is absent from cysteine proteases having, instead of the aspartate residue, a neutral asparagine which is hydrogen bonded to the catalytic imidazole group.

All the above differences appear to be connected to the different transition-state stabilization associated with the two protease groups.[226] In both cases, the THI is the highest energy *covalent* species formed during the catalysis. The study of this species most resembling the transition state may help understanding transition-state stabilization. There is a remarkable difference between the formation of the THIs in serine and cysteine protease catalyses. The serine protease catalysis involves an energetically unfavorable charge separation, i.e., the formation of a negatively charged tetrahedral adduct and a positively charged imidazolium ion. On the other hand, in the catalytically competent form of cysteine proteases, the active-site thiol forms an ion-pair with a neighboring imidazole group. Therefore, upon going from the ground state to the transition state, and further to the THI, the ion-pair is merely rearranged from the thiolate-imidazolium into the tetrahedral adduct-imidazolium couple. Thus, due to the pre-existent ion-pair, there is no need for a net charge separation. This may account for the lack of charge stabilization by the aspartate and possibly by the oxyanion binding site in the cysteine protease catalysis. In the serine protease catalysis, however, the charge separation is apparently facilitated by the aspartate and the oxyanion binding site (Chapter 3, Section VI). The importance of oxyanion binding in serine proteases relative to that of cysteine enzymes is consistent with the greater rate of acid hydrolysis always observed for esters as compared to thiolesters (Chapter 1, Section X). This implies that protonation of the carbonyl oxygen does not facilitate significantly the acid-catalyzed hydrolysis of thiolesters. Consequently, possible partial protonation by hydrogen bonds from the oxyanion binding site may not represent a great contribution to the catalysis by cysteine proteases.

Charge separation occurring in the free cysteine proteases but not in the serine enzymes is consistent with the different chemical properties of oxygen and sulfur. Thiol compounds are substantially more acidic than their hydroxyl counterparts. Thus, the pK_a values of the nucleophile and histidine are much closer in cysteine than in serine proteases, which facilitates the formation of the thiolate-imidazolium ion-pair. However, an ordinary thiol group ($pK_a \sim 8.5$) is still more basic than an imidazole ($pK_a \sim 7$). Therefore, the ion-pair needs some additional stabilization by its microenvironment, such as the effect of a helix dipole[227] or other through-space electrostatic effects. Because of their capability to stabilize the ion-pair even in the ground state, cysteine proteases can be regarded as "activated" enzymes.[226]

Finally, it is worth noting that due to the existence of an ion-pair in the ground state of the cysteine protease reaction, general base catalysis in acylation and general acid catalysis in deacylation are not operative. The lack of these catalytic steps renders the stereochemistry of cysteine enzymes less complicated than that of the serine enzymes. The less stringent stereochemistry permits more freedom of motion in the transition state. This implies that a gain in enthalpy at the expense of entropy contributes to the driving force in the catalysis by serine proteases, whereas the opposite is true for the cysteine protease catalysis. The simpler mechanism of cysteine proteases may serve as a model, when relatively simple catalysts with enzymic activity are to be designed. This is a great challenge to the present day organic chemists. Furthermore, one may speculate whether the simple mechanism was realized earlier during evolution.

REFERENCES

1. **Glazer, A. N. and Smith, E. L.,** Papain and other plant sulfhydryl proteolytic enzymes, in *The Enzymes,* Vol. 3, 3rd ed., Boyer, P. D., Ed., Academic Press, New York, 1971, 501.
2. **Brocklehurst, K., Baines, B. S., and Kierstan, M. P. J.,** Papain and other constituents of Carica papaya L, *Top. Enzyme Ferment. Biotechnol.,* 5, 262, 1981.
2a. **Brocklehurst, K., Willenbrock, F., and Salih, E.,** Cysteine proteinases, in *New Comprehensive Biochemistry,* Vol. 16, Neuberger, A. and Brocklehurst, K., Eds., Elsevier, Amsterdam, 1987, 39.
3. **Drenth, J., Jansonius, J. N., Koekoek, R., and Wolthers, B. G.,** Papain, X-ray structure, in *The Enzymes,* Vol. 3, 3rd ed., Boyer, P. D., Ed., Academic Press, New York, 1971, 485.
4. **Drenth, J., Jansonius, J. N., Koekoek, R., and Wolthers, B. G.,** The structure of papain, *Adv. Protein Chem.,* 25, 79, 1971.
5. **Lowe, G.,** The cysteine proteinases, *Tetrahedron,* 32, 291, 1976.
6. **Polgár, L.,** The mechanism of action of thiolenzymes, *Int. J. Biochem.,* 8, 171, 1977.
7. **Polgár, L. and Halász, P.,** Current problems in mechanistic studies of serine and cysteine proteinases, *Biochem. J.,* 207, 1, 1982.
7a. **Baker, E. N. and Drenth, J.,** The thiol proteases: structure and mechanism, in *Biological Macromolecules and Assemblies,* Vol. 3, Jurnak, F. A. and McPherson, A., Eds., John Wiley & Sons, 1987, 313.
8. **Carne, A. and Moore, C. H.,** The amino acid sequence of the tryptic peptides from actinidin, a proteolytic enzyme from the fruit of *Actinidia chinensis, Biochem. J.,* 173, 73, 1978.
9. **Goto, K., Murachi, T., and Takahashi, N.,** Structural studies on stem bromelain. Isolation, characterization and alignment of the cyanogen bromide fragments, *FEBS Lett.,* 62, 93, 1976.
10. **Goto, K., Takahashi, N., and Murachi, T.,** Structural studies on stem bromelain. Cyanogen bromide cleavage and amino acid sequence of carboxyl-terminal half of the molecule, *Int. J. Pept. Protein Res.,* 15, 335, 1980.
11. **Lynn, K. R., Brockbank, W. J., and Clevette, N. A.,** Multiple forms of asclepains. Cysteinyl proteases from milkweed, *Biochim. Biophys. Acta,* 612, 119, 1980.
12. **Baker, E. N.,** Structure of actinidin: details of the polypeptide chain conformation and active site from an electron density map at 2.8 Å resolution, *J. Mol. Biol.,* 115, 263, 1977.
13. **Baker, E. N.,** Structure of actinidin, after refinement at 1.7 Å resolution, *J. Mol. Biol.,* 141, 441, 1980.
14. **Heinemann, U., Pal, G. P., Hilgenfeld, R., and Saenger, W.,** Crystal and molecular structure of the sulfhydryl protease calotropin DI at 3.2 Å resolution, *J. Mol. Biol.,* 161, 591, 1982.
15. **Smith, E. L., Kimmel, J. R., Brown, D. M., and Thompson, E. D. P.,** Isolation and properties of a crystalline mercury derivative of a lysozyme from papaya latex, *J. Biol. Chem.,* 215, 67, 1955.
16. **Balls, A. K., Lineweaver, H., and Thompson, R. R.,** Crystalline papain, *Science,* 86, 379, 1937.
17. **Husain, S. S. and Lowe, G.,** Completion of the amino acid sequence of papain, *Biochem. J.,* 114, 279, 1969.
18. **Mitchell, R. E. J., Chaiken, I. M., and Smith, E.,** The complete amino acid sequence of papain, *J. Biol. Chem.,* 245, 3485, 1970.
19. **Kimmel, J. R. and Smith, E. L.,** Crystalline papain. I. Preparation, specificity, and activation, *J. Biol. Chem.,* 207, 515, 1954.
20. **Kimmel, J. R. and Smith, E. L.,** Crystalline papain and benzoyl-L-argininamide, *Biochem. Prep.,* 6, 61, 1958.
21. **Finkle, B. J. and Smith, E. L.,** Crystalline papain: number and reactivity of thiol groups; chromatographic behavior, *J. Biol. Chem.,* 230, 669, 1958.
22. **Sluyterman, L. A. AE.,** The activation reaction of papain, *Biochim. Biophys. Acta,* 139, 430, 1967.

23. **Glazer, A. N. and Smith, E. L.,** The sulfur distribution of papain, *J. Biol. Chem.*, 240, 201, 1965.
24. **Klein, I. B. and Kirsch, J. F.,** The mechanism of activation of papain, *Biochem. Biophys. Res. Commun.*, 34, 575, 1969.
25. **Blumberg, S., Schechter, I., and Berger, A.,** The purification of papain by affinity chromatography, *Eur. J. Biochem.*, 15, 97, 1970.
26. **Burke, D. E., Lewis, S. D., and Shafer, J. A.,** A two-step procedure for purification of papain from extract of papaya latex, *Arch. Biochem. Biophys.*, 164, 30, 1974.
27. **Funk, M. O., Nakagawa, Y., Skochdopole, J., and Kaiser, E. T.,** Affinity chromatographic purification of papain. A reinvestigation, *Int. J. Pept. Protein Res.*, 13, 296, 1979.
28. **Schack, P. and Kaarsholm, N. C.,** Subsite differences between the active centres of papaya peptidase A and papain as revealed by affinity chromatography. Purification of papaya peptidase A by ionic-strength-dependent affinity adsorption on an immobilized peptide inhibitor of papain, *Biochem. J.*, 219, 727, 1984.
29. **Sluyterman, L. A. AE. and Wijdense, J.,** An agarose mercurial column for the separation of mercaptopapain and nonmercaptopapain, *Biochim. Biophys. Acta*, 200, 593, 1970.
30. **Brocklehurst, K., Carlsson, J., Kierstan, M. P. J., and Crook, E. M.,** Covalent chromatography. Preparation of fully active papain from dried papaya latex, *Biochem J.*, 133, 573, 1973.
31. **Brocklehurst, K.,** Specific covalent modification of thiols: applications in the study of enzymes and other biomolecules, *Int. J. Biochem.*, 10, 259, 1979.
32. **Asbóth, B. and Polgár, L.,** Hydrolysis of alkyl ester and amide substrates by papain, *Acta Biochim. Biophys. Acad. Sci. Hung.*, 12, 329, 1977.
33. **Polgár, L.,** Deuterium isotope effects on papain acylation. Evidence for lack of general base catalysis and for enzyme-leaving group interaction, *Eur. J. Biochem.*, 98, 369, 1979.
34. **Asbóth, B. and Polgár, L.,** On the enhanced catalytic activity of papain towards amide substrates, *Acta Biochim. Biophys. Acad. Sci. Hung.*, 12, 223, 1977.
35. **Schechter, I. and Berger, A.,** On the size of the active site of proteinases. I. Papain, *Biochem. Biophys. Res. Commun.*, 27, 157, 1967.
36. **Lowbridge, J. and Fruton, J. S.,** Studies on the extended active site of papain, *J. Biol. Chem.*, 249, 6754, 1974.
37. **Mattis, J. A. and Fruton, J. S.,** Kinetics of the action of papain on fluorescent peptide substrates, *Biochemistry*, 15, 2191, 1976.
38. **Alecio, M. R., Dann, M. L., and Lowe, G.,** The specificity of the S_1' subsite of papain, *Biochem. J.*, 141, 495, 1974.
39. **Kaarsholm, N. C. and Schack, P.,** Characterization of papaya peptidase A as an enzyme of extreme basicity, *Acta Chem. Scand.*, B37, 607, 1983.
40. **Schack, P.,** Fractionation of proteolytic enzymes of dried papaya latex. Isolation and preliminary characterization of a new proteolytic enzyme, *C. R. Trav. Lab. Carlsberg*, 36, 67, 1967.
41. **Robinson, G. W.,** Isolation and characterization of papaya peptidase A from commercial chymopapain, *Biochemistry*, 14, 3695, 1975.
42. **Buttle, D. J. and Barrett, A. J.,** Chymopapain. Chromatographic purification and immunological characterization, *Biochem. J.*, 223, 81, 1984.
43. **Kóródi, I., Asbóth, B., and Polgár, L.,** Disulfide bond formation between the active-site thiol and one of the several free thiol groups of chymopapain, *Biochemistry*, 25, 6895, 1986.
44. **Polgár, L.,** Problems of classification of papaya latex proteinases, *Biochem. J.*, 221, 555, 1984.
45. **Brocklehurst, K., Salih, E., McKee, R., and Smith, H.,** Fresh non-fruit latex of *Carica papaya* contains, papain, multiple forms of chymopapain A and papaya proteinase Ω, *Biochem. J.*, 228, 525, 1985.
46. **Barrett, A. J. and Buttle, D. J.,** Names and numbers of papaya proteinases, *Biochem. J.*, 228, 527, 1985.
47. **McKee, R. A., Adams, S., Matthews, J. A., Smith, C. J., and Smith, H.,** Molecular cloning of two cysteine proteinases from paw-paw *(Carica papaya), Biochem. J.*, 237, 105, 1986.
48. **Polgár, L.,** Isolation of highly active papaya peptidases A and B from commercial chymopapain, *Biochim. Biophys. Acta*, 658, 262, 1981.
49. **Baines, B. S., Brocklehurst, K., Carey, P. R., Jarvis, M., Salih, E., and Storer, A. C.,** Chymopapain A. Purification and investigation by covalent chromatography and characterization by two-protonic-state reactivity-probe kinetics, steady-state kinetics and resonance Raman spectroscopy of some dithioacyl derivatives, *Biochem. J.*, 233, 119, 1986.
50. **Asbóth, B., Stokum, É., Khan, I. U., and Polgár, L.,** Mechanism of action of cysteine proteinases; oxyanion binding site is not essential in the hydrolysis of specific substrates, *Biochemistry*, 24, 606, 1985.
51. **Smith, L. and Brown, J. E.,** Treatment of lumbar intervertebral disc leasions by direct injection of chymopapain, *J. Bone Jt. Surg.*, 49B, 502, 1967.
52. **Onofrio, B. M.,** Injection of chymopapain into intervertebral discs. Preliminary report on 72 patients with symptoms of disc disease, *J. Neurosurg.*, 42, 384, 1975.

53. **Reddy, M. N., Keim, P. S., Heinrickson, R. L., and Kézdy, F. J.,** Primary structural analysis of sulfhydryl protease inhibitors from pineapple stem, *J. Biol. Chem.,* 250, 1741, 1975.
54. **Katunuma, N. and Kominami, E.,** Structures and functions of lysosomal thiol proteinases and their endogenous inhibitor, in *Current Topics in Cellular Regulation,* Vol. 22, Horecker, B. L. and Stadtman, E. R., Eds., Academic Press, New York, 1983, 71.
55. **Barrett, A. J. and Kirschke, H.,** Cathepsin B, cathepsin H, and cathepsin L, *Methods Enzymol.,* 80, 535, 1981.
56. **Sloane, B. F., Rozhin, J., Johnson, K., Taylor, H., Crissman, J. D., and Hohn, K. V.,** Cathepsin B: association with plasma membreane in metastatic tumors, *Proc. Natl. Acad. Sci. U.S.A.,* 83, 2483, 1986.
57. **Takio, K., Towatari, T., Katunuma, N., Teller, D. C., and Titani, K.,** Homology of amino acid sequences of rat cathepsins B and H with that of papain, *Proc. Natl. Acad. Sci. U.S.A.,* 80, 3666, 1983.
58. **Otto, K.,** Über ein neues Kathepsin. Reinigung aus Rindermilz, Eigenschaften, sowie Vergleich mit Kathepsin B, *Hoppe-Seiler's Z. Physiol. Chem.,* 348, 1449, 1967.
59. **McDonald, J. K. and Ellis, S.,** On the substrate specificity of cathepsins B1 and B2 including a new flourogenic substrate for cathepsin B1, *Life Sci.,* 17, 1269, 1975.
60. **Willenbrock, F. and Brocklehurst, K.,** Preparation of cathepsins B and H by covalent chromatography and characterization of their catalytic sites by reaction with a thiol-specific two-proteonic-state reactivity probe. Kinetic study of cathepsins B and H extending into alkaline media and a rapid spectroscopic titration of cathepsin H at pH 3-4, *Biochem. J.,* 227, 511, 1985.
61. **Takahashi, T., Dehdarani, A. H., Schmidt, P. G., and Tang, J.,** Cathepsins B an H from porcine spleen. Purification and polypepetide chain arrangements, and carbohydrate content, *J. Biol. Chem.,* 259, 9874, 1984.
62. **Polgár, L. and Csoma, C.,** Dissociation of ionizing groups in the binding cleft invesely controls the endo- and exopeptidase activities of cathepsin B, *J. Biol. Chem.,* 262, 14448, 1987.
63. **Barrett, A. J.,** Human cathepsin B1. Purification and some properties of the enzyme, *Biochem. J.,* 131, 809, 1973.
64. **Towatari, T., Kawabata, Y., and Katunuma, N.,** Crystallization and properties of cathepsin B from rat liver, *Eur. J. Biochem.,* 102, 279, 1979.
65. **McGregor, R. R., Hamilton, J. W., Shofstall, R. E., and Cohn, D. V.,** Isolation and characterization of porcine parathyroid cathepsin B, *J. Biol. Chem.,* 254, 4423, 1979.
66. **Zvonar, T., Kregar, I., and Turk, V.,** Isolation of cathepsin B and α-N-benzoylarginine-β-naphtylamide hydrolase by covalent chromatography on activated thiol Sepharose, *Croat. Chem. Acta,* 52, 411, 1979.
67. **Mort, J. S., Recklies, A. D., and Poole, A. R.,** Characterization of a thiol proteinase secreted by malignant human breast tumours, *Biochim. Biophys. Acta,* 614, 134, 1980.
68. **Suhar, A. and Marks, N.,** Purification and properties of brain cathepsin B. Evidence for cleavage of pituitary lipotropins, *Eur. J. Biochem.,* 101, 23, 1979.
69. **Pohl, J., Baduys, M., Tomasek, V., and Kostka, V.,** Identification of the active site cysteine and of the disulfide bonds in the N-terminal part of the molecule of bovine spleen cathepsin B, *FEBS Lett.,* 142, 23, 1982.
70. **Kirschke, H., Langner, J., Wiederanders, B., Ansorge, S., Bohley, P., and Hanson, H.,** Intrazellulärer Proteinabbau. VII. Kathepsin L und H: zwei neue Proteinasen aus Rattenleberlysosomen, *Acta. Biol. Med. Ger.,* 35, 285, 1976.
71. **Kirschke, H., Langner, J., Wiederanders, B., Ansorge, S., Bohley, P., and Hanson, H.,** Cathepsin H: an endo-aminopeptidase from rat-liver lysosomes, *Acta Biol. Med. Ger.,* 36, 185, 1977.
72. **Kirschke, H., Langner, J., Wiederanders, B., Ansorge, S., and Bohley, P.,** Cathepsin L. A new proteinase from rat-liver lysosomes, *Eur. J. Biochem.,* 74, 293, 1977.
73. **Evans, B. and Shaw, E.,** Inactivation of cathepsin B by active site-directed disulfide exchange. Application in covalent affinity chromatography, *J. Biol. Chem.,* 258, 10227, 1983.
74. **Rich, D. H., Brown, M. A., and Barrett, A. J.,** Purification of cathepsin B by a new form of affinity chromatography *Biochem. J.,* 235, 731, 1986.
75. **Aronson, N. N., Jr. and Barrett, A. J.,** The specificity of cathepsin B. Hydrolysis of glucagon at the C-terminus by a peptidyldipeptidase mechanism, *Biochem. J.,* 171, 759, 1978.
76. **Bond, J. S. and Barrett, A. J.,** Degradation of fructose-1,6-bisphophate aldolase by cathepsin B. A further example of peptidyldipeptidase activity of this proteinase, *Biochem. J.,* 189, 17, 1980.
77. **Singh, H. and Kalnitsky, G.,** α-N-Benzoylarginine-β-naphthylamide hydrolase, an aminopeptidase from rabbit lung, *J. Biol. Chem.,* 255, 369, 1980.
78. **Schwartz, W. N. and Barrett, A. J.,** Human cathepsin H, *Biochem. J.,* 191, 487, 1980.
79. **Mason, R. W., Green, G. D. J., and Barrett, A. J.,** Human liver cathepsin L, *Biochem. J.,* 226, 233, 1985.
80. **Kirschke, H., Kembhavi, A. A., Bohley, P., and Barrett, A. J.,** Action of rat liver cathepsin L on collagen and other substrates, *Biochem. J.,* 201, 367, 1982.

81. **Mason, R. W., Johnson, D. A., Barrett, A. J., and Chapman, H. A.,** Elastinolytic activity of human cathepsin L, *Biochem. J.,* 233, 925, 1986.

82. **Ritonja, A., Popovic, T., Turk, V., Wiedenmann, K., and Machleidt, W.,** Amino acid sequence of human liver cathepsin B, *FEBS Lett.,* 181, 169, 1985.

83. **Meloun, B., Pohl, J., and Kostka, V.,** Tentative amino acid sequence of bovine spleen cathepsin B, in *Proc. Int. Symp. Cysteine Preoteinases and Their Inhibitors,* Turk, V., Ed., Walter de Gruyter, Berlin, 1986, 19.

84. **Segundo, B. S., Chan, S. J., and Steiner, D. F.,** Identification of cDNA clones encoding a precursor of rat liver cathepsin B, *Proc. Natl. Acad. Sci. U.S.A.,* 82, 2320, 1985.

85. **Docherty, K., Carroll, R., and Steiner, D. F.,** Indentification of a 31,500 molecular weight islet cell protease as cathepsin B, *Proc. Natl. Acad. Sci. U.S.A.,* 80, 3245, 1983.

86. **Katunuma, N., Towatari, T., Kominami, E., Hashida, S., and Takio, K.,** Rat liver thiol proteinases: cathepsin B, cathepsin H and cathepsin L, *Acta Biol. Med. Ger.,* 40, 1419, 1981.

87. **Kamphuis, I. G., Drenth, J., and Baker, E. N.,** Thiol proteases. Comparative studies based on the high-resolution structures of papain and actinidin, and on amino acid sequence information for cathepsins B and H, and stem bromelain, *J. Mol. Biol.,* 182, 317, 1985.

87a. **Chan, S. J., Segundo, B. S., McCormick, M. B., and Steiner, D. F.,** Nucleotide and predicted amino acid sequences of cloned human and mouse preprocathepsin B cDNAs, *Proc. Natl. Acad. Sci. U.S.A.,* 83, 7721, 1986.

87b. **Ishidoh, K., Imajoh, S., Emori, Y., Ohno, S., Kawasaki, H., Minami, Y., Kominami, E., Katunuma, N., and Suzuki, K.,** Molecular cloning and sequencing of cDNA for rat cathepsin H. Homology in pro-peptide regions of cysteine proteinases, *FEBS Lett.,* 226, 33, 1987.

87c. **Ishidoh, K., Towatari T., Imajoh, S., Kawasaki, H., Kominami, E., Katunuma, N., and Suzuki, K.,** Molecular cloning and sequencing of cDNA for rat cathepsin L, *FEBS Lett.,* 223, 69, 1987.

87d. **Cohen, L. W., Coghlan, V. M., and Dihel, L. C.,** Cloning and sequencing of papain-encoding cDNA, *Gene,* 48, 219, 1986.

88. **Williams, J. G., North, M. J., and Mahbubani, H.,** A developmentally regulated cysteine proteinase in *Dictyostelium discoideum, EMBO J.,* 4, 999, 1985.

89. **Pears, C. J., Mahbubani, H. M., and Williams, J. G.,** Characterization of two highly diverged but developmentally co-regulated cysteine proteinase genes in *Dictyostelium discoideum, Nucleic Acids Res.,* 13, 8853, 1985.

90. **Rogers, J. C., Dean, D., and Heck, G. R.,** Aleurain; a barley thiol protease closely related to mammalian cathepsin H, *Proc. Natl. Acad. Sci. U.S.A.,* 82, 6512, 1985.

91. **North, M. J.,** Homology within the N-terminal extension of cysteine proteinases, *Biochem. J.,* 238, 623, 1986.

92. **Leary, R. and Shaw, E.,** Inactivation of cathepsin B_1 by diazomethyl ketones, *Biochem. Biophys. Res. Commun.,* 79, 926, 1977.

93. **Shaw, E. and Green, G. D. J.,** Inactivation of thiol proteases with peptidyl diazomethyl ketones, *Methods Enzymol.,* 80, 820, 1981.

94. **Shaw, E., Wikstrom, P., and Ruscica, J.,** An exploration of the primary specificity site of cathepsin B, *Arch. Biochem. Biophys.,* 222, 424, 1983.

95. **Müller-Esterl, W., Fritz, H., Lottspeich, F., Kellermann, J., Machleidt, W., and Turk, V.,** Genealogy of mammalian cysteine proteinase inhibitors. Common evolutionary origin of stefins, cystatins, and kininogens, *FEBS Lett.,* 191, 221, 1985.

96. **Järvinen, M. and Rinne, A.,** Human spleen cysteine proteinase inhibitor, *Biochim. Biophys. Acta,* 708, 210, 1982.

97. **Anastasi, A., Brown, M. A., Kembhavi, A. A., Nicklin, M. J. H., Sayers, C. A., Sunter, D. C., and Barrett, A. J.,** Cystatin, a protein inhibitor of cysteine proteinases. Improved purification from egg white, characterization, and detection in chicken serum, *Biochem. J.,* 211, 129, 1983.

98. **Machleidt, W., Borchart, U., Fritz, H., Brzin, J., Ritonja, A., and Turk, V.,** Protein inhibitors of cysteine proteinases. II. Primary structure of stefin, a cytosolic protein inhibitor of cysteine proteinases from human polymorphonuclear granulocytes, *Hoppe-Seyler's Z. Physiol. Chem.,* 364, 1481, 1983.

99. **Ritonja, A., Machleidt, W., and Barrett, A. J.,** Amino acid sequence of the intracellular cysteine proteinase inhibitor, cystatin B from human liver, *Biochem. Biophys. Res. Commun.,* 131, 1187, 1985.

100. **Takio, K., Kominami, E., Bando, Y., Katunuma, N., and Titani, K.,** Amino acid sequence of rat epidermal thiol proteinase inhibitor, *Biochem. Biophys. Res. Commun.,* 121, 149, 1984.

101. **Takio, I., Kominami, E., Wakamatsu, N., Katunuma, N., and Titani, K.,** Amino acid sequence of rat liver thiol proteinase inhibitor, *Biochem. Biophys. Res. Commun.,* 115, 902, 1983.

102. **Fossum, K. and Whitaker, J. R.,** Ficin and papain inhibitor from chicken egg white, *Arch. Biochem. Biophys.,* 125, 367, 1968.

103. **Keilova, H. and Tomasek, V.,** Effect of papain inhibitor from chicken egg white on cathepsin Bl, *Biochim. Biophys. Acta,* 334, 179, 1974.

104. **Turk, V., Brzin, J., Longer, M., Ritonja, A., Eropkin, M., Borchart, U., and Machleidt, W.,** Protein inhibitors of cysteine proteinases. III. Amino acid sequence of cystatin from chicken egg white, *Hoppe-Seyler's Z. Physiol. Chem.,* 364, 1487, 1983.

105. **Schwabe, C., Anastasi, A., Crow, H., McDonald, J. K., and Barrett, A. J.,** Cystatin. Amino acid sequence and possible secondary structure, *Biochem. J.,* 217, 813, 1984.

106. **Grubb, A. and Löfberg, H.,** Human γ-trace, a basic microprotein: amino acid sequence and presence in the adenohypophysis, *Proc. Natl. Acad. Sci. U.S.A.,* 79, 3024, 1982.

107. **Isemura, S., Saitoh, E., and Sanada, K.,** Isolation and amino acid sequence of SAP-1, an acidic protein of human whole saliva, and sequence homology with human γ-trace, *J. Biochem. (Tokyo),* 96, 489, 1984.

108. **Hirado, M., Tsunasawa, S., Sakiyama, F., Niinobe, M., and Fujii, S.,** Complete amino acid sequence of bovine colostrum low-M_r cysteine proteinase inhibitor, *FEBS Lett.,* 186, 41, 1985.

109. **Ghiso, J., Jensson, O., and Frangione, B.,** Amyloid fibrils in hereditary cerebral hemorrhage with amyloidosis of Icelandic type is a variant of γ-trace basic protein (cystatin C), *Proc. Natl. Acad. Sci. U.S.A.,* 83, 2974, 1986.

110. **Kato, H., Nagasawa, S., and Iwanaga, S.,** HMW and LMW kininogens, *Methods Enzymol.,* 80, 172, 1981.

111. **Lottspeich, F., Kellerman, J., Henschen, A., Foertsch, B., and Muller-Esterl, W.,** The amino acid sequence of the light chain of human high-molecular-mass kininogen, *Eur. J. Biochem.,* 152, 307, 1985.

112. **Sasaki, M., Minakata, K., Yamamoto, H., Niwa, M., Kato, T., and Ito, N.,** A new serum component which specifically inhibits thiol proteinases, *Biochem. Biophys. Res. Commun.,* 76, 917, 1977.

113. **Ryley, H. C.,** Isolation and partial characterization of a thiol proteinase inhibitor from human plasma, *Biochem. Biophys. Res. Commun.,* 89, 871, 1979.

114. **Gounaris, A. D., Brown, M. A., and Barrett, A. J.,** Human plasma alpha-cysteine proteinase inhibitor. Purification by affinity chromatography, characterization and isolation of an active fragment, *Biochem. J.,* 221, 445, 1984.

115. **Travis, J. and Salvesen, G. S.,** Human plasma proteinase inhibitors, *Annu. Rev. Biochem.,* 52, 655, 1983.

116. **Ohkubo, I., Kurachi, K., Takasawa, T., Shiokawa, H., and Sasaki, M.,** Isolation of a human cDNA for α_2-thiol proteinase inhibitor and its identity with low molecular weight kininogen, *Biochemistry,* 23, 5691, 1984.

117. **Müller-Esterl, W., Fritz, H., Machleidt, W., Ritonja, A., Brzin, J., Kotnik, M., Turk, V., Kellerman, J., and Lottspeich, F.,** Human plasma kininogens are identical with α-cysteine proteinase inhibitors. Evidence from immunological, enzymological and sequence data, *FEBS Lett.,* 182, 310, 1985.

118. **Sueyoshi, T., Enjyoji, K., Shimada, T., Kato, H., Iwanaga, S., Bando, Y., Kominami, E., and Katunuma, N.,** A new function of kininogens as thiol-proteinase inhibitors: inhibition of papain, and cathepsins B, H and L by bovine, rat and human plasma kininogens, *FEBS Lett.,* 182, 193, 1985.

119. **Takagaki, Y., Kitamura, N., and Nakanishi, S.,** Cloning and sequence analysis of cDNAs for human high molecular weight and low molecular weight prekininogens. Primary structures of two human prekininogens, *J. Biol. Chem.,* 260, 8601, 1985.

120. **Kitamura, N., Kitagawa, H., Fukushima, D., Takagaki, Y., Miyata, T., and Nakanishi, S.,** Structural organisation of the human kininogen gene and a model for its evolution, *J. Biol. Chem.,* 260, 8610, 1985.

121. **Kitamura, N., Takagaki, Y., Furuto, S., Tanaka, T., Nawa, H., and Nakanishi, S.,** A single gene for bovine high molecular weight and low molecular weight kininogens, *Nature (London),* 305, 545, 1983.

122. **Cole, T., Inglis, A. S., Roxburgh, C. M., Howlett, G. J., and Schreiber, G.,** Major acute phase α_1-protein of the rat is homologus to bovine kininogen and contains the sequence for bradykinin: its synthesis is regulated at the mRNA level, *FEBS Lett.,* 182, 57, 1985.

123. **Salvesen, G., Parkes, C., Abrahamson, M., Grubb, A., and Barrett, A. J.,** Human low-M_r kininogen contains three copies of a cystatin sequence that are divergent in structure and in inhibitory activity for cysteine proteinases, *Biochem. J.,* 234, 429, 1986.

124. **Bode, W., Brzin, J., and Turk, V.,** Crystallization of chicken egg white cystatin, a low molecular weight protein inhibitor of cysteine proteinases, and preliminary X-ray diffraction data, *J. Mol. Biol.,* 181, 331, 1985.

125. **Murachi, T.,** Calpain and calpastatin, *Trends Biochem. Sci.,* 8, 167, 1983.

125a. **Suzuki, K., Imajoh, S., Emori, Y., Kawasaki, H., Minami, Y., and Ohno, S.,** Calcium-activated neutral protease and its endogenous inhibitor. Activation at the cell membrane and biological function, *FEBS Lett.,* 220, 271, 1987.

126. **Suzuki, K., Tsuji, S., Kubota, S., Kimura, Y., and Imahori, K.,** Limited autolysis of Ca^{2+}-activated neutral protease (CANP) changes its sensitivity to Ca^{2+} ions, *J. Biochem. (Tokyo),* 90, 275, 1981.

127. Nomenclature committee of the international union of biochemistry (NC-IUB), enzyme nomenclature. Recommendations 1978. Supplement 2: corrections and additions, *Eur. J. Biochem.,* 116, 423, 1981.

128. **Kitahara, A., Ohtsuki, H., Kirihata, Y., Yamagata, Y., Takano, E., Kannagi, R., and Murachi, T.,** Selective localization of calpain I (the low Ca $^{2+}$-requiring form of Ca^{2+}-dependent cysteine proteinase) in B-cells of human pancreatic islets, *FEBS Lett.,* 184, 120, 1985.

129. **Wolheim, C. B. and Sharp, G. W. G.,** Regulation of insulin release by calcium, *Physiol. Rev.,* 61, 914, 1981.

130. **Sasaki, T., Kikuchi, T., Yumoto, M., Yoshimura, N., and Murachi, T.,** Comparative specificity and kinetic studies on porcine calpain I and calpain II with naturally occurring peptides and synthetic fluorogenic substrates, *J. Biol. Chem.,* 259, 12489, 1984.

131. **Suzuki, K., Tsuji, S., and Ishiura, S.,** Effect of Ca^{2+} on the inhibition of calcium-activated neutral protease by leupeptin, antipain and epoxysuccinate derivatives, *FEBS Lett.,* 136, 119, 1981.

132. **Parkes, C., Kembhavi, A. A., and Barrett, A. J.,** Calpain inhibition by peptide epoxides, *Biochem. J.,* 230, 509, 1985.

133. **Waxman, L. and Krebs, E. G.,** Identification of two protease inhibitors from bovine cardiac muscle, *J. Biol. Chem.,* 253, 5888, 1978.

134. **Nishiura, I., Tanaka, K., Yamamoto, S., and Murachi, T.,** The occurrence of an inhibitor of Ca^{2+}-dependent neutral protease in rat liver, *J. Biochem. (Tokyo),* 84, 1657, 1978.

134a. **Emori, Y., Kawasaki, H., Imajoh, S., Imahori, K., and Suzuki, K.,** Endogenous inhibitor for calcium-dependent cysteine protease contains four internal repeats that could be responsible for its multiple reactive sites, *Proc. Natl. Acad. Sci. U.S.A.,* 84, 3590, 1987.

134b. **Maki, M., Takano, E., Mori, H., Sato, A., Murachi, T., and Hatanaka, M.,** All four internally repetitive domains of pig calpastatin possess inhibitory activities against calpains I and II, *FEBS Lett.,* 223, 174, 1987.

135. **Karlsson, J.-O., Gustavsson, S., Hall, C., and Nilsson, E.,** A simple one step procedure for the separation of calpain I and calpain II and calpastatin, *Biochem. J.,* 231, 201, 1985.

136. **Ohno, S., Emori, Y., Imajoh, S., Kawasaki, H., Kisaragi, M., and Suzuki, K.,** Evolutionary origin of a calcium-dependent protease by fusion of genes for a thiol protease and a calcium-binding protein?, *Nature (London),* 312, 566, 1984.

137. **Aoki, K., Imajoh, S., Ohno, S., Emori, Y., Koike, M., Kosaki, G., and Suzuki, K.,** Complete amino acid sequence of the large subunit of the low- Ca^{2+}-requiring form of human Ca^{2+}-activated neutral protease (μCANP) deduced from its cDNA sequence, *FEBS Lett.,* 205, 313, 1986.

138. **Kretsinger, R. H., and Nockolds, C. E.,** Carp muscle calcium-binding protein. II. Structure determination and general description, *J. Biol. Chem.,* 248, 3313, 1973.

139. **Kakiuchi, S. and Sobue, K.,** Control of the cytoskeleton by calmodulin and calmodulin-binding proteins, *Trends Biochem. Sci.,* 8, 59, 1983.

140. **Emory, Y., Ohno, S., Tobita, M., and Suzuki, K.,** Gene structure of calcium-dependent protease retains the ancestral organization of the calcium-binding protein gene, *FEBS Lett.,* 194, 249, 1986.

141. **Sakihama, T., Kakidani, H., Zenitja, K., Yumoto, N., Kikuchi, T., Sasaki, T., Kannagi, R., Nakanishi, S., Ohmori, M., Takio, K., Titani, K., and Murachi, T.,** A putative Ca^{2+}-binding protein: structure of the light subunit of porcine calpain elucidated by molecular cloning and protein sequence analysis, *Proc. Natl. Acad. Sci. U.S.A.,* 82, 6075, 1985.

142. **Emori, Y., Kawasaki, H., Imajoh, S., Kawashima, S., and Suzuki, K.,** Isolation and sequence analysis of cDNA clones for the small subunit of rabbit calcium-dependent protease, *J. Biol. Chem.,* 261, 9472, 1986.

143. **Liu, T.-Y. and Elliott, S. D.,** Streptococcal proteinase, in *The Enzymes,* Vol. 3, 3rd ed., Boyer, P. D., Ed., Academic Press, New York, 1971, 609.

144. **Elliot, S. D. and Liu, T.-Y.,** Streptococcal proteinase, *Methods Enzymol.,* 19, 252, 1970.

145. **Liu, T.-Y. and Elliott, S. D.,** Activation of streptococcal proteinase and its zymogen by bacterial cell walls, *Nature (London),* 206, 33, 1965.

146. **Elliott, S. D.,** The crystallization and serological differentiation of a streptococcal proteinase and its precursor, *J. Exp. Med.,* 92, 201, 1950.

147. **Lo, S.-S., Fraser, B. A., and Liu, T.-Y,** The mixed disulfide in the zymogen of streptococcal proteinase. Characterization and implication for its biosynthesis, *J. Biol. Chem.,* 259, 11041, 1984.

148. **Gerwin, B. I., Stein, W. H., and Moore, S.,** On the specificity of streptococcal proteinase, *J. Biol. Chem.,* 241, 3331, 1966.

149. **Tai, J. Y., Kortt, A. A., Liu, T.-Y., and Elliott, S. D.,** Primary structure of streptococcal proteinase. III. Isolation of cyanogen bromide peptides; complete covalent structure of the polypeptide chain, *J. Biol. Chem.,* 251, 1955, 1976.

150. **Mitchel, W. M. and Harrington, W. F.,** Clostripain, in *The Enzymes,* Vol. 3, 3rd ed., Boyer, P. D., Ed., Academic Press, New York, 1971, 699.

151. **Mitchell, W. M. and Harrington, W. F.,** Purification and properties of clostridiopeptidase (clostripain), *J. Biol. Chem.,* 243, 4683, 1968.

152. **Mitchell, W. M. and Harrington, W. F.**, Clostripain, *Methods Enzymol.*, 19, 635, 1970.
153. **Gilles, A.-M., Imhoff, J.-M., and Keil, B.**, α-Clostripain. Chemical characterization, activity, and thiol content of the highly active form of clostripain, *J. Biol. Chem.*, 254, 1462, 1979.
154. **Gilles, A.-M., Lecroisey, A., and Keil, B.**, Primary structure of α-clostripain light chain, *Eur. J. Biochem.*, 145, 469, 1984.
155. **Argos, P., Kamer, G., Nicklin, M. J. H., and Wimmer, E.**, Similarity in gene organization and homology between proteins of animal picornaviruses and a plant comovirus suggest common ancestry of these virus families, *Nucleic Acids Res.*, 12, 7251, 1984.
156. **Ivanoff, L. A., Towatari, T., Ray, J., Korant, B. D., and Petteway, S. R., Jr.**, Expression and site-specific mutagenesis of the poliovirus 3C protease in *Escherichia coli*, *Proc. Natl. Acad. Sci. U.S.A.*, 83, 5392, 1986.
157. **Kitamura, N., Semler, B. L., Rothberg, P. G., Larsen, G. R., Adler, C. J., Dorner, A. J., Emini, E. A., Hanecak, R., Lee, J. J., van der Werf, S., Anderson, C. W., and Wimmer, E.**, Primary structure, gene organization and polypeptide expression of poliovirus RNA, *Nature (London)*, 291, 547, 1981.
158. **Lowe, G. and Williams, A.**, Papain-catalysed hydrolysis of some hippuric esters: A new mechanism for papain-catalysed hydrolyses, *Biochem. J.*, 96, 199, 1965.
159. **Kirsch, J. F. and Igelström, M.**, The kinetics of the papain-catalyzed hydrolysis of esters of carbobenzoxyglycine. Evidence for an acyl-enzyme intermediate, *Biochemistry*, 5, 783, 1966.
160. **Lucas, E. C. and Williams, A.**, The pH-dependencies of individual rate constants in papain-catalyzed reactions, *Biochemistry*, 8, 5125, 1969.
161. **Smith, E. L. and Parker, M. J.**, Kinetics of papain action. III. Hydrolysis of benzoyl-L-arginine ethyl ester, *J. Biol. Chem.*, 233, 1387, 1958.
162. **Smith, E. L.**, Active site of papain and covalent "high energy" bonds of proteins, *J. Biol. Chem.*, 233, 1392, 1958.
163. **Whitaker, J. R. and Bender, M. L.**, Kinetics of papain-catalyzed hydrolysis of α-N-benzoyl-L-arginine ethyl ester and α-N-benzoyl-L-arginine amide, *J. Am. Chem. Soc.*, 87, 2728, 1965.
164. **Sluyterman, L. A. AE.**, The rate-limiting reaction in papain action as derived from the reaction of the enzyme with chloroacetic acid, *Biochim. Biophys. Acta*, 151, 178, 1968.
165. **Brocklehurst, K., Crook, E. M., and Wharton, C. W.**, The kinetic analysis of hydrolytic enzyme catalyses: consequences of non-productive binding, *FEBS Lett.*, 2, 69, 1968.
166. **Glick, B. R. and Brubacher, L. J.**, Evidence for nonproductive binding subsites within the active site of papain, *Can. J. Biochem.*, 52, 877, 1974.
167. **Whitaker, J. R.**, Ficin-and papain-catalyzed reactions. Changes in reactivity of the essential sulfhydryl group in the presence of substrates and competitive inhibitors, *Biochemistry*, 8, 4591, 1969.
168. **Lowe, G. and Williams, A.**, Direct evidence for an acylated thiol as an intermediate in papain- and ficin-catalyzed hydrolyses, *Biochem. J.*, 96, 189, 1965.
169. **Brubacher, L. J. and Bender, M. L.**, The preparation and properties of *trans*-cinnamoyl-papain, *J. Am. Chem. Soc.*, 88, 5871, 1966.
170. **Smith, E. L. and Kimmel, J. R.**, *The Enzymes*, Vol. 4., 2nd ed., Boyer, P. D., Lardy, H., and Myrbäck, K., Eds., Academic Press, New York, 1960, 133.
171. **Zannis, V. I. and Kirsch, J. F.**, Effects of substituents on the rates of deacylation of substituted benzoyl papains. Role of a carboxylate residue in the catalytic mechanism, *Biochemistry*, 17, 2669, 1978.
172. **Angelides, K. J. and Fink, A. L.**, Mechanism of action of papain with a specific anilide substrate, *Biochemistry*, 18, 2355, 1979.
173. **Johnson, F. A., Lewis, S. D., and Shafer, J. A.**, Perturbations in the free energy and enthalpy of ionization of histidine-159 at the active site of papain as determined by fluoresence spectroscopy, *Biochemistry*, 20, 52, 1981.
174. **Polgár, L.**, On the mode of activation of the catalytically essential sulfhydryl group of papain, *Eur. J. Biochem.*, 33, 104, 1973.
175. **Sluyterman, L. A. AE. and Wijdenes, J.**, Benzoylamidoacetonitrile as an inhibitor of papain, *Biochim. Biophys. Acta*, 302, 95, 1973.
176. **Lewis, S. D., Johnson, F. A., Ohno, A. K., and Shafer, J. A.**, Dependence of the catalytic activity of papain on the ionization of two acidic groups, *J. Biol. Chem.*, 253, 5080, 1978.
177. **Allen, K. G. D., Stewart, J. A., Johnson, P. E., and Wettlaufer, D. G.**, Identification of the functional ionic groups of papain by pH/rate profile analysis, *Eur. J. Biochem.*, 87, 575, 1978.
178. **Schack, P. and Kaarsholm, N. C.**, Absence in papaya peptidase A catalyzed hydrolyses of a $pK_a \sim 4$ present in papain-catalyzed hydrolyses, *Biochemistry*, 23, 631, 1984.
179. **Bendall, M. R. and Lowe, G.**, Cooperative ionisation of aspartic-acid-158 and histidine-159 in papain. Evidence from ^{19}F nuclear-magnetic-resonance and fluorescence spectroscopy, *Eur J. Biochem.*, 65, 481, 1976.

180. **Bendall, M. R. and Lowe, G.,** A spectroscopic investigation of S-trifluoroethylthiopapain. An investigation of the active site of papain, *Eur. J. Biochem.,* 65, 493, 1976.

181. **Shipton, M., Kierstan, M. P. J., Malthouse, J. P. G., Stuchbury, T., and Brocklehurst, K.,** The case for assigning a value of approximately 4 to pKa_1 of the essential histidine-cysteine interactive systems of papain, bromelain and ficin, *FEBS Lett.,* 50, 365, 1975.

182. **Kamphuis I. G., Kalk, K. H., Swarte, M. B. A., and Drenth, J.,** Structure of papain refined at 1.65 Å resolution, *J. Mol. Biol.,* 179, 233, 1984.

183. **Drenth, J., Kalk, K. H., and Swen, H. M.,** Binding of chloromethyl ketone substrate analogs to crystalline papain, *Biochemistry,* 15, 3731, 1976.

184. **Lowe, G. and Yuthavong, Y.,** Kinetic specificity in papain-catalysed hydrolyses, *Biochem. J.,* 124, 107, 1971.

185. **Barrett, A. J., Nicklin, M. J. H., and Rawlings, N. D.,** The papain superfamily of cysteine proteinases and their inhibitors, in *Proc. Biol. Hung.,* Vol. 25, Elödi, P., Ed., Akadémiai Kiadó, Budapest, 1984, 203.

186. **Brocklehurst, K. and Little, G.,** Reactivities of the various protonic states in the reaction of papain and L-cysteine with 2,2'- and 4,4'-dipyridyl disulphide: evidence for nucleophilic reactivity in the unionized thiol group of cysteine-25 residue of papain occasioned by its interaction with the His-159-Asn-175 hydrogen-bonded system, *Biochem. J.,* 128, 471, 1972.

187. **Campbell, P. and Kaiser, E. T.,** Reaction of 6-membered cyclic sulfonate ester, β-(2-hydroxy-3,5-dinitrophenyl)ethanesulfonic acid sultone, with the active site of papain, *J. Am. Chem. Soc.,* 95, 3735, 1973.

188. **Polgár, L.,** On the mode of activation of the catalytically essential sulfhydryl group of papain, *Eur. J. Biochem.,* 33, 104, 1973.

189. **Polgár, L.,** Deuterium isotope effects on papain acylation. Evidence for lack of general base catalysis and for enzyme-leaving group interaction, *Eur. J. Biochem.,* 98, 369, 1979.

190. **Polgár, L.,** Spectrophotometric determination of mercaptide ion, an activated form of SH-group in thiol enzymes, *FEBS Lett.,* 38, 187, 1974.

191. **Polgár, L.,** Mercaptide-imidazolium ion-pair: the reactive nucleophile in papain catalysis, *FEBS Lett.,* 47, 15, 1974.

192. **Polgár, L.,** Ion-pair formation as a source of enhanced reactivity of D-glyceraldehyde-3-phosphate dehydrogenase, *Eur. J. Biochem.,* 51, 63, 1975.

193. **Sluyterman, L. A. AE. and De Graaf, M. J. M.,** The fluorescence of papain, *Biochim. Biophy. Acta,* 200, 595, 1970.

194. **Sluyterman, L. A. AE. and Wijdenes, J.,** Proton equilibria in the binding of Zn^{2+} and of methylmercuric iodide to papain, *Eur. J. Biochem.,* 71, 383, 1976.,

195. **Migliorini, M. and Creighton, D. J.,** Active site ionizations of papain. An evaluation of the potentiometric difference titration method, *Eur. J. Biochem.,* 156, 189, 1986.

196. **Johnson, F. A., Lewis, S. D., and Shafer, J. A.,** Determination of a low pK_a for histidine-159 in the S-methyltihio derivative of papain by proton nuclear magnetic resonance spectroscopy, *Biochemistry,* 20, 44, 1981.

197. **Jordan, F. and Polgár, L.,** Proton nuclear magnetic resonance evidence for the absence of a stable hydrogen bond between the active sites aspartate and histidine residues of native subtilisins and for its presence in thiolsubtilisins, *Biochemistry,* 20, 6366, 1981.

198. **Lewis, S. D., Johnson, F. A., and Shafer, J. A.,** Potentiometric determination of ionizations at the active site of papain, *Biochemistry,* 15, 5009, 1976.

199. **Chaiken, I. M. and Smith, E. L.,** Reaction of the sulfhydryl group of papain with chloroacetic acid, *J. Biol. Chem.,* 244, 5095, 1969.

200. **Wallenfels, K. and Eisele, B.,** Stereospecific alkylation with asymmetric reagents, *Eur. J. Biochem.,* 3, 267, 1968.

201. **Jolley, C. J. and Yankeelov, J. A., Jr.,** Reaction of papain with α-bromo-β-(5-imidazoyl) propionic acid, *Biochemistry,* 11, 164, 1972.

202. **Halász, P. and Polgár, L.,** Negatively charged reactants as probes in the study of the essential mercaptide-imidazolium ion-pair of thiolenzymes, *Eur. J. Biochem.,* 79, 491, 1977.

203. **Brocklehurst, K., Mushiri, S. M., Patel, G., and Willenbrock F.,** Evidence for a close similarity in the catalytic sites of papain and ficin in near-neutral media despite differences in acidic and alkaline media. Kinetics of the reactions of papain and ficin with chloroacetate, *Biochem. J.,* 201, 101, 1982.

204. **Gerwin, B. I.,** Properties of the single sulfhydryl group of streptococcal proteinase. A comparison of the rates of alkylation by chloroacetic acid and chloroacetamide, *J. Biol. Chem.,* 242, 451, 1967.

205. **Csoma, C. and Polgár, L.,** Proteinase from germinating bean cotyledons. Evidence for involvement of a thiol group in catalysis, *Biochem. J.,* 222, 769, 1984.

155

206. **Bone, R. and Wolfenden, R.**, Solvent isotope effects on formation of protease complexes with inhibitory aldehydes, with an appendix on the determination of deuterium fractionation factors by NMR, *J. Am. Chem. Soc.*, 107, 4772, 1985.
207. **Lowe, G. and Yuthavong, Y.**, pH Dependence and structure-activity relationship in the papain-catalysed hydrolysis of anilides, *Biochem. J.*, 124, 117, 1971.
208. **O'Leary, M. H., Urberg, M., and Young, A. P.**, Nitrogen isotope effects on the papain-catalyzed hydrolysis of N-benzoyl-L-argininamide, *Biochemistry*, 13, 2077, 1974.
209. **Wolthers, B. G., Drenth, J., Jansonius, J. N., Koekoek, R., and Swen, H. M.**, The three-dimensional structure of papain, *Proc. Int. Symp. Structure-Function Relationships of Proteolytic Enzymes*, Desnuelle, P., Neurath, H., and Ottesen, M., Eds., Munksgaard, Copenhagen, 1970, 272.
210. **Asbóth, B. and Polgár, L.**, Transition-state stabilization at the oxyanion binding sites of serine and thiol proteinases: hydrolyses of thiono and oxygen esters, *Biochemistry*, 22, 117, 1983.
211. **Storer, A. C. and Carey, P. R.**, Comparison of the kinetics and mechanism of the papain-catalyzed hydrolysis of esters and thiono esters, *Biochemistry*, 24, 6808, 1985.
212. **Gamcsik, M. P., Malthouse, J. P. G., Primrose, W. V., Mackenzie, N. E., Boyd, A. S. F., Russell, R. A., and Scott, A. I.**, Structure and stereochemistry of tetrahedral inhibitor complexes of papain by direct NMR observation, *J. Am. Chem. Soc.*, 105, 6324, 1983.
213. **Polgár, L. and Bender, M. L.**, A new enzyme containing a synthetically formed active site. Thiolsubtilisin, *J. Am. Chem. Soc.*, 88, 3153, 1966.
214. **Neet, K. E. and Koshland, D. E., Jr.**, The conversion of serine at the active site of stubtilisin to cysteine: a "chemical mutation", *Proc. Natl. Acad. Sci. U.S.A.*, 56, 1606, 1966.
215. **Polgár, L.**, Modified preparation of thiolsubtilisins and their purification on agarose-mercurial column, *Acta Biochim. Biophys. Acad. Sci. Hung.*, 11, 81, 1976.
216. **Polgár, L. and Sajgó, M.**, Peptic peptide of thiolsubtilisin. Analytical evidence for the chemical transformation of the essential serine-221 to cysteine-221, *Biochim. Biophys. Acta*, 667, 351, 1981.
217. **Polgár, L. and Bender, M. L.**, Simulated mutation at the active site of biologically active proteins, *Adv. Enzymol. Relat. Areas Mol. Biol.*, 33, 381, 1970.
218. **Philipp, M. and Bender, M. L.**, Kinetics of subtilisin and thiolsubtilsin, *Mol. Cell. Biochem.*, 51, 5, 1983.
219. **Polgár, L. and Bender, M. L.**, The reactivity of thiolsubtilisin, an enzyme containing a synthetic functional group, *Biochemistry*, 6, 610, 1967.
220. **Tsai, I.-H., and Bender, M. L.**, Conformation of the active site of thiolsubtilisin: reaction with specific chloromethyl ketones and arylacryloylimidazoles, *Biochemistry*, 18, 3764, 1979.
221. **Philipp, M., Tsai, I.-H, and Bender, M. L.**, Comparison of the kinetic specificity of subtilisin and thiolsubtilisin toward n-alkyl p-nitrophenyl esters, *Biochemistry*, 18, 3769, 1979.
222. **Polgár, L.**, Conversion of the serine residue at the active site of alkalase to a cysteine side chain, *Acta Biochim. Biophys. Acad. Sci. Hung.*, 3, 397, 1968.
223. **Polgár, L.**, Transformation of a serine protease of *Aspergillus oryzae* into a thiol-enzyme, *Acta Biochim. Biophys. Acad. Sci. Hung.*, 5, 53, 1970.
224. **Yokosawa, H., Ojima, S., and Ishii, S.**, Thioltrypsin. Chemical transformation of the active-site serine residue of *Streptomyces griseus* trypsin to a cysteine residue, *J. Biochem.*, 82, 869, 1977.
225. **Polgár, L. and Fejes, J.**, Mechanism-controlled stereospecificity. Acylation of subtilisin with enantiomeric alkyl and nitrophenyl ester substrates, *Eur. J. Biochem. (Tokyo)*, 102, 531, 1979.
226. **Polgár, L. and Asbóth, B.**, The basic difference in catalyses by serine and cysteine proteinases resides in charge stabilization in the transition state, *J. Theor. Biol.*, 121, 323, 1986.
227. **Van Duijnen, P.Th., Thole, B.Th., and Hol, W. G. J.**, On the role of the active site helix in papain, an ab initio molecular orbital study, *Biophys. Chem.*, 9, 273, 1979.
228. **Hilgenfeld, R.**, personal communication.
229. **Asbóth, B. and Polgár, L.**, unpublished observations.

Chapter 5

ASPARTIC PROTEASES

I. THE MAJOR GROUPS OF ASPARTIC PROTEASES

The aspartic proteases were previously called acid proteases because of their optimal activities at pH 1.5 to 5. Then with the recognition that two carboxyl groups were essential for their activity, they were renamed carboxyl proteases. When the carboxyl groups were shown to belong to aspartyl residues in several members of this family, the preferred terminology was changed to aspartic or aspartyl proteases. The term aspartic proteases seems to be more consistent with the names of the other protease groups (serine, cysteine proteases) as compared to the previously used term acid proteases. Furthermore, the new term is more appropriate because some enzymes of this group act optimally on their substrates near pH 7.

As compared to the serine proteases, there are relatively few aspartic proteases. These include the gastric proteases (pepsin, gastricsin, chymosin), renin, cathepsin D, and proteases isolated from numerous fungi, some of which have been studied by X-ray crystallography, as will be discussed later in this chapter. Comprehensive reviews on aspartic proteases have been published.[1-3a]

A. Gastric Proteases

As the first enzyme to be discovered and the second enzyme to be crystallized, pepsin has a long history. The term pepsin includes several gastric proteases that arise from pepsinogens. Excellent reviews concerning the general features of pepsin and other gastric proteases are available.[1,3,4]

Chromatographic analysis[5] of the extracts of porcine gastric mucosa has shown that the main component, pepsin A, is accompanied by pepsin C, also denoted gastricsin, as well as by minor components, pepsin B and pepsin D.

The amino acid sequence of pig pepsin A[6] contains 327 residues, and the relative molecular mass calculated from this sequence is 34,644. The enzyme possesses only three strongly basic amino acid residues (one Lys, and two Arg) and an overwhelming predominance of side chain carboxyl groups. This is consistent with the low pI of pepsin A, which is between pH 2 and 3 as determined by the isoelectric focusing technique.[7,8] Pig pepsin A contains one phosphoryl group per molecule,[9] which is linked to Ser 68. This phosphoryl group is absent from both pig pepsin D and its zymogen,[10] which are actually dephosphorylated pepsin A and pepsinogen A, respectively.

When activated, pig pepsinogen A releases a 44-residue N-terminal peptide. The amino acid sequence of this fragment has been determined.[11] The peptide contains most of the Lys and Arg residues of the zymogen, which accounts for its higher isoelectric point relative to that of the active enzyme. In addition to pig pepsinogen A, the corresponding zymogens of man,[12] monkey,[13] and chicken[11] have also been sequenced; that of human pepsinogen A was determined through nucleotide sequence analysis. The homology between the human, monkey, and pig zymogens is high (>82%), but it is lower between the mammalian and the chicken zymogens (>62%).[13] It is noteworthy that chicken pepsin has more Lys and Arg residues than the mammalian enzymes. Thus, its higher pI may account for its stability above pH 6, whereas the pig enzyme is rapidly denatured.[14] Another distinctive feature of chicken pepsin is its glycoprotein nature. The protein molecule contains three mannose and seven glucosamine units.[15]

The zymogen of pepsin C or gastricsin, which is the currently preferred name, has also been sequenced.[16] Figure 1 shows the sequence data in comparison with four pepsinogens A of different species and with bovine prochymosin. The phylogenic tree (Chapter 2, Section II) illustrated in Figure 2 indicates that the divergence of progastricsin from the ancestral gastric proenzymes occurred prior to the divergence of prochymosin and pepsinogen A.

Prochymosin, the zymogen of chymosin, is encountered in the gastric juice of fetal and newborn mammals. Chymosin is also called rennin because it is the main protease component of rennet, the calf stomach extract used in cheese making.[17] Unfortunately, rennin may be confused with renin, another aspartic protease (see below and Chapter 2, Section III, Figure 7). Therefore, the term chymosin is preferred currently.[4] The amino acid sequence of prochymosin has been determined[18] (Figure 1). In contrast to the mammalian pepsin A, chymosin contains more Lys and Arg residues.

Because of its involvement in the manufacture of cheese, chymosin is of great practical importance. Its production in *Escherichia coli* has recently been accomplished using recombinant plasmids containing the preprochymosin coding sequence.[19] Chymosin is responsible for the limited proteolysis of κ-casein, a glycoprotein fraction of casein. The selective cleavage removes a soluble glycopeptide required to stabilize the casein micelle.[20]

B. Renin and Cathepsin D

Renin is a highly specific protease that acts on the α-globulin angiotensinogen, cleaving the N-terminal decapeptide angiotensin I [1] from its large substrate.[21,22] Angiotensin I is then hydrolyzed by a "converting enzyme" to yield the C-terminal dipeptide and the octapeptide angiotensin II, which is one of the most potent vasoconstrictors known.

$$\text{Asp-Arg-Val-Tyr-Ile-His-Pro-Phe-His-Leu}$$

$$[1]$$

Renin functions in the blood plasma, but is purified primarily from kidney and the submaxillary gland where it is most abundant. The kidney renin is glycosylated and thermostable, whereas the submaxillary gland renin is unglycosylated and thermolabile. Chromatography using pepstatin as an affinity ligand[22-23a] is the key step in most of the procedures for renin isolation. Its pH optimum is around neutrality, i.e., higher than that of most aspartic proteases.[22]

The amino acid sequence of human and mouse renin has recently been determined.[24-26] A considerable homology with pepsin was found. The entire human[27,28] and mouse[28a] renin genes have also been cloned and their organizations have been established.

Renins are species specific. Subprimate renin is incapable of cleaving human angiotensinogen although human renin can act on substrates from a variety of sources. The scissile bond is Leu[10]-Val[11] in the human enzyme and Leu[10]-Leu[11] in equine renin. The N-terminal tetradecapeptide of angiotensinogen is a good substrate of renin. The smallest peptide portion still hydrolyzed by the protease at an appreciable rate is the octapeptide containing residues 6 to 13 although the cleavage is slow enough to allow the octapeptide to function as a weak ($IC_{50} = 0.2$ mM) competitive inhibitor of renin.[29] Reduction of the scissile -CONH- peptide bond to -CH_2NH- in the octapeptide and in its analogs resulted in the formation of potent inhibitors of renin that may be used in the control of hypertension. An analog containing the reduced Leu[10]-Val[11] scissile peptide bond is highly specific for human renin ($IC_{50} = 10$ nM), being 1000 times more effective than against the dog enzyme ($IC_{50} = 10$ μM).[30] It is possible that the reduced peptide linkage assumes a transition-state conformation and is therefore bound to the enzyme more tightly than the substrate. Transition-state analogs containing a hydroxyethylene group (-$CHOHCH_2$-) instead of the scissile peptide bond,[30a] and those having an unusual amino acid (statine, see Section III.C) at the S_1 subsite[30a,30b]

```
                 1
Monkey   A   IIYKVPLVRK KSLRRNLSEH GLLKDFLKKH NLNPASKYFP QAEAPTLIDE QPLENYLDVE YFGTIGIGTP AQDFTVIFDT GSSHLWVPSV YCSSLACTNH
                                                                                                                    100
Human    A   -M-------- ------T--R ---------- ---------- ---W---I-- ---M------ ---------- ----------V ---------- -S--------
Porcine  A   .LV------- ---Q-IKD-- -K---T-KH- --------PY E-.-A-+GD- E--T------ ---------- -------Q--S ---------I -K-S--S-
Chicken  A   S-HR----- ---KQ-KO-- -E------- ---------- VLT.ATESY E-MT-M-AS -Y--S---- ----------- --------- -K-N-K-

Bovine   Prochy  AE-TRI--YKG --KA-K----- ---E--Q-QYGIS --SG --FG...EVAS V--T---SQ -K-YL---- P-E--L---- ----------DF -K-N-K-
Monkey   Progas  AVV-----KKF --I-ETMK-K --GE-RT--K --KYD-W-HF GD...LSVSY E-MA-M-AA -E-S----- -----P-N-L-L ---------- -Q-Q--S-

                                                                          150
         NLFNPQDSST YQSTSGTLSI TYGTGSMTGI LGYDTVQVGG ISDTNQIFGL SETEPGSFLY YAPFDGILGL AYPSISSSGA TPVFDNIWDQ GLVSQCLFSV
MA
H    -R---E---- ----E-V---                                                                   -N-
P    -Q---D----                            FEA-QE---                          -A---- L---        -H-
C    KR-D-RK-- -V-NE-VY- -A----- -------A-SS -DVQ---- ----------F- -CN------ --LA-EYS I--- -MMS--H-
                                                                                                          200
B    QR-D-RK-- --F-NLGKP- ---H----Q---                          -T-SN-V-IQ-TV- -TQ-DVFT -E---M----- -LA-EYS I------MMNR H-
MP   SR---SE--- --STNGQ-F-L Q-S--L--F F---LT-QS -QVP--E--- --N-TNFV -Q---M--- -TL-VD- -TAMQGMVQE -ALTSPI-

                                                    250
                                                      i                                                       300
         YLSADDQ SGS VVIFGGIDSS YYTGSLNHVP VSVEGYWQIS VDSITMDGEA IA CAEGCQAI VDTGTSLLTG PTSPIANIQS DIGASENSDG EMVVSCSAIS
MA
H    --SN-0--- ----LL---                -T----- --T------- --N------ ---SG------                                 -0-----S-0
P    -K-GE T-- --F-L---PH- -T-KGIY--- -L-A-T--- -T M-RV-VGHKY V--FFT -A---------                            -.-.I--DD-
C                                               VM -QGAYNR-IK -L-V-.-- -...-
                                                                                     *
B    -MDR-G- E.- MLTL-A-P-- ----H----T-QQ-FT -V-IS-VV V--EG--- L---K-V--S-D-L-Q A-TQ-QY- -FDID-DHL-
MP   -DQQGS--G A--V--V--- L---QIY-A--TQ-L--- G IEEFLIG-Q- SGW-S---- --V-QQYMSALLQ AT-Q-DEY- QFL-N-NS-Q

                                     iii               350
         SLPDIVFTIN GIQYPVPPSA YILQSQGSCT SGF QGHDVPT ESGELWLGD VFIRQYFTVF DRANNQVGLA PVA
MA                                                                              373
H    -----V---- ---E--I--- ---NL------- -- S---------- -Y------- -K--------
P    K--VT-H-- -HAFTL-A- -V-NED-M L-- -EN-GT--L-Q E-YVI---- --K--S -LS
C    YM-TV-E-- -KH-LT--- ---TS-D-F- -SEN...- N-.QK------ E-YS------- ----L--- KAI
B    N--TLT-I-- -VEF-L---S ----NNN-V-- V-VPTYELSAQNS QPLYY---- --L-S-YS-Y -LS-R-F- TA-
MP
```

FIGURE 1.   Comparison of the amino acid sequences of the Japanese monkey, human, porcine, chicken pepsinogen A, bovine prochymosin, and Japanese monkey progastricsin. The numbering employed here is based on the sequence of monkey pepsinogen A. (In an alternate numbering system the propeptide and the active enzyme are numbered separately, and the numbers for the propeptide are marked with a letter P.) The residues common to monkey pepsinogen A are shown by bars. The dots indicate deletions. The asterisks designate the active site aspartic residues. The vertical dashed lines show the boundaries between the propeptides and the active enzymes. In monkey progastricsin, five residues are inserted as compared with monkey pepsinogen A, and these positions are designated with letter i. (From Kageyama, T. and Takahashi, K., *J. Biol. Chem.*, 261, 4406, 1986. With permission.)

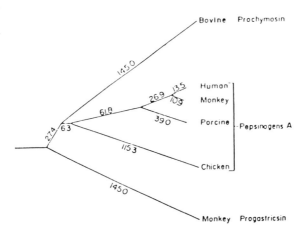

FIGURE 2.   A phylogenic tree of gastric aspartic proteases. The branch lengths are proportional to the values shown. (From Kageyama, T. and Takahashi, K., *J. Biol. Chem.*, 261, 4406, 1986. With permission.)

also proved to be powerful inhibitors. The problems of designing inhibitors to renin has recently been discussed.[30d]

Whereas renin functions in the plasma, cathepsin D exerts its action intracellularly. It is a lysosomal enzyme involved in the breakdown of proteins.[2] Cathepsin D is readily isolated from mammalian spleen, in which it is most abundant,[31,32] but it can also be purified from other sources, e.g., brain.[33,34] As a lysosomal enzyme, cathepsin D is a glycoprotein.[2,31] It contains mannose-6-phosphate residues which are required for the selection of proteases to be packaged in the lysosomes following their biosynthesis.[35,36] Because of its oligosaccharide content, cathepsin D binds to a concanavalin A Sepharose column,[32] which can be utilized in the purification procedure.

Cathepsin D is composed of several isoenzymes.[31,34] Like the cysteine protease cathepsins (Chapter 4), it occurs in single-chain and two-chain forms, the latter having a light (15 kdaltons) and a heavy chain (35 kdaltons).[2,31] It had been suggested that the polypeptide chain of cathepsin D is larger than that of pepsin and other aspartic proteases, the former containing a unique hydrophobic tail made up of about 100 residues added to the C-terminal portion.[2,31] However, the amino acid sequence of porcine spleen cathepsin D[37] and the sequence of the cDNA for the human enzyme[38] have shown a high degree of homology between cathepsin D and other aspartic proteases. The amino acid sequence predicted from the cDNA sequence shows that human cathepsin D consists of 412 amino acid residues, with 20 and 44 amino acids in the pre- and prosegment, respectively. The mature protein region shows an 87% amino acid identity with porcine cathepsin D, but differs in having nine additional amino acids. Two of these are at the C-terminus; the other seven are positioned at residues 98 to 104, i.e., between the heavy and light chains of the porcine enzyme. The sequence predicts N-glycosylation sites (-Asn-Xaa-Thr-) at positions 70 and 199, both of which have been found to be glycosylated.[39,40]

C. Microbial Proteases

Microbial aspartic proteases have been isolated from a number of fungi.[3] Their primary structures are highly homologous to those of the mammalian enzymes, indicating a common evolutionary origin.[40a-40c] Penicillopepsin, produced by *Penicillium janthinellum* has been sequenced and its three-dimensional structure has been determined by X-ray crystallography.[41] Proteases from *Endothia parasitica* and *Rhisopus chinensis* have also been analyzed by X-ray crystallography.[42] In all cases, the peptide chains are folded in the same way as found with pepsin.[43] The steric structures of aspartic proteases will be discussed in detail in Section V.

There is a considerable interest in the industrial application of microbial aspartic proteases. Increasing cheese production with a concomitant decrease in slaughtering milk-fed calves has led to a shortage of rennet and a search for rennet substitutes. Many acid proteases are capable of coagulating milk, but most are unsuitable as rennet because they are less specific than chymosin, thereby producing defects in the flavor and texture of the cheese produced. At present, only porcine and bovine pepsins, and proteases from *Mucor miehel, M. pusillus* and *Endothia parasitica* are used in cheese making.[44] Microbial proteases[45] and their substrate specificities[46] have been reviewed.

The most spectacular recent discovery concerns the aspartic proteases encoded by viral genome.[46a,46b] Retroviruses, such as the AIDS virus, code for a polyprotein precursor. The protease releases itself from this polyprotein autocatalytically and cleaves the polyprotein at specific sites, namely between a hydrophobic residue and a proline residue,[46c] to provide viral structural proteins. Retroviral proteases are much smaller than is typical for proteases of living cells. It appears that the amino acid sequence of a viral protease corresponds to a single domain of the bilobal aspartic proteases (Section V.A), and the active enzyme may be a dimer.[46a]

II. ZYMOGEN ACTIVATION

In common with the other extracellular proteases of vertebrates, gastric proteases are secreted in a zymogen form. However, the activation of gastric proenzymes is substantially different from the activation of the pancreatic zymogens. We have discussed in Chapter 3 (Section II.B) that the pancreatic proenzymes have negligible activity, and full activity is generated by limited proteolysis, brought about by other proteases. By contrast, fully active enzymes are formed from gastric zymogens at low pH, which elicit a conformational change followed by the removal of the prosegment.[1-4,17]

In his pioneering work on the activation of pig pepsinogen, Herriott[47] demonstrated that at pH 4.0 to 4.6 the reaction is autocatalytic, i.e., pepsin catalyzes its own production. He also observed that the activation is much faster at about pH 2. Detailed investigations have shown that the activation is intramolecular, and the rate is independent of the pepsinogen concentration.[48-51] Under these conditions, the carboxyl groups of the protein become protonated, and the basic amino terminal of the zymogen is released from the surface, leading to conformational change and enzymic activity. This is followed by limited proteolysis at sites which depend on the amino acid sequence of the zymogen and the specificity of the enzyme. In pig pepsinogen, for example, the cleavage takes place between Leu[16] and Ile.[17] (Figure 1 in Section I.A., three deletions may be noted in the prosegment of the porcine zymogen.) The resulting intermediate is active and has been named pseudopepsin. The intramolecular formation of pseudopepsin is rapidly followed by an intermolecular cleavage between Leu 44 and Ile 45, yielding the final product.[51] This mechanism which occurs at pH 2 is probably favored under physiological conditions.[52,53] Notably, pepsinogen is secreted together with HCl in the narrow lumen of the oxyntic glands,[4] providing the appropriate low pH for rapid activation. The pH in the mammalian stomach may rise to about 4 after a meal which permits only a much slower activation.

The occurrence of a conformational change with the conversion of pepsinogen to pepsin is supported by fluorescence studies.[54-56] Thus, upon activation of pepsinogen at pH 2.35 in the presence of Mns-Phe-Phe-OP4P (Mns = mansyl [2]; OP4P = 3-(4-pyridyl)propyl-1-oxy [3]), the mansyl fluorescence greatly increases, indicating that the newly formed binding site becomes accessible to the fluorescent probe. The change in mansyl fluorescence is abolished by the addition of pepstatin, a strong inhibitor binding at the active site.[54]

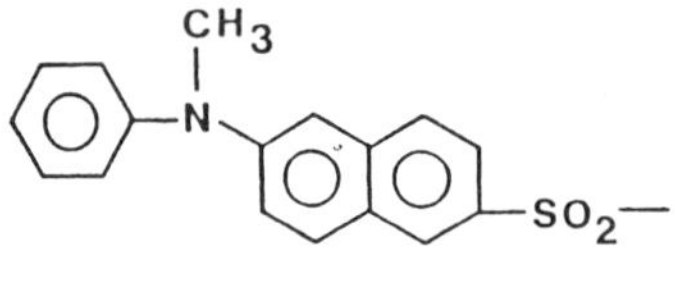

[2]    [3]

It should be noted that the mechanisms of activation of the gastric proteases are not identical. The activation of calf prochymosin at pH 2 occurs predominantly as an intermolecular, second-order reaction.[57] This reaction leads to pseudochymosin through cleavage of the bond linking Phe 27 and Leu 28. (Figure 1 in Section I.A; the addition of one amino acid residue to the N-terminus may be noticed.) No other bond is hydrolyzed at this pH. If the reaction is conducted near pH 5, activation is still autocatalytic but occurs at a different site and yields an N-terminal Gly (see Figure 1 in Section I.A). The formation of N-terminal Gly is surprising in light of the known specificity of the aspartic proteases (see Section III) which leads to a preferential splitting of peptide bonds that are flanked by hydrophobic residues. However, pig pepsin and porcine chymosin differ considerably in their secondary specificities, thus accounting, in part, for the unusual cleavage site in prochymosin at pH 5.[3] It is interesting that the activation process with chicken pepsinogen resembles that for prochymosin rather than that for pig pepsinogen.[58]

The aspartic protease renin, which is active at neutral pH (Section I.B), has evidently evolved a different mode for the activation of its zymogen. The amino acid sequence of prorenin shows that the cleavage site for zymogen activation occurs after a pair of basic amino acid residues (Arg-Arg).[26] This is the recognition signal for the activation cleavage of many peptide hormones (Chapter 2, Section III). Hence, the activation of prorenin is probably carried out by a specialized protease under physiological conditions.

Pig pepsin has been shown to be inhibited by the peptides liberated from the amino terminus of its zymogen.[59] Synthetic peptides derived from the prosegment of mouse submaxillary renin[60] and human renal renin[61] inhibited these enzymes in the micromolar range. This phenomenon, however, is not common to all aspartic proteases. Thus, calf chymosin is not inhibited by its prosegment peptides.[57]

III. SUBSTRATE AND INHIBITOR BINDING

A. Primary Specificity

The comprehensive investigation of pepsin specificity has been summarized by Fruton.[1,3,62,63] In contrast to the specificity of serine and cysteine proteases discussed in the preceding chapters, the aspartic proteases express their primary specificity on both sides of the scissile bond. Hydrophobic residues, in particular aromatic residues, are preferred as the P_1 and P_1' amino acids. This was known from early studies of the action of pepsin on small substrates such as acetyl-Phe-Tyr or acetyl-Phe-Phe.[64] The pH optimum for the hydrolysis of the acetyl dipeptides is near pH 2, and it is shifted into the range between 3 and 4.5 upon esterification of the free carboxyl group.[65] The neutral substrates are sparingly soluble in aqueous solution, so that it is difficult to carry out reliable kinetic studies under these conditions. Addition of organic solvents, although increasing peptide solubility, inhibits the action of pepsin.[66]

To overcome solubility problems, peptides containing cationic groups have been employed. One such type of substrate has an His residue at the P_2 position,[67-70] as in Z-His-Phe-Phe-OMe, where Z is a benzyloxycarbonyl group. In another type, exemplified by Z-Phe-Phe-OP4P, the C-terminal carboxyl group is esterified by a pyridylpropyloxy group, OP4P, shown in Structure [3] in the Section II.[71,72] The *N*-methyl pyridinium group can also be used in place of the pyridinium group.[73]

A simple method for monitoring hydrolysis by pepsin was introduced by using substrates containing a p-nitro-L-phenylalanine residue, Phe(NO$_2$) instead of Phe at the P$_1$ position.[68] The release of the acidic product, e.g., Z-His-Phe(NO$_2$)-OH, can be followed spectrophotometrically at 310 nm. This is more convenient than performing the ninhydrin or fluorescamine reactions subsequent to the liberation of the amine product.

Variation of the residues in the peptide substrates discussed above, has shown that pepsin exhibits an absolute requirement for L-amino acids at positions P$_1$ and P$_2$.[68] Substrates possessing D-Phe at either position are competitive inhibitors with k$_i$ values similar to the K$_m$ for the reaction of the corresponding L-compound. This indicates that K$_m$ approximates the dissociation constant of the enzyme-substrate complex (Chapter 2, Section VIII.A).

It was also shown that substrates with Val or Ile at the P$_1$ position are completely resistant to the action of pepsin,[69] indicating steric hindrance by the branched side chain. On the other hand, the P$_1'$ position is not sensitive to those residues with branching at the β-carbon.[68,69] Moreover, the S$_1'$ site can accept the large Trp ring which cannot easily be accommodated at the S$_1$ subsite.[68,69]

Besides peptide bonds, pepsin is capable of hydrolyzing ester bonds.[68] Thus, replacement of Phe in Z-His-Phe(NO$_2$)-Phe-OMe by a β-phenyl-L-lactyl (Pla) residue gives a depsipeptide, which is hydrolyzed at the Phe(NO$_2$)-Pla bond, more rapidly than the parent peptide substrate. The enantiomeric specificity is also important in the case of the ester bond cleavage; L-Pla cannot be replaced by D-Pla.

Several organic sulfites are also hydrolyzed by pepsin.[74-76] The kinetics of the hydrolysis of bis-p-nitrophenylsulfite have been investigated.[76]

The above specificity studies are concerned with porcine pepsin A, and significant differences have been found with other proteases of the pepsin family. Thus, gastricsin has a preference for Tyr at the P$_1$ position,[77] and exhibits poor activity against acetyl-L-phenylalanyl-L-diiodotyrosine, Ac-Phe-Tyr(I$_2$), which is a good substrate of pepsin A.[78] Chicken pepsin also fails to hydrolyze Ac-Phe-Tyr(I$_2$) at an appreciable rate.[15] Chymosin was shown to hydrolyze Z-Glu-Tyr, a dipeptide containing an acidic residue at the P$_1$ position.[79] On the other hand, the microbial aspartic proteases[46] can cleave substrates with the basic Lys at the same P$_1$ position.[80-82]

B. Secondary Specificity

Interactions of polypeptide substrates with the secondary binding sites of the enzyme can effectively promote the hydrolysis by aspartic proteases.[3,63] Similar effects have been observed with certain serine proteases, such as elastase, but not with trypsin (Chapter 3, Section III). Whereas with elastase and the related serine proteases, the S subsites appear to be more important than are the S′ subsites, in the catalysis by aspartic proteases, the S and S′ subsites are equally important. Since in the interaction between oligopeptides and proteases nonproductive binding may also occur, k$_{cat}$/K$_m$, which is independent of this effect, is the most meaningful rate constant when the specificity data are compared (Chapter 2, Section V).

A few characteristic results of the comprehensive investigations on the secondary specificity of pepsin are shown in Table 1 and 2. Table 1 shows the contribution by the P residues at constant P′ residues;[83,84] Table 2 demonstrates the effects of variation in the P′ residues at constant P residues.[70,85] It is seen that there are considerable rate increases upon addition of residues on either side of the scissile bond. For example, the k$_{cat}$/K$_m$ value for the hydrolysis of Phe-Gly-His-Phe(NO$_2$)-Phe-Val-Leu-OMe is about 6000 times greater than that for the cleavage of Phe-Gly-His-Phe(NO$_2$)-Phe-OMe (Table 2). It is of particular interest that the K$_m$, which in this case corresponds to the dissociation constant, K$_s$, varies much less than does the k$_{cat}$. It has been suggested that these secondary interactions affect the catalytic efficiency of pepsin by promoting the better positioning of the catalytic groups of the enzyme in relation to the sensitive peptide bond.[70] In effect, the binding energy arising from the

Table 1
SECONDARY INTERACTIONS AT THE S
SUBSITES OF PEPSIN[a83,84]

P₄, P₃, P₂	k_{cat} (sec^{-1})	K_m (m*M*)	k_{cat}/K_m (m*M*$^{-1}$ sec^{-1})
Z-Gly-	3.1	0.36	8.6
Z-Gly-Gly-	71.8	0.42	179
Z-Gly-Gly-Gly-	4.5	0.40	10.1
Z-Gly-Ala-	409	0.11	3720
Z-Ala-Ala-	282	0.04	7050

[a] The P_1-P_1' residues are constant: Phe-Phe-OP4P. OP4P stands for 3-(4-pyridinium)propyl-1-oxy and Z represents benzyloxycarbonyl.

Table 2
SECONDARY INTERACTIONS AT
THE S′ SUBSITES OF PEPSIN[a70,85]

P₂′, P₃′	k_{cat} (sec^{-1})	K_m (m*M*)	k_{cat}/K_m (m*M*$^{-1}$)sec^{-1}
-OMe	0.12	0.4	0.3
-Ala-Ala-OMe	28	0.16	175
-Val-Leu-OMe	62	0.04	1540

[a] The P_4-P_3-P_2-P_1-P_1' residues are constant: Phe-Gly-His-Phe(NO₂)-Phe. Phe(NO₂) is *p*-nitrophenylalanyl.

additional residues is utilized to decrease the activation energy of the reaction (Chapter 2, Section VII). Furthermore, the results have also shown that the binding site of pig pepsin has at least seven subsites; this extended active site has been estimated to be about 2.5 nm long.[84]

The secondary interactions are important in the catalysis by other aspartic proteases as well.[86] For example, chymosin hydrolyzes κ-casein (Section I.A); it cleaves specifically the Phe-Met bond in the sequence Leu-Ser-Phe-Met-Ala-Ile. Removal of the P_3 Leu or the P_3' Ile from the hexapeptide decreases considerably the value of k_{cat}/K_m.[87]

C. Pepstatin, a Transition-State Analog Inhibitor

As compared to the ubiquitous distribution of serine and cysteine protease inhibitors in various biological systems, naturally occurring inhibitors of the aspartic proteases are relatively uncommon. There is, however, an exceedingly potent aspartic protease inhibitor discovered by Umezawa and his associates.[88-90] This inhibitor, named pepstatin, has been utilized not only as an effective inhibitor but also as a transition-state analog in mechanistic studies. Also, it greatly facilitates the isolation of aspartic proteases as a specific adsorbent in affinity chromatography.

Pepstatin and its variants are produced by Actinomycetaceae and have the general form of acyl-Val-Val-Sta-Ala-Sta, where the acyl group is isovaleryl (as in the common pepstatin) or acetyl, and Sta stands for statine, a hydroxyamino acid, 3*S*-hydroxy-4*S*-amino-6-methylheptanoic acid (Structure [4]).

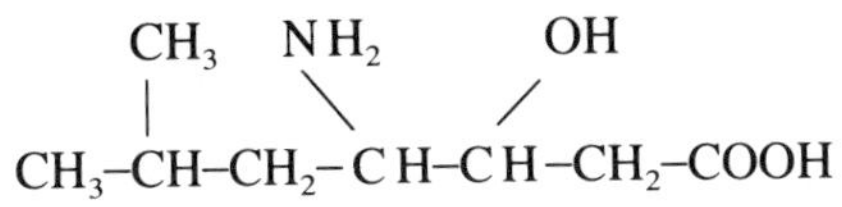

$$CH_3 \quad NH_2 \quad OH$$
$$CH_3-CH-CH_2-CH-CH-CH_2-COOH$$

[4]

Pepstatins are very effective inhibitors of most aspartic proteases. They exhibit the highest inhibitory activity against pepsin, with a K_i of about $10^{-10}M$.[91] Only aspartic proteases of *Scytalidium lignicolum* are known to be insensitive to pepstatin.[92] The extraordinarily tight binding between pepsin and pepstatin has led to the suggestion that the inhibitor mimics the tetrahedral intermediate for the peptide-bond hydrolysis.[93] Studies with a series of analogs of pepstatin[94-96] indicate that the C-terminal statine residue is relatively unimportant, and that the Val-Sta-Ala portion together with the 3*S*, 4*S* configuration is of primary importance in the inhibitory action. For example, dideoxypepstatin lacking the 3*S* hydroxyl groups on the two statine residues was found to be more than 4000 times weaker an inhibitor than pepstatin.[94] Upon binding of pepstatin, pig pepsin undergoes conformational changes as detected by NMR spectroscopy.[97]

As anticipated from its structure, pepstatin is very poorly soluble in aqueous solution. By introducing a hydrophilic lactyl as the acylation group, the resulting lactyl-Val-Sta-Ala-Sta tetrapeptide analog becomes much more soluble.[98] Lactyl-pepstatin retains activity against pig pepsin and gastricsin, but is less effective than isovaleryl- or acetyl-pepstatin against human gastricsin and calf chymosin.

The structural features of the pepstatin-protease complex will be discussed in Section V.

IV. THE CATALYTIC GROUPS

In mechanistic investigations of chymotrypsin, pH-rate profile studies were of great value in identifying the histidine residue as a catalytically competent group (Chapter 3, Section I.B). Most pH-dependence studies of pepsin catalysis have been carried out with protein substrates which were not suitable for such mechanistic investigations because the substrates themselves also changed with pH. Studies with small neutral substrates, e.g., Ac-Phe-Tyr-OEt, have shown bell-shaped pH dependences for k_{cat}/K_m.[99-102] The two pK_a values calculated from the curves were near 1 and 4, indicating that pepsin has two catalytically competent groups, probably two carboxyl groups, one acting as an acid (nondissociated form), the other as a base (dissociated form). It is interesting to note that the presence of a cationic group in the substrate (as in Z-His-Phe-Phe-OEt) altered the pH-rate profile so that the rate was pH independent between 1 and 3 and increased between 3 and 4.5.[103] This may not be a consequence of a dissociation of the imidazolium group since this is not expected to occur significantly over the pH range studied.

More direct evidence in favor of participation of one or more carboxyl groups in pepsin catalysis was obtained from chemical modification studies. The carboxyl groups were modified with several nonspecific reagents, which resulted in loss of enzymic activity.[1] Reactions with specific diazoketones, such as Structure [5], or diazoacetamido compounds, such as Structure [6], have shown that pepsin can be esterified on a single carboxyl group, resulting in complete inactivation of the enzyme.[93,94] The rate of this reaction is promoted by a cupric ion with both types of diazocompounds.[104] The L-enantiomer of Tos-Phe-CHN$_2$ (Structure [5]) reacted with pepsin much more rapidly than the D-compound. With ^{14}C-labeled reagent the incorporation of the tosyl-L-phenylalanyl group into pepsin was stoichiometric, but no incorporation was observed with pepsinogen.[104] The use of diazoacetyl-DL-norleucine methyl ester (Structure [6]), often abbreviated as DAN, permitted the determination of the incorporated norleucine by amino acid analysis.[105] In contrast to Compound [5], Compound [6] does not react stereospecifically, the D and L forms showing the same rate.[104] Compound [6] reacts with most aspartic proteases and therefore, is considered as a standard inhibitor of this group of enzymes.[3] The site of modification by the above diazocompounds was localized in the amino acid sequence Ile-Val-*Asp*-Thr-Gly-Thr-Ser,[106-108] the catalytic group later being identified as Asp 215 of pig pepsin A.[109]

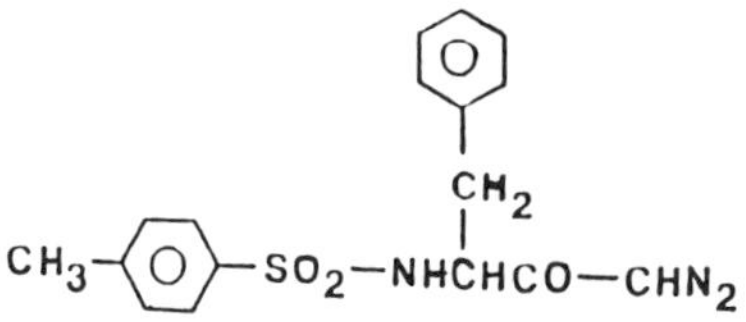

$$CH_3 - \langle O \rangle - SO_2-NHCHCO-CHN_2$$

with CH$_2$ substituent bearing a phenyl ring

[5]

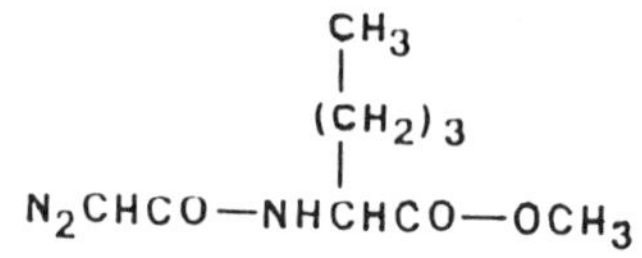

$$N_2CHCO-NHCHCO-OCH_3$$

with (CH$_2$)$_3$ chain terminating in CH$_3$

[6]

$$O_2N - \langle O \rangle - O-CH_2-CH-CH_2$$ (epoxide)

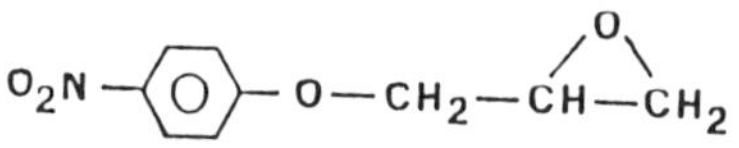

[7]

Epoxides represent another type of esterifying reagents. In particular, 1,2-epoxy-3(4-nitrophenoxy)propane (Structure [7], abbreviated EPNP) has been used as an inactivator of pepsin[110,111] and other aspartic proteases.[92,112] This compound is apparently less specific than the diazoacetyl and diazoketone derivatives because two peptides could be isolated with a labeled Asp residue. One included Asp 32, the other Asp 215.[109] These results are in good agreement with the concept of the two catalytically competent carboxyl groups.

V. TERTIARY STRUCTURE

A. Molecular Architecture

Although pepsin was the first enzyme for which an X-ray diffraction pattern was obtained,[113] the folding of the peptide chain was reported relatively recently.[43] At the same time, the three-dimensional structures of three microbial aspartic proteases were also published. These include penicillopepsin,[41] the *Rhizopus chinensis* protease,[42] and the *Endothia parasitica* protease,[42] all determined at medium resolution. High resolution structures were later published for penicillopepsin (0.18 nm),[114] *Rhizopus* pepsin (0.18 nm),[114a] and pepsin (0.20 nm).[115] The tertiary structures of pepsin and the microbial enzymes show a striking similarity to each other.

The structural features of aspartic proteases are illustrated in Figure 3. It is seen from a scheme of penicillopepsin that the molecule is bilobal with two domains separated by a pronounced cleft. The lobes are nearly equal in size. The polypeptide chain folds first into one lobe (N-terminal domain), and then into the other lobe (C-terminal domain). The two lobes are linked through a single piece of polypeptide chain and by considerable hydrophobic interactions. The entire molecule is predominantly composed of β-structures with little helical content (Figure 3). A general feature is the presence of several hairpin loops connecting antiparallel β-strands.

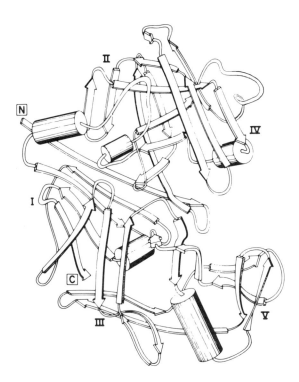

FIGURE 3. A scheme of the structure of penicillopepsin. Arrows represent β-strands, barrels depict the helical conformation of the chain. The amino (N) and carboxyl (C) termini are labeled. The five prominent regions of antiparallel β-sheet are designated with Roman numerals. Sheet I is the large interdomain structure comprising six antiparallel strands. Sheets II and III are related by the interdomain dyad. The two intradomain dyad axes pass between the central parts of these sheets. Sheet IV is the so-called flap which lies across the entrance of the active site cleft. High temperature factors of the atoms in the flap provide evidence for the mobility of this region of the molecule. Sheet V, consisting of four strands in the C-terminal domain, is extremely twisted. This sheet does not have a corresponding structure in the N-terminal domain. (From James, M. N. G. and Sielecki, A. R., *J. Mol. Biol.*, 163, 299, 1983. With permission.)

The most interesting feature of the molecular architecture is the twofold symmetry of the lobes. In pepsin and penicillopepsin, 62 and 70 pairs of residues, respectively, are in topologically equivalent position.[114,115] It has been suggested that the primordial gene for this two-domain structure had arisen by gene duplication.[116] Through evolution the gene duplication has been blurred in most parts of the primary structure, but maintained in the tertiary structure. However, a reminiscence of the gene duplication is still seen from the sequences around the two catalytically competent Asp residues (Figure 4). This homology had suggested a role for gene duplication in the evolution of aspartic proteases even before their steric structures became known.[109] This evolutionary mechanism has also received support from the determination of the nucleotide sequences of human[28] and mouse[28a] renin genes organized in two homologous clusters of four exons.

In addition to the interdomain symmetry, an intradomain twofold symmetry axis has been also observed in each domain.[117,118] It is, therefore, possible that the first step in the evolution of aspartic proteases was the duplication and fusion of a small gene coding for a sequence of about 80 amino acid residues. A second gene duplication and fusion have led to the modern proteases containing about 320 to 330 amino acid residues.

B. The Active Site Region

The substrate binding cleft is large enough to accommodate polypeptides of about seven

```
Pig pepsin A        Ile-Phe-Asp-Thr-Gly-Ser-Ser-Asn-Leu

Calf chymosin       Leu-Phe-Asp-Thr-Gly-Ser-Ser-Asp-Phe

Penicillopepsin     Asp-Phe-Asp-Thr-Gly-Ser-Ala-Asp-Leu
                                  32
```

```
Pig pepsin A        Ile-Val-Asp-Thr-Gly-Thr-Ser-Leu-Leu

Calf Chymosin       Ile-Leu-Asp-Thr-Gly-Thr-Ser-Lys-Leu

Penicillopepsin     Ile-Ala-Asp-Thr-Gly-Thr-Thr-Leu-Leu
                                 215
```

FIGURE 4.   Comparison of the amino acid sequences around the two catalytically important Asp residues of aspartic proteases (Asp 32 in the N-terminal domain, and Asp 215 in the C-terminal domain, when using the pepsin numbering). The homology between the two segments is evident.

amino acid residues. This is consistent with the known specificity of aspartic proteases (Section III.B). The catalytic aspartyl residues provided by the N-terminal lobe and the C-terminal lobe, respectively, are located deep in the center of the cleft, in close proximity to each other and accessible to the solvent. The electron density maps suggest that the two carboxyl groups are hydrogen bonded. In addition, the carboxyl groups are involved in a number of other hydrogen-bonding interactions.[114,119] These interactions should help to stabilize the net negative charge associated with this system.[114] A water molecule, an oxonium ion, or an ammonium ion is bound to the two carboxyl groups in penicillopepsin[114] and *Endothia* pepsin.[119]

Several hydrophobic pockets are located around the catalytic groups. One of them in the C-terminal domain is probably formed by Ile 211(213), Phe 295(299), and Ile 297(301), both in penicillopepsin and pepsin, the numbering of pepsin residues being shown in parenthesis. This pocket, described as the S_1' binding site in penicillopepsin,[120] has some special features in pepsin.[115] Specifically, there is a segment above this site involving residues 292 to 298 which is absent from the fungal asparic proteases. This segment restricts the position of the Tyr 189 side chain that lines one wall of the pocket.

The S_1 binding site is located in the N-terminal domain and may include Tyr 75(75), Phe 112(111), and Leu 121(120) in penicillopepsin.[120] In pepsin, it contains additional hydrophobic residues, namely Phe 117 and Ile 130 (pepsin numbering), the latter being in close contact with Ile 120.[115]

Most of the aspartic proteases of fungal origin show specificity for a lysine residue at the P_1 position. Because the K_m for the hydrolysis of lysine peptides increases considerably at low pH where the carboxyl groups become protonated, it seems to be probable that the binding of lysine residues is controlled by negatively charged carboxyl groups(s).[120] These groups have been tentatively identified in penicillopepsin as Asp 115(114) and Glu 16(13).[120]

In the absence of X-ray crystallographic data, a model of the three-dimensional structure of renin has been constructed by using interactive computer graphics.[121] The model is based on the three-dimensional structure of *Endothia* pepsin and the primary structure of mouse submaxillary renin. The amino acids which make up the specificity pockets, in particular at S_1 and S_2, are mainly hydrophobic in renin as in the homologous enzymes.

As in the case of serine or cysteine proteases, the mode of substrate binding was inferred from inhibitor binding studies. Thus, X-ray crystallographic analysis of the binding of pepstatin (see Section III.C) to *Rhisopus* pepsin has clearly shown that the inhibitor binds in the cleft with an extended conformation.[122] The structure of the shorter pepstatin analog, isovaleryl-Val-Val-Sta-OEt, in a complex with penicillopepsin has also been determined.[123] Both complexes indicated conformational changes in the protein structure involving a flexible loop called a flap (Trp 71 to Gly 83 in penicillopepsin). It appears that accessibility of the cleft to pepstatin requires some opening of the flap to allow the central statine (P_1 residue) to bind, which is followed by a subsequent closing of the flap. This movement may play a role in forming an adjustable hydrophobic pocket which can be finely tuned to a particular P_1 residue, such as Leu, Phe, Tyr, or Trp. Hydrogen-bonding interactions between the P_2 residue and the flap (Asp 7 of penicillopepsin) have also been noted.[123] These bonds may be important for positioning the scissile bond properly with respect to the catalytic groups.

A different binding mode to pepsin was found with the substrate-like dipeptide, Phe-TyrI$_2$-OMe.[115] Although the diiodotyrosine ring was localized in the S_1' binding pocket, as expected, the phenylalanine pocket was different from the S_1 pocket observed in the fungal enzymes. This may be the consequence of an electrostatic interaction between the protonated free amino group of the dipeptide and the active site carboxylate group.

C. Pepsinogen

In Section II, we already discussed the extensive studies on zymogen activation of aspartic proteases. The most recent X-ray crystallographic investigations of the pepsinogen molecule have revealed the reasons for the catalytic incompetence of the intact zymogen.[124]

The structure of the pepsin portion of pepsinogen is similar to that of the native pepsin, except for the N-terminal 12 residues. In the active pepsin, the N-terminal 6 residues form a strand of a large six-stranded antiparallel β-sheet, a common feature of aspartic proteases (Figure 3). In pepsinogen, the first six residues of the propeptide occupy the same position, and the N-terminal Ile of pepsin moves about 4 nm to the opposite side of the molecule where most of the prosegment is concentrated. The N-terminal residues of pepsin and the prosegment are intimately associated, forming a disk-like inactivation domain which blocks access to the catalytically important aspartyl residues. The prosegment of pig pepsinogen consists of 44 residues (Leu 1P-Leu 44P) and many of them are involved in the formation of three short helices. Lys 36P has a pivotal role in the inhibition of the intact zymogen at high pH. The positively charged amino group of the side chain of this residue interacts electrostatically and through hydrogen bonding with the aspartate diad at the active site. In addition to this interaction, several other salt bridges contribute to the binding energy between the prosegment and the pepsin molecule.

In the inactivation domain, a cluster of charged groups is of particular note. The positively charged guanidinium groups of Arg 8P and Arg 308 stack with their planes approximately parallel at a van der Waals distance. This unusual interaction is stabilized by the close proximity of the carboxylate groups of Glu 13 and Asp 304. On lowering the pH of the pepsinogen solution, the carboxylate groups become protonated, which leads to the destabilization of the stacked guanidinium groups. As a consequence, the two guanidinium groups are electrostatically repelled, triggering a conformational change in the prosegment of the pepsinogen molecule. Activation then occurs with the intramolecular cleavage of the peptide bond Leu 16P-Ile 17P. This scissile bond is part of the first helix of the prosegment, and residues Leu 16P and Ile 17P must move about 1.8 nm to reach the catalytic aspartate residues. The activation is completed by an intermolecular cleavage step in which the peptide bond Leu 44P-Ile 1 is hydrolyzed by another active enzyme molecule.

VI. THE MECHANISM OF CATALYTIC ACTION

A. The Acyly- and Amino-Enzyme Hypotheses

It is apparent from the preceding sections that two carboxyl groups are implicated in the catalysis by aspartic proteases. It seems to be appropriate to recapitulate the most important pieces of evidence. (1) The bell-shaped pH dependence of k_{cat}/K_m for the pepsin-catalyzed reactions indicate that two groups, one with an apparent pK_a of about 1 and another with pK_a between 4 and 5, participate in the catalytic reactions. (2) The Asp 32 and Asp 215 residues of pepsin can be labeled with specific reagents such as diazoacetyl-DL-norleucine methyl ester or 1,2-epoxy-3-(*p*-nitrophenoxy)propane. (3) The carboxyl groups of Asp 32 and Asp 215 are located in the active site cleft, within a hydrogen bond distance.

The mode of action of the two carboxyl groups was a major issue for a long time and important questions are still to be solved. From the pH-dependence studies, it is clear that the two carboxyl groups must react in different dissociated forms: one in the ionized, and the other in the un-ionized form.[99] The low pK_a (about 1) associated with the latter species is readily explicable in terms of the proximity of the groups. The acid strengthening, effect is well-known in the case of small dicarboxylic acids,[125] e.g., oxalic or malonic acid. This effect arises from the dipole of the un-ionized carboxyl group, which stabilizes the negative charge of the carboxylate anion [8]. A similar effect in the stabilization of the negatively charged tetrahedral intermediate in the catalysis by serine proteases has already been discussed (Chapter 3, Section VI.A).

$$\overleftarrow{HO\overset{+}{O}C} \text{ mmm } COO^-$$

[8]

The second pK_a of a dicarboxylic acid is generally higher than that of a simple acid, such as acetic acid ($pK_a = 4.74$). This effect is associated with the fact that the negative charge on one carboxylate ion makes the other carboxylate ion more basic.[125] However, a pK_a higher than 4.8 was observed only in a few limited cases.[99] Moreover, the bell-shaped curve was rarely found to be regular. Indeed, the details of the pH-dependence studies are not firmly established and may be affected by the dissociation of other carboxyl groups located in the active site cleft. In this respect, pepsin represents a more difficult case than that of the serine proteases where more definitive data have been obtained (Chapter 3, Section I.B). A slight deviation from the normal dissociation curve has also been observed for papain (Chapter 4, Section II).

An additional question concerns which of the two carboxyl groups reacts as the anion and which as the protonated group. Medium resolution X-ray analyses have indicated that Asp 32 was extensively hydrogen bonded and less accessible than Asp 215. Thus, Asp 32 was assumed to be ionized when the enzyme was active.[41] However, completion of the refinement at high resolution has shown that the environments of the two groups are equivalent and related by a local twofold axis,[114,119] which does not permit preferential assignment of the proton to either carboxy group.

As for the catalytic role of the aspartic residues, transpeptidation reactions originally appeared to provide a clue. Unfortunately, the experimental data were misinterpreted, as we shall see later. We have already seen that chymotrypsin can catalyze transfer reactions not only to water (hydrolysis), but also to other acceptor molecules such as alcohols or amines (Chapter 3, Section I.A). Such transfer reactions are generally called transpeptidation or transamidation, and may be of two types: (1) acyl-transfer, in which the P residues of the substrate are transferred to the acceptor and (2) amino-transfer, in which the P′ residues are transferred to the acceptor molecule.

Acyl-transfer reactions have been shown to occur with chymotrypsin (Chapter 3, Section I.A). These reactions involve a nucleophilic attack by the acceptor molecule on the acyl-enzyme. The first example of an enzyme-catalyzed amino-transfer was demonstrated in pepsin catalysis.[126] Notably, the incubation of a substrate such as benzyloxycarbonyl-Glu-

Tyr with pepsin led to the formation of a significant amount of Tyr-Tyr. The amide derivative of the same substrate resulted in Tyr-NH$_2$ as the only ninhydrin-positive component. The results were interpreted in terms of the formation of an intermediate amino-enzyme through a series of reactions,[126] as shown by Equations 1a to 1c.

$$E + \text{Z-Glu-Try-OH} \rightleftharpoons E.\text{Tyr-OH} + \text{Z-Glu-OH} \tag{1a}$$

$$E.\text{Tyr-OH} + \text{Z-Glu-Tyr-OH} \rightleftharpoons E + \text{Z-Glu-Tyr-Tyr-OH} \tag{1b}$$

$$\text{Z-Glu-Tyr-Tyr-OH} \xrightarrow{E} \text{Z-Glu-OH} + \text{H-Tyr-Tyr-OH} \tag{1c}$$

The finding that acyl-dipeptides undergo transpeptidation reactions has been confirmed in numerous reports.[127-131] However, with some substrates no transpeptidation could be found.[130] The failure of transpeptidation was explained by a much higher rate of hydrolysis of the substrate relative to that of transpeptidation.[131]

Besides amino-enzyme formation, which involves a peptide linkage between the enzyme and substrate, acyl-enzyme formation through an acid anhydride bond was also considered. The demonstration of ^{18}O exchange upon incubation of pepsin and acylamino acids, such as acetyl-Phe, with H$_2$^{18}O[132,133] suggested an acyl-enzyme mechanism analogous to that operative in the case of chymotrypsin. However, the presumed acyl-enzyme could not be trapped with ^{14}C-labeled methanol.[134]

The acyl-enzyme hypothesis, although discounted in favor of the amino-enzyme intermediate,[99] received support from the studies showing that pepsin and penicillopepsin catalyze the transfer of the N-terminal Leu of Leu-Tyr-Leu according to Equations 2a to 2c.[135-137]

$$E + \text{Leu-Tyr-Leu} \rightleftharpoons E.\text{Leu} + \text{Tyr-Leu} \tag{2a}$$

$$E.\text{Leu} + \text{Leu-Tyr-Leu} \rightleftharpoons E + \text{Leu-Leu-Tyr-Leu} \tag{2b}$$

$$\text{Leu-Leu-Tyr-Leu} \xrightarrow{E} \text{Leu-Leu} + \text{Tyr-Leu} \tag{2c}$$

The above series of reactions was confirmed by using labeled tripeptide substrate that contained ^{14}C in the N-terminal residue and ^{3}H in the C-terminal residue.[138] The finding that 80 to 90% of the resulting Leu-Leu dipeptide was [^{14}C]Leu-[^{14}C]Leu is consistent with the formation of [^{14}C]Leu-pepsin that can react with another tripeptide substrate to form the tetrapeptide (Equation 2b), which is then hydrolyzed by a free pepsin molecule (Equation 2c). The remainder of the labeled Leu-Leu dipeptide was labeled with ^{3}H in both residues, which suggested the Tyr-Leu bond cleavage followed by an amino-transfer.[138] However, one cannot conclude from these data that the aspartic proteases operate through two alternative mechanisms involving either an acyl-enzyme or an amino-enzyme. This dilemma will be discussed in the next section.

B. Noncovalent Intermediate

The concurrent acyl- and amino-transfer reactions discussed in the preceding section were reinterpreted in terms of differential product release, implying that after cleavage of the peptide link, either half of the substrate may leave first,[138] avoiding the problem of a particular covalent intermediate. Indeed, the formation of covalent intermediates in protease catalysis was generally favored in that time. Even recently considerable efforts have been made to prove that all enzyme reactions, including pepsin catalysis, proceed through covalent catalysis.[139] Fruton was the first who questioned the formation of a covalent intermediate in the catalysis by aspartic proteases. He stated: "As regards the bond-breaking step in the cleavage of peptide substrates by pepsin, it seems necessary to critically re-examine the status of both the amino-enzyme and the acyl-enzyme hypotheses and to consider the possibility that no detectable covalent intermediate is involved in pepsin catalysis".[63] Indeed, experimental data inconsistent with a simple acyl- or amino-enzyme intermediate have accumulated. For example, it has been found that the ratio of the rates of transpeptidation and hydrolysis is not independent of the acyl portion (X-Phe) of the substrate, as would be expected on the basis of Equation 3.[140] Specifically, breakage of the Phe-Trp bond of X-Phe-Trp (X = acetyl, acetyl-Gly-Gly, benzyloxycarbonyl-His, benzyloxycarbonyl-Ala-His) would yield the amino-enzyme, E-Trp, which then would undergo either hydrolysis or trapping by radioactive acetyl-Phe (Ac-Phe*) to yield the transpeptidation product, Ac-Phe*-Trp. It is clear that the decomposition of such an amino-enzyme would be independent of the liberated acyl-portion of the substrate, which is contrary to the experimental finding.

$$
\begin{array}{c}
\text{E + Ac-Phe}^*\text{-Trp} \\
\uparrow \text{ Ac-Phe}^* \\
\text{X-Phe-Trp + E} \rightarrow \text{E-Trp + X-Phe} \qquad\qquad (3) \\
\downarrow \text{ H}_2\text{O} \\
\text{E + Trp}
\end{array}
$$

Additionally, no supporting evidence for the existence of covalent intermediates has been obtained, in spite of much effort in many laboratories. Thus, the putative acyl-enzyme intermediate could not be trapped by methanol[134] although this method proved to be effective in detecting acyl-enzyme formation in the case of chymotrypsin (see the added nucleophile method in Chapter 2, Section VIII.B). The observation of burst release of the acyl or amino product was also unsuccessful even by using stopped-flow[82,141] or subzero temperature spectrophotometric methods.[142,143]

Negative results, of course, cannot be considered as conclusive evidence against the existence of covalent intermediates. More convincing evidence was provided by ^{18}O incorporation studies by Antonov and his colleagues.[144,145] They proposed that if transpeptidation reactions are carried out in $H_2^{18}O$ enriched water, covalent and noncovalent intermediates can be distinguished by determining the amount of ^{18}O incorporated into the transpeptidation products. Equations 4 and 5 illustrate the two reaction pathways for covalent and the noncovalent catalysis, respectively. In these pathways ^{18}O is designated by ●.

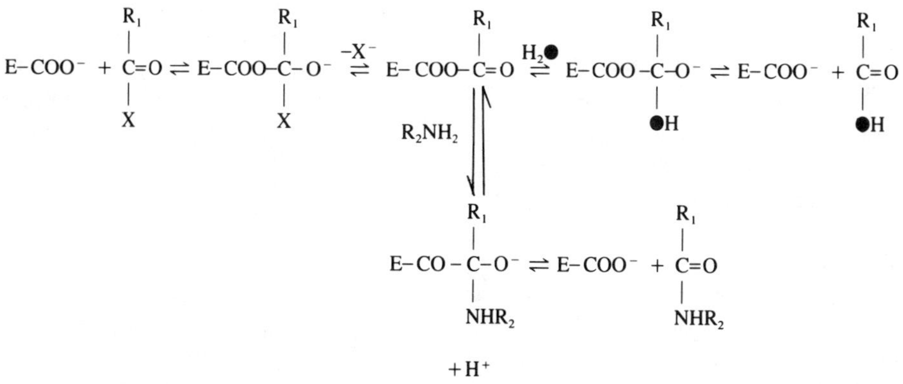

$$(4)$$

$$E-COO^- + H-\bullet + \underset{\underset{H}{|}}{\overset{\overset{R_1}{|}}{C}}=O \rightleftharpoons E-COOH.H\bullet-\underset{\underset{X}{|}}{\overset{\overset{R_1}{|}}{C}}-O^- \overset{-X^-}{\rightleftharpoons} E-COO^-.\underset{\underset{OH}{|}}{\overset{\overset{R_1}{|}}{C}}=\bullet \rightleftharpoons E-COO^- + \underset{\underset{OH}{|}}{\overset{\overset{R_1}{|}}{C}}=\bullet$$

$$R_2NH_2$$

$$E-COOH.^-\bullet-\underset{\underset{OH}{|}}{\overset{\overset{R_1}{|}}{C}}-NHR_2 \rightleftharpoons E-COO^- + \underset{\underset{NHR_2}{|}}{\overset{\overset{R_1}{|}}{C}}=\bullet \qquad (5)$$

The two mechanisms differ in the step that involves the water attack. In the covalent catalysis (Equation 4), water enters into the reaction after the formation of the acyl-enzyme, whereas in the other mechanism (Equation 5), which describes a general base catalysis by the enzymic carboxylate, the water attack occurs at the first step of the chemical reaction. It is clear that ^{18}O incorporation can take place only in the second case, where water attack precedes transpeptidation. This was observed experimentally.[144,145] The possibility that the transpeptidation product becomes labeled by a covalent acyl-enzyme that has previously been labeled through reversible hydration-dehydration seems to be improbable.[114,145]

Additional evidence against covalent intermediate formation during catalysis by the aspartate proteases has come from ^2H and ^{18}O isotope effects on ^{13}C NMR spectra of the enzyme-bound ketone analog of pepstatin.[146] From the observed chemical shifts it can be concluded that water, and not an enzymic nucleophile, adds to the peptide carbonyl to yield a tetrahedral diol adduct in the enzyme catalyzed reaction.

The noncovalent mechanism (Equation 5) requires a relatively stable complex between the enzyme and the substrate fragment that combines with the acceptor to form the transpeptidation product. This relatively stable complex transforms into a less stable one before dissociation into free enzyme and product.[145] Such a stepwise formation and decomposition of enzyme-substrate or enzyme-product complexes has been noted in several cases.[63,147-149] In fact, conformational flexibility of an aspartic protease has also been demonstrated by binding of pepstatin analog to penicillopepsin.[123] Probably the acyl fragment of the split substrate (P_1 and P_2 residues) is kept preferentially in place by new interactions made with some residues of the flap after its conformational change. The low yield of amino transpeptidation with penicillopepsin compared with the corresponding pepsin reaction may be attributed to the four-residue insertion in pepsin after Asn 290 relative to penicillopepsin. These residues form a second flap which could interact preferentially with the residues P_1', P_2' and P_3' of a substrate, thus providing an explanation for the enhanced efficiency of amino transpeptidation with pepsin.[123] This is an attractive structural description of the concept of ordered release of the split products in the absence of covalent intermediates.[63]

Further argument against the existence of covalent intermediates comes from X-ray crystallographic studies which suggest that for steric reasons the formation of covalent acyl- or amino-enzyme is improbable.[122,123] However, there is not universal agreement concerning the general base mechanism. It has recently been suggested that the available experimental data are not sufficient to exclude covalent catalysis.[150] Moreover, it has been speculated that an amino-enzyme may be formed from the acyl-enzyme if this anhydride intermediate is attacked by the amine product before it is released from the enzyme.[151] However, in the light of the tertiary structure of the active site of aspartic proteases, one argument against a covalent intermediate, and in particular against an amino-enzyme, may be put forth. The basic feature of the function of aspartic proteases is the concerted action of two enzymic carboxyl groups. Apparently, no other groups are directly involved in the catalytic reaction. If a peptide substrate were converted into an amino-enzyme, the resulting peptide bond would not offer a chemical advantage. Moreover, one carboxyl group would be lost for the catalysis. In other words, for the substrate hydrolysis, two carboxyl groups can be operative in the general base mechanism, whereas for the hydrolysis of the peptide bond of the amino-enzyme, only one carboxyl group is available, the other being engaged in the formation of the intermediate. In essence, the original catalytic potential of aspartic proteases would be seriously reduced by the formation of an amino-enzyme intermediate. The acyl-enzyme as an anhydride intermediate would be more reactive than the amino-enzyme, but here again, only one carboxyl group could be operative during the hydrolysis of the intermediate.

C. The Current Status

Numerous mechanistic pathways for the action of aspartic proteases have been suggested so far. A few inconsistent examples have recently been summarized.[150] The earlier proposals could not explain the stereochemical aspects of the catalysis, and even in the present-day mechanisms the limitations imposed by the stereochemistry are often disregarded. The mechanistic pathway to be described next involves a noncovalent, general base mechanism, which is based on the highly refined crystal structures of penicillopepsin[114] and of its complex with a pepstatin fragment at 0.18 nm resolution.[123]

We have seen in the case of serine protease catalysis (Chapter 3) that the cleavage of the peptide bond can be promoted by the enzyme groups at three sites. (1) Attack by a nucleophilic component at the carbonyl carbon of the substrate. This component is a serine hydroxyl group in the serine protease catalysis, and it is a water molecule in the hydrolysis by aspartic proteases. The nucleophilic attack is facilitated by general base catalysis in both cases. (2) Participation of an electrophilic component to enhance the polarization of the carbonyl bond of the substrate. This is attained by hydrogen bonding from the oxyanion binding site in the case of serine proteases and by general acid catalysis by a nondissociated carboxyl group in the case of aspartic proteases. (3) Proton donation to the nitrogen atom to make it a good leaving group. This is achieved by the protonated imidazole group of serine proteases, and presumably by a protonated carboxyl group of aspartic enzymes (see later discussion).

The binding of substrate as deduced from the binding of pepstatin and a pepstatin analog provides the clue to place the scissile peptide bond with respect to the catalytic groups, Asp 32 and Asp 215 (pepsin numbering).[120,122,123,152] The position of the scissile bond indicates that it is Asp 32 that catalyzes the nucleophilic attack by the water molecule (Figure 5A). Thus, in the enzyme-substrate complex, Asp 32 bears the negative charge and Asp 215 is protonated. A water molecule hydrogen bonded in the native enzyme to Asp 32 is ideally positioned for nucleophilic attack on the substrate carbonyl carbon atom.[152] This solvent molecule is part of a hydrogen-bonded network of several water molecules, which is disrupted on binding the substrate. Thus, it is not necessary that the water in the complex be exactly in the same position as in the native enzyme.

The general base-catalyzed nucleophilic attack by the water molecule leads to the formation of a tetrahedral intermediate, the process of which is facilitated by general acid catalysis by Asp 215. Indeed, the carbonyl oxygen of the scissile peptide bond is in a good position to accept the proton from the carboxyl group of Asp 215. As for the details of the formation of the tetrahedral intermediate, it was proposed[152] that the reaction is initiated by the protonation of the carbonyl oxygen of the substrate, and the subsequent nucleophilic attack by an OH^- ion leads to the formation of the tetrahedral adduct. However, it is not probable that the proton associated with the two carboxyl groups and held by the negative charge on them, would be transferred to the less basic neutral peptide carbonyl oxygen atom in the absence of a nucleophilic attack on the carbonyl carbon. It is even more unlikely that an OH^- ion would be generated in the acidic media where aspartic proteases usually operate. This stepwise mechanism would even be inconsistent with the general base catalysis, which requires the reactants and the catalyst to be involved in the *same* transition state (Chapter 1, Section V).

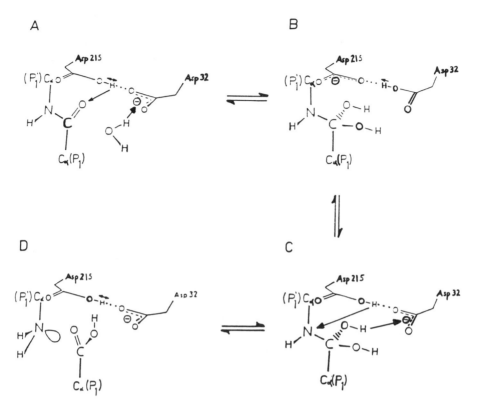

FIGURE 5. A proposed pathway for the catalysis by aspartic proteases. Proton transfers are indicated by arrows. Schemes B and C differ only in the position of the diad proton. This mechanism is a modification of that proposed in Reference 152.

The breakdown of the tetrahedral intermediate is even less clear than its formation. A major problem is the mode of protonation of the peptide nitrogen atom, which seems to be mandatory in order to render the amino part of the substrate a good leaving group. In the case of serine and cysteine proteases, protonation of the leaving group is accomplished by general acid catalysis by an imidazolium ion. As for the aspartic protease catalysis, it was originally thought that the proton donor was the phenolic hydroxyl group of Tyr 75.[153] Inhibitor binding studies with penicillopepsin[123] and with *Rhisopus* pepsin[122] have ruled out this possibility. It was also suggested that the backbone carbonyl of Gly 34 would be implicated in the proton transfer.[122] However, it is improbable for chemical reasons that a peptide carbonyl group would function as a general acid catalyst. Additionally, analysis of substrate binding based on highly refined coordinates indicated that the carbonyl oxygen of Gly 34 is most probably involved with substrate binding as a hydrogen bond acceptor from the -NH- group of P_2'.[120]

Two alternative mechanisms for protonation of the leaving nitrogen atom have recently been proposed.[152] One involves a proton transfer mediated by Asp 215 from the protonated carbonyl oxygen atom of the tetrahedral intermediate, the other involves protonation from the bulk water. In the former case, proton transfer is a two-step reaction: in the first step the proton is transferred to the carboxylate ion of Asp 215 with the resultant formation of a negative tetrahedral adduct; the second step is a general acid catalysis involving the proton transfer from Asp 215 to the P_1' nitrogen atom. The validity of this mechanism can be tested by considering the reverse reaction, i.e., the synthesis of the peptide bond. This reaction would imply general base catalysis by Asp 215 and the formation of a negatively charged tetrahedral intermediate, which would be followed by the proton transfer from Asp 215 to the negative oxygen atom. It is difficult to understand in such a mechanism why the proton hydrogen bonded between the two carboxyl groups is not transferred to the carbonyl oxygen while the covalent bond is formed. This proton must be released concurrently with the general base catalysis, otherwise, two protons would be associated with the diad which above pH 2 is inconsistent with the pK_a values of the carboxyl groups.

The other source for the protonation of the leaving group nitrogen may be the bulk solvent.[152] However, enzymes rarely utilize specific acid or specific base catalysis in their reactions. Proton uptake from the bulk water is a typical specific acid catalysis, apparently not aided by the enzyme. One may argue that aspartic proteases operate at low pH where the concentration of the oxonium ion is sufficiently high for an effective specific acid catalysis. However, this argument cannot be valid in the case of renin which has a pH optimum at neutrality. Furthermore, the bell-shaped pH-rate profile obtained for pepsin catalysis indicates the importance of the two carboxyl groups rather than the importance of specific acid catalysis. In addition, the principle of microscopic reversibility (Chapter 1, Section XIV) prescribes that specific base catalysis, the reverse of specific acid catalysis, must facilitate peptide synthesis. The reaction of hydroxide ions, however, could hardly be important at acidic pH.

The difficulties occasioned by these two alternatives for the protonation of the leaving nitrogen atom can be overcome by assuming different origins and timing for the proton transfers within the framework of the same heavy atom movements.[154] It is proposed that formation and decomposition of the tetrahedral intermediate are symmetrical processes. The symmetrical mechanism is consistent with the extraordinarily symmetrical structural details around the two carboxyl groups.[114,119] Due to this symmetry, the diad proton may be bound covalently to either carboxyl group with equal probability. The basic feature of the proposed mechanism involves two proton transfers concerted with bond making (formation of the tetrahedral intermediate) and bond breaking (decomposition of the tetrahedral intermediate). The concurrent proton transfers are indicated by arrows in Figure 5. In the formation of the tetrahedral intermediate, the proton of the nucleophile is accepted by the diad carboxylate ion, while the diad proton is donated to the carbonyl oxygen of the substrate (Figure 5A and 5B). This is essentially a "push-pull" general acid-base catalysis (Chapter 1, Section XIII). The breakdown of the resultant adduct proceeds by the same mechanism (Figure 5C and 5D). The proton from the protonated oxyanion is transferred to the diad, while the diad proton protonates the leaving nitrogen. In this mechanism the protonation of the nitrogen atom takes place not from the protonated oxyanion, as has been previously proposed,[152] but the proton stems from the nucleophile and is conveyed to the leaving group by the diad. This mechanism is similar to that of chymotrypsin, where protonation of the leaving group also takes place from the nucleophile with the help of a histidine residue. However, in serine protease catalysis, a single proton is transferred, whereas in the case of aspartic proteases, two concurrent proton transfers promote the formation of the tetrahedral adduct, as well as its breakdown.

In a recent paper, an alternative stereochemistry was proposed. This suggests that the attacking water molecule is positioned almost symmetrically with respect to the two carboxyl groups.[155] The push-pull catalysis is possible also in this case although the authors do not consider that a proton transfer would occur from the carboxyl group to the oxyanion of the tetrahedral intermediate. The proposed mechanism involves a hydrogen-bonded rather than a protonated transition state, in common with the serine protease catalysis. However, there is a principal difference in the chemistry of the catalyses by aspartic and serine proteases; namely the acidic carboxyl can readily protonate the oxyanion, whereas the more basic peptide -NH- group cannot.

REFERENCES

1. **Fruton, J. S.,** Pepsin, in *The Enzymes,* Vol. 3, 3rd ed., Boyer, P. D., Ed., Academic Press, New York, 1971, 119.
2. **Tang, J.,** Evolution in the structure and function of carboxyl proteases, *Mol. Cell. Biochem.,* 26, 93, 1979.
3. **Fruton, J. S.,** Aspartic proteinases, in *New Comprehensive Biochemistry,* Neuberger, A. and Brocklehurst, K., Eds., Elsevier, Amsterdam, 1987, chap. 1.
3a. **James, M. N. G. and Sielecki, A. R.,** Aspartic proteinases and their catalytic pathway, in *Biological Macromolecules and Assemblies* Vol. 3, Jurnak, F. A. and McPherson, A., Eds., John Wiley & Sons, New York, 1987, 413.
4. **Foltman, B.,** Gastric proteinases, — structure, function, evolution and mechanism of action, *Essays Biochem.,* 17, 52, 1981.
5. **Ryle, A. P.,** The porcine pepsins and pepsinogens, *Methods Enzymol.,* 19, 316, 1970.
6. **Sepulveda, P., Marciniszyn, J., Jr., Liu, D., and Tang, J.,** Primary structure of porcine pepsin. III. Amino acid sequence of a cyanogen bromide fragment, CB2A, and the complete structure of porcine pepsin, *J. Biol. Chem.,* 250, 5082, 1975.
7. **Jonsson, M.,** Isoelectric spectra of native and base denatured crystallized swine pepsin, *Acta Chem. Scand.,* 26, 3435, 1972.
8. **Vesterberg, O.,** Isoelectric focusing of acidic proteins. Studies on pepsin, *Acta Chem. Scand.,* 27, 2415, 1973.
9. **Perlmann, G. E.,** The phosphorus in pepsin and pepsinogen, *J. Am. Chem. Soc.,* 74, 6308, 1952.
10. **Lee, D. and Ryle, A. P.,** Pepsinogen D, *Biochem. J.,* 104, 735, 1967.
11. **Pederson, V. B. and Foltman, B.,** The amino acid sequence of a hitherto unobserved segment from porcine pepsinogen preceding the N-terminus of pepsin, *FEBS Lett.,* 35, 255, 1973.
12. **Sogawa, K., Fujii-Kuriyama, Y., Mizukami, Y., Ichihara, Y., and Takahashi, K.,** Primary structure of human pepsinogen gene, *J. Biol. Chem.,* 258, 5306, 1983.
13. **Kageyama, T. and Takahashi, K.,** The complete amino acid sequence of monkey pepsinogen A, *J. Biol. Chem.,* 261, 4395, 1986.
14. **Green, M. L. and Llewellin, J. M.,** The purification and properties of a single chicken pepsinogen fraction and the pepsin derived from it, *Biochem. J.,* 133, 105, 1973.
15. **Bohak, Z.,** Purification and characterization of chicken pepsinogen and chicken pepsin, *J. Biol. Chem.,* 244, 4638, 1969.
16. **Kageyama, T. and Takahashi, K.,** The complete amino acid sequence of monkey progastricsin, *J. Biol. Chem.,* 261, 4406, 1986.
17. **Foltmann, B.,** A review on prorennin and rennin, *C. R. Trav. Lab. Carlsberg,* 35, 143, 1966.
18. **Foltmann, B., Pedersen, V. B., Jacobsen, H., Kauffman, D., and Wybrandt, G.,** The complete amino acid sequence of prochymosin, *Proc. Natl. Acad. Sci. U.S.A.,* 74, 2321, 1977.
19. **Emtage, J. S., Angal, S., Doel, M. T., Harris, T. J. R., Jenkins, B., Lilley, G., and Lowe, P. A.,** Synthesis of calf prochymosin (prorennin) in *Escherichia coli, Proc. Natl. Acad. Sci. U.S.A.,* 80, 3671, 1983.
20. **Armstrong, C. E., Mackinlay, A. G., Hill, R. J., and Wake, R. G.,** The action of rennin on κ-casein: the heterogeneity and origin of the soluble product, *Biochim. Biophys. Acta,* 140, 123, 1967.
21. **Ondetti, M. A. and Cushman, D. W.,** Enzymes of the renin-angiotensin system and their inhibitors, *Annu. Rev. Biochem.,* 51, 283, 1982.
22. **Slater, E. E.,** Renin, *Methods Enzymol.,* 80, 427, 1981.
23. **Inagami, T. and Murakami, K.,** Pure renin. Isolation from hog kidney and characterization, *J. Biol. Chem.,* 252, 2978, 1977.
23a. **Do, Y.-S., Shinagawa, T., Tam, H., Inagami, T., and Hsueh, W. A.,** Characterization of pure human renal renin. Evidence for a subunit structure, *J. Biol. Chem.,* 262, 1037, 1987.
24. **Imai, T., Miyazaki, H., Hirose, S., Hori, H., Hayashi, T., Kageyama, R., Ohkubo, H., Nakanishi, S., and Murakami, K.,** Cloning and sequence analysis of cDNA for human renin precursor *Proc. Natl. Acad. Sci. U.S.A.,* 80, 7405, 1983.
25. **Misono, K. S., Chang, J.-J., and Inagami, T.,** Amino acid sequence of mouse submaxillary gland renin *Proc. Natl. Acad. Sci. U.S.A.,* 79, 4858, 1982.
26. **Panthier, J.-J., Foote, S., Chambraud, B., Strosberg, A. D., Corvol, P., and Rougeon, F.,** Complete amino acid sequence and maturation of the mouse submaxillary gland renin precursor, *Nature (London),* 298, 90, 1982.
27. **Hobart, P. M., Fogliano, M., O'Connor, B. A., Schaefer, I. M., and Chirgwin, J. M.,** Human renin gene: structure and sequence analysis, *Proc. Natl. Acad. Sci. U.S.A.,* 81, 5026, 1984.
28. **Miyazaki, H., Fukamizu, A., Hirose, S., Hayashi, T., Hori, H., Ohkubo, H., Nakanishi, S., and Murakami, K.,** Structure of the human renin gene, *Proc. Natl. Acad. Sci. U.S.A.,* 81, 5999, 1984.

28a. **Holm, I., Ollo, R., Panthier, J.-J., and Rougeon, F.,** Evolution of aspartyl proteases by gene duplication: the mouse renin gene is organized in two homologous clusters of four exons, *EMBO J.,* 3, 557, 1984.

29. **Skeggs, L. T., Lentz, K. E., Hochstrasser, H., and Kahn, J. R.,** Kinetics of the reaction of renin with nine synthetic peptide substrates, *J. Exp. Med.,* 128, 13, 1968.

30. **Szelke, M., Leckie, B., Hallett, A., Jones, D. M., Sueiras, J., Atrash, B., and Lever, A. F.,** Potent new inhibitors of human renin, *Nature (London),* 299, 555, 1982.

30a. **Kati, W. M., Pals, D. T., and Thaisrivongs, S.,** Kinetics of the inhibition of human renin by an inhibitor containing a hydroxy-ethylene dipeptide isostere, *Biochemistry,* 26, 7621, 1987.

30b. **Boger, J., Lohr, N. S., Ulm, E. H., Poe, M., Blaine, E. H., Fanelli, G. M., Lin, T.-Y., Payne, L. S., Schorn, T. W., LaMont, B. I., Vassil, T. C., Stabilito, I. I., Veber, D. F., Rich, D. H., and Bopari, A. S.,** Novel renin inhibitors containing the amino acid statine, *Nature (London),* 303, 81, 1983.

30c. **Cumin, F., Nisato, D., Gagnol, J.-P., and Corvol, P.,** A potent radiolabeled human renin inhibitor [^{3}H]SR42128: enzymatic, kinetic, and binding studies to renin and other aspartic proteases, *Biochemistry,* 26, 7615, 1987.

30d. **Blundell, T. L., Cooper, J., Foundling, S. I., Jones, D. M., Atrash, B., and Szelke, M.,** On the rational design of renin inhibitors: X-ray studies of aspartic proteinases complexed with transition-state analogues, *Biochemistry,* 26, 5585, 1987.

31. **Takahashi, T. and Tang, J.,** Cathepsin D from porcine and bovine spleen, *Methods Enzymol.,* 80, 565, 1981.

32. **Huang, J. S., Huang, S. S., and Tang, J.,** Cathepsin D isozymes from porcine spleens. Large scale purification and polypeptide chain arrangements, *J. Biol. Chem.,* 254, 11405, 1979.

33. **Hackenthal, E., Hackenthal, R., and Hilgenfeldt, U.,** Purification and partial characterization of rat brain acid proteinase (isorenin), *Biochim. Biophys. Acta,* 522, 561, 1978.

34. **Azaryan, A., Akopyan, T., and Buniatian, H.,** Cathepsin D from human brain: purification and multiple forms, *Biomed. Biochim. Acta,* 42, 1237, 1983.

35. **Hickman, S., Shapiro, L. J., and Neufeld, E. F.,** A recognition marker required for uptake of a lysosomal enzyme by cultured fibroblasts, *Biochem. Biophys. Res. Commun.,* 57, 55, 1974.

36. **Hasilik, A., Klein, U., Waheed, A., Strecker, G., and von Figura, K.,** Phosphorylated oligosaccharides in lysosomal enzymes: identification of α-N-acetylglucosamine(1)phospho(6)mannose diester groups, *Proc. Natl. Acad. Sci. U.S.A.,* 77, 7074, 1980.

37. **Shewale, J. G. and Tang, J.,** Amino acid sequence of porcine spleen cathepsin D, *Proc. Natl. Acad. Sci. U.S.A.,* 81, 3703, 1984.

38. **Faust, P. L., Kornfeld, S., and Chirgwin, J. M.,** Cloning and sequence analysis of cDNA for human cathepsin D, *Proc. Natl. Acad. Sci. U.S.A.,* 82, 4910, 1985.

39. **Hasilik, A. and von Figura, K.,** Oligosaccharides in lysosomal enzymes. Distribution of high-mannose and complex oligosaccharides in cathepsin D and β-hexoaminidase, *Eur. J. Biochem.,* 121, 125, 1981.

40. **Takahashi, T., Schmidt, P., and Tang, J.,** Oligosaccharide units of lysosomal cathepsin D from porcine spleen. Amino acid sequence and carbohydrate structure of the glycopeptides, *J. Biol. Chem.,* 258, 2819, 1983.

40a. **Barkholt, V.,** Amino acid sequence of endothiapepsin. Complete primary structure of the aspartic protease from *Endothia parasitica, Eur. J. Biochem.,* 167, 327, 1987.

40b. **Delaney, R., Wong, R. N. S., Meng, G., Wu, N., and Tang, J.,** Amino acid sequence of rhizopuspepsin isozyme pI 5, *J. Biol. Chem.,* 262, 1461, 1987.

40c. **Takahashi, K.,** The amino acid sequence of rhizopus-pepsin, and aspartic proteinase from *Rhizopus chinensis, J. Biol. Chem.,* 262, 1468, 1987.

41. **Hsu, I.-N., Delbaere, L. T. J., James, M. N. G., and Hofmann, T.,** Penicillopepsin from *Penicillium janthinellum* crystal structure at 2.8 Å and sequence homology with porcine pepsin, *Nature (London),* 266, 140, 1977.

42. **Subramanian, E., Swan, I. D. A., Liu, M., Davies, D. R., Jenkins, J. A., Tickle, I. J., and Blundell, T. L.,** Homology among acid proteases: comparison of crystal structures at 3 Å resolution of acid proteases from *Rhisopus chinensis* and *Endothia parasitica, Proc. Natl. Acad. Sci. U.S.A.,* 74, 556, 1977.

43. **Andreeva, N. S., Gustchina, A. E., Fedorov, A. A., Shutzkever, N. E., and Volnova, T. V.,** X-Ray crystallographic studies of pepsin, *Adv. Exp. Med. Biol.,* 95, 23, 1977.

44. **Green, M. L.,** Review of the progress of dairy science: milk coagulant, *J. Dairy Res.,* 44, 159, 1977.

45. **Matsubara, H. and Feder, J.,** Other bacterial, mold and yeast proteases, in *The Enzymes,* Vol. 3, 3rd ed., Boyer, P. D., Ed., Academic Press, New York, 1971, 723.

46. **Morihara, K.,** Comparative specificity of microbial proteinases, *Adv. Enzymol. Relat. Areas Mol. Biol.,* 41, 179, 1974.

46a. **Pearl, L. H. and Taylor, W. R.,** A structural model for the retroviral proteases, *Nature (London),* 329, 351, 1987.

46b. **Katoh, I., Yasunaga, T., Ikawa, Y., and Yoshinaka, Y.,** Inhibition of retroviral protease activity by an aspartyl proteinase inhibitor, *Nature (London),* 329, 654, 1987.

46c. **Pearl, L. H. and Taylor, W. R.,** Sequence specificity of retroviral proteases, *Nature (London)*, 328, 482, 1987.

47. **Herriott, R. M.,** Kinetics of the formation of pepsin from swine pepsinogen and identification of an intermediate compound, *J. Gen. Physiol.*, 22, 65, 1939.

48. **Bustin, M. and Conway-Jacobs, A.,** Intramolecular activation of porcine pepsinogen, *J. Biol. Chem.*, 246, 615, 1971.

49. **McPhie, P.,** A spectrophotometric investigation of the pepsinogen-pepsin conversion, *J. Biol. Chem.*, 247, 4277, 1972.

50. **Dykes, C. W. and Kay, J.,** Conversion of pepsinogen into pepsin is not a one-step process, *Biochem. J.*, 153, 141, 1976.

51. **Christensen, K. A., Pedersen, V. B., and Foltmann, B.,** Identification of an enzymatically active intermediate in the activation of porcine pepsinogen, *FEBS Lett.*, 76, 214, 1977.

52. **Al-Janabi, J., Hartsuck, J. A., and Tang, J.,** Kinetics and mechanism of pepsinogen activation, *J. Biol. Chem.*, 247, 4628, 1972.

53. **Marciniszyn, J., Jr., Huang, J. S., Hartsuck, J. A., and Tang, J.,** Mechanism of intramolecular activation of pepsinogen. Evidence for an intermediate δ and the involvement of the active site of pepsin in the intramolecular activation of pepsinogen, *J. Biol. Chem.*, 251, 7095, 1976.

54. **Sachdev, G. P., Brownstein, A. D., and Fruton, J. S.,** N-methyl-2-anilinonaphthalene-6-sulfonyl peptides as fluorescent probes for pepsin-substrate interaction, *J. Biol. Chem.*, 248, 6292, 1973.

55. **Auer, H. E. and Glick, D. M.,** Early events of pepsinogen activation, *Biochemistry*, 23, 2735, 1984.

56. **Glick, D. M., Auer, H. M., Rich, D. H., Kawai, M., and Kamath, A.,** Pepsinogen activation: genesis of the binding site, *Biochemistry*, 25, 1858, 1986.

57. **Pedersen, V. B., Christensen, K. A., and Foltmann, B.,** Investigations on the activation of bovine prochymosin, *Eur. J. Biochem.*, 94, 573, 1979.

58. **Keilova, H., Kostka, V., and Kay, J.,** The first step in the activation of chicken pepsinogen is similar to that of prochymosin, *Biochem. J.*, 167, 855, 1977.

59. **Dunn, B. M., Deyrup, C., Moesching, W. C., Gilbert, W. A., Nolan, R. J., and Trach, M. L.,** Inhibition of pepsin by zymogen activation fragments. Spectrum of peptides released from pepsinogen NH₂ terminus and solid phase synthesis of two inhibitory peptide sequences, *J. Biol. Chem.*, 253, 7269, 1978.

60. **Evin, G., Devin, J., Castro, B., Menard, J., and Corvol, P.,** Synthesis of peptides related to the prosegment of mouse submaxillary gland renin precursor: an approach to renin inhibitors, *Proc. Natl. Acad. Sci. U.S.A.*, 81, 48, 1984.

61. **Cumin, F., Evin, G., Fehrentz, J.-A., Seyer, R., Castro, B., Menard, J., and Corvol, P.,** Inhibition of human renin by synthetic peptides derived from its prosegment, *J. Biol. Chem.*, 260, 9154, 1985.

62. **Fruton, J. S.,** Specificity and mechanism of pepsin action, *Adv. Enzymol. Relat. Areas Mol. Biol.*, 33, 401, 1970.

63. **Fruton, J. S.,** The mechanism of the catalytic action of pepsin and related acid proteinases, *Adv. Enzymol. Relat. Areas Mol. Biol.*, 44, 1, 1976.

64. **Baker, L. E.,** New synthetic substrates for pepsin, *J. Biol. Chem.*, 193, 809, 1951.

65. **Clement, G. E., Rooney, J., Zakheim, D., and Eastman, J.,** The pH dependence of the dephosphorylated pepsin-catalyzed hydrolysis of N-acetyl-L-phenylalanyl-L-tyrosine methyl ester, *J. Am. Chem. Soc.*, 92, 186, 1970.

66. **Tang, J.,** Competitive inhibition of pepsin by aliphatic alcohols, *J. Biol. Chem.*, 240, 3810, 1965.

67. **Inouye, K., Voynick, I. M., Delpierre, G. R., and Fruton, J. S.,** New synthetic substrates for pepsin, *Biochemistry*, 5, 2473, 1966

68. **Inouye, K. and Fruton, J. S.,** Studies on the specificity of pepsin, *Biochemistry*, 6, 1765, 1967.

69. **Trout, G. E. and Fruton, J. S.,** The side-chain specificity of pepsin, *Biochemistry*, 8, 4183, 1969.

70. **Medzihradszky, K., Voynick, I. M., Medzihradszky-Schweiger, H., and Fruton, J. S.,** Effect of secondary enzyme-substrate interactions on the cleavage of synthetic peptides by pepsin, *Biochemistry*, 9, 1154, 1970.

71. **Sachdev, G. P. and Fruton, J. S.,** Pyridyl esters of peptides as synthetic substrates of pepsin, *Biochemistry*, 8, 4231, 1969.

72. **Sachdev, G. P., Johnston, M. A., and Fruton, J. S.,** Fluorescent studies on the interaction of pepsin with its substrates, *Biochemistry*, 11, 1080, 1972.

73. **Deyrup, C. and Dunn, B. M.,** A new substrate for porcine pepsin possessing cryptic fluorescence properties, *Anal. Biochem.*, 129, 502, 1983.

74. **Reid, T. W. and Fahrney, D.,** The pepsin-catalyzed hydrolysis of sulfite esters, *J. Am. Chem. Soc.*, 89, 3941, 1967.

75. **Reid, T. W., Stein, T. P., and Fahrney, D.,** The pepsin-catalyzed hydrolysis of sulfite esters. II. Resolution of alkyl phenyl sulfites, *J. Am. Chem. Soc.*, 89, 7125, 1967.

76. **May, S. W. and Kaiser, E. T.,** The pepsin-catalyzed hydrolysis of bis-p-nitrophenyl sulfite and its inhibition by diphenyl sulfite at pH 2, *J. Am. Chem. Soc.*, 91, 6491, 1969.

77. **Auffret, C. A. and Ryle, A. P.,** The catalytic activity of pig pepsin C towards small synthetic substrates, *Biochem. J.,* 179, 239, 1979.

78. **Tang, J.,** Gastricsin and pepsin, *Methods Enzymol.,* 19, 406, 1970.

79. **Fish, J. C.,** Activity and specificity of rennin, *Nature (London),* 180, 345, 1957.

80. **Morihara, K. and Oka, T.,** Comparative specificity of microbial acid proteinases for synthetic peptides. III. Relationship with their trypsinogen activating ability, *Arch. Biochem. Biophys.,* 157, 561, 1973.

81. **Oka, T. and Morihara, K.,** Comparative specificity of microbial acid proteinases for synthetic peptides. Primary specificity with Z-tetrapeptides, *Arch. Biochem. Biophys.,* 165, 65, 1974.

82. **Hofmann, T. and Hodges, R. S.,** A new chromophoric substrate for penicillopepsin and other fungal aspartic proteinases, *Biochem. J.,* 203, 603, 1982.

83. **Sachdev, G. P. and Fruton, J. S.,** Secondary enzyme-substrate interactions and the specificity of pepsin, *Biochemistry,* 9, 4465, 1970.

84. **Sampath-Kumar, P. S. and Fruton, J. S.,** Studies on the extended active sites of acid proteinases, *Proc. Natl. Acad. Sci. U.S.A.,* 71, 1070, 1974.

85. **Ferguson, J. B., Andrews, J. R., Voynick, I. M., and Fruton, J. S.,** The specificity of cathepsin D, *J. Biol. Chem.,* 248, 6701, 1973.

86. **Dunn, B. M., Jimenez, M., Parten, B. F., Valler, M. J., Rolph, C. E., and Kay, J.,** A systematic series of synthetic chromophoric substrates for aspartic proteinases, *Biochem. J.,* 237, 899, 1986.

87. **Visser, S., van Rooijen, P. J., Schattenkerk, C., and Kerling, K. E. T.,** Peptide substrates for chymosin (rennin). Kinetic studies with peptides of different chain length including parts of the sequence 101-112 of bovine κ-casein, *Biochem. Biophys. Acta,* 438, 265, 1976.

88. **Umezawa, H., Aoyagi, T., Morishima, H., Matsuzaki, M., Hamada, M., and Takeuchi, T.,** Pepstatin, a new pepsin inhibitor produced by actinomycetes, *J. Antibiot.,* 23, 259, 1970.

89. **Umezawa, H.,** Structures and activities of protease inhibitors of microbial origin, *Methods Enzymol.,* 45, 678, 1976.

90. **Umezawa, H.,** Low-molecular-weight enzyme inhibitors of microbial origin, *Annu. Rev. Microbiol.,* 36, 75, 1982.

91. **Workman, R. J. and Burkitt, D. W.,** Pepsin inhibition by a high specific activity radioiodinated derivative of pepstatin, *Arch. Biochem. Biophys.,* 194, 157, 1979.

92. **Takahashi, K. and Chang, W. J.,** The structure and function of acid proteases. V. Comparative studies on the specific inhibition of acid proteases by diazoacetyl-DL-norleucine methyl ester, 1,2-epoxy-3-(p-nitrophenoxy) propane and pepstatin, *J. Biochem. (Tokyo),* 80, 497, 1976.

93. **Marciniszyn, J., Jr., Hartsuck, J. A., and Tang, J.,** Mode in inhibition of acid proteases by pepstatin, *J. Biol. Chem.,* 251, 7088, 1976.

94. **Rich, D. H., Sun, E., and Singh, J.,** Synthesis of dideoxypepstatin. Mechanism of inhibition of porcine pepsin, *Biochem. Biophys. Res. Commun.,* 74, 762, 1977.

95. **Rich, D. H. and Sun, E. T. O.,** Synthesis of analogues of the carboxyl protease inhibitor pepstatin. Effect of structure on inhibition of pepsin and renin, *J. Med. Chem.,* 23, 27, 1980.

96. **Rich, D. H. and Bernatowicz, M. S.,** Synthesis of analogues of the carboxyl protease inhibitor pepstatin. Effect of structure in subsite P_3 on inhibition of pepsin, *J. Med. Chem.,* 25, 791, 1982.

97. **Schmidt, P. G., Bernatowicz, M. S., and Rich, D. H.,** Pepstatin binding to pepsin. Enzyme conformation changes monitored by nuclear magnetic resonance, *Biochemistry,* 21, 6710, 1982.

98. **Kay, J., Afting, E. G., Aoyagi, T., and Dunn, B. M.,** The effects of lactoyl-pepstatin and the pepsin inhibitor peptide on pig cathepsin D, *Biochem. J.,* 203, 795, 1982.

99. **Clement, G. E.,** Catalytic activity of pepsin, *Prog. Bioorg. Chem.,* 2, 177, 1973.

100. **Denburg, J. L., Nelson, R., and Silver, M. S.,** The effect of pH on the rates of hydrolysis of three acylated dipeptides by pepsin, *J. Am. Chem. Soc.,* 90, 479, 1968.

101. **Clement, G. E., Snyder, S. L., Price, H., and Cartmell, R.,** The pH dependence of the pepsin-catalyzed hydrolysis of neutral dipeptides, *J. Am. Chem. Soc.,* 90, 5603, 1968.

102. **Cornish-Bowden, A. J. and Knowles, J. R.,** The pH- dependence of pepsin-catalyzed reactions, *Biochem. J.,* 113, 353, 1969.

103. **Hollands, T. R. and Fruton, J. S.,** Kinetics of the hydrolysis of synthetic substrates by pepsin and by acetyl-pepsin, *Biochemistry,* 7, 2045, 1968.

104. **Delpierre, G. R. and Fruton, J. S.,** Inactivation of pepsin by diphenyldiazomethane, *Proc. Natl. Acad. Sci. U.S.A.,* 54, 1161, 1965.

105. **Rajagopalan, T. G., Stein, W. H., and Moore, S.,** The inactivation of pepsin by diazoacetyl norleucine methyl ester, *J. Biol. Chem.,* 241, 4295, 1966.

106. **Bayliss, R. S., Knowles, J. R., and Wybrandt, G. B.,** An aspartic acid residue at the active site of pepsin. The isolation and sequence of the heptapeptide, *Biochem. J.,* 113, 377, 1969.

107. **Fry, K. T., Kim, O.-K., Spona, J., and Hamiliton, G. A.,** A reactive aspartyl residue of pepsin, *Biochem. Biophys. Res. Commun.,* 30, 489, 1968.

108. **Stephanov, V. M. and Vaganova, T. I.,** Identification of the carboxyl group of pepsin reacting with diazoacetamide derivatives, *Biochem. Biophys. Res. Commun.,* 31, 825, 1968.

109. **Tang, J., Sepulveda, P., Marciniszyn, J., Jr., Chen, K. C. S., Huang, W.-Y., Too, N., Liu, D., and Lanier, J. P.,** Amino-acid sequence of procine pepsin, *Proc. Natl. Acad. Sci. U.S.A.,* 70, 3437, 1973.

110. **Tang, J.,** Specific and irreversible inactivation of pepsin by substrate-like epoxides, *J. Biol. Chem.,* 246, 4510, 1971.

111. **Hartsuck, J. A. and Tang, J.,** The carboxylate ion in the active center of pepsin, *J. Biol. Chem.,* 247, 2575, 1972.

112. **Chang, W.-J. and Takahashi, K.,** The structure and function of acid proteases. II. Inactivation of bovine rennin by acid protease-specific inhibitors, *J. Biochem. (Tokyo),* 74, 231, 1973.

113. **Bernal, J. D. and Crowfoot, D.,** X-Ray photographs of crystalline pepsin, *Nature (London),* 133, 794, 1934.

114. **James, M. N. G. and Sielecki, A. R.,** Structure and refinement of penicillopepsin at 1.8 Å resolution, *J. Mol. Biol.,* 163, 299, 1983.

114a. **Suguna, K., Bott, R. R., Padlan, E. A., Subramanian, E., Sheriff, S., Cohen, G. H., and Davies, D. R.,** Structure and refinement at 1.8 Å resolution of the aspartic proteinase from *Rhisopus chinensis, J. Mol. Biol.,* 196, 877, 1987.

115. **Andreeva, N. S., Zdanov, A. S., Gustchina, A. E., and Fedorov, A. A.,** Structure of ethanol-inhibited porcine pepsin at 2-Å resolution and binding of the methyl ester of phenylalanyl-diiodotyrosine to the enzyme, *J. Biol. Chem.,* 259, 11353, 1984.

116. **Tang, J., James, M. N. G., Hsu, I. N., Jenkins, J. A., and Blundell, T. L.,** Structural evidence for gene duplication in the evolution of the acid proteases, *Nature (London),* 271, 618, 1978.

117. **Andreeva, N. S. and Gustchina, A. E.,** On the super-secondary structure of acid proteases, *Biochem. Biophys. Res. Commun.,* 87, 32, 1979.

118. **Blundell, T. L., Sewell, B. T., and McLachlan, A. D.,** Four-fold structural repeat in the acid proteases, *Biochim. Biophys. Acta,* 580, 24, 1979.

119. **Pearl, L. and Blundell, T.,** The active site of aspartic proteinases, *FEBS Lett.,* 174, 96, 1984.

120. **Hofmann, T., Hodges, R. S., and James, M. N. G.,** Effect of pH on the activities of penicillopepsin and *Rhisopus* pepsin and a proposal for the productive substrate binding mode in penicillopepsin, *Biochemistry,* 23, 635, 1984.

121. **Blundell, T., Sibanda, B. L., and Pearl, L.,** Three- dimensional structure, specificity and catalytic mechanism of renin, *Nature (London),* 304, 273, 1983.

122. **Bott, R., Subramanian, E., and Davies, D. R.,** Three-dimensional structure of the complex of the *Rhisopus chinensis* carboxyl proteinase and pepstatin at 2.5 Å resolution, *Biochemistry,* 21, 6956, 1982.

123. **James, M. N. G., Seilecki, A., Salituro, F., Rich, D. H., and Hofmann, T.,** Conformational flexibility in the active sites of aspartyl proteinases revealed by a pepstatin fragment binding to penicillopepsin, *Proc. Natl. Acad. Sci. U.S.A.,* 79, 6137, 1982.

124. **James, M. N. G. and Sielecki, A. R.,** Molecular structure of an aspartic proteinase zymogen, porcine pepsinogen, at 1.8 Å resolution, *Nature (London),* 319, 33, 1986.

125. **Streitwieser, A., Jr. and Heathcock, C. H.,** *Introduction to Organic Chemistry,* 2nd., ed., Macmillan, New York, 1981, 864.

126. **Neumann, H., Levin, Y., Berger, A., and Katchalski, E.,** Pepsin-catalysed transpeptidation of the aminotransfer type, *Biochem. J.,* 73, 33, 1959.

127. **Fruton, J. S., Fujii, S., and Knappenberger, M. H.,** The mechanism of pepsin action, *Proc. Natl. Acad. Sci. U.S.A.,* 47, 759, 1961.

128. **Maltsev, N. I., Ginodman, L. M., Orekhovich, V. N., Valueva, T. A., and Akimova, L. N.,** A study of pepsin specificity in transpeptidation reactions, *Biokhimiya,* 31, 983, 1966.

129. **Kitson, T. M. and Knowles, J. R.,** The pathway of pepsin-catalysed transpeptidation. Evidence for the reactive species being the anion of the acceptor molecule, *Biochem. J.,* 122, 249, 1971.

130. **Silver, M. S. and Stoddard, M.,** Amino-enzyme intermediates in pepsin-catalyzed reactions, *Biochemistry,* 11, 191, 1972.

131. **Antonov, V. K., Rumsh, L. D., and Tikhodeeva, A. G.,** Kinetics of pepsin-catalysed transpeptidation: evidence for the ''amino-enzyme'' intermediate, *FEBS Lett.,* 46, 29, 1974.

132. **Sharon, N., Grisaro, V., and Neumann, H.,** Pepsin-catalyzed exchange of oxygen atoms between water and carboxylic acids, *Arch. Biochem. Biophys.,* 97, 219, 1962.

133. **Kozlov, L. V.,** pH-Dependence of pepsin-catalyzed isotopic exchange of oxygen in N-acetyl-L-phenylalanine, *Biokhimiya,* 39, 512, 1974.

134. **Cornish-Bowden, A. J., Greenwell, P., and Knowles, J. R.,** The rate-determining step in pepsin-catalysed reactions, and evidence against an acyl-enzyme intermediate, *Biochem. J.* 113, 369, 1969.

135. **Takahashi, M., Wang, T. T., and Hofmann, T.,** Acyl intermediates in pepsin and penicillopepsin catalyzed reactions, *Biochem. Biophys. Res. Commun.,* 57, 39, 1974.

136. **Takahashi, M. and Hofmann, T.,** Acyl intermediates in penicillopepsin-catalysed reactions and a discussion of the mechanism of action of pepsins, *Biochem. J.,* 147, 549, 1975.
137. **Wang, T. T. and Hofmann, T.,** Acyl and amino intermediates in penicillopepsin-catalysed reactions and activation by nonsubstrate peptides, *Can. J. Biochem.* 55, 286, 1977.
138. **Newmark, A. K. and Knowles, J. R.,** Acyl- and amino-transfer routes in pepsin-catalyzed reactions, *J. Am. Chem. Soc.,* 97, 3557, 1975.
139. **Spector, L. B.,** *Covalent Catalysis by Enzymes,* Springer-Verlag, Berlin, 1982, 137.
140. **Silver, M. S., Stoddard, M., and Kelleher, M. H.,** Nature of amino-enzyme intermediate in pepsin-catalyzed reactions, *J. Am. Chem. Soc.,* 98, 6684, 1976.
141. **Fruton, J. S.,** Fluorescence studies on the active sites of proteinases, *Mol. Cell. Biochem.,* 32, 105, 1980.
142. **Dunn, B. M. and Fink, A. L.,** Cryoenzymology of porcine pepsin, *Biochemistry,* 23, 5241, 1984.
143. **Hofmann, T. and Fink, A. L.,** Cryoenzymology of penicillopepsin, *Biochemistry,* 23, 5247, 1984.
144. **Antonov, V. K., Ginodman, L. M., Kapitannikov, Yu. V., Barshevskaya, T. N., Gurova, A. G., and Rumsh, L. D.,** Mechanism of pepsin catalysis: general base catalysis by the active-site carboxylate ion, *FEBS Lett.,* 88, 87, 1978.
145. **Antonov, V. K., Ginodman, L. M., Rumsh, L. D., Kapitannikov, Yu. V., Barshevskaya, T. N., Yavashev, L. P., Gurova, A. G., and Volkova, L. I.,** Studies on the mechanism of action of proteolytic enzymes using heavy oxygen exchange, *Eur. J. Biochem.,* 177, 195, 1981.
146. **Schmidt, P. G., Holladay, M. W., Salituro, F. G., and Rich, D. H.,** Identification of oxygen nucleophiles in tetrahedral intermediates: ^{2}H and ^{18}O induced isotope shifts in ^{13}C NMR spectra of pepsin-bound peptide ketone pseudosubstrates, *Biochem. Biophys. Res. Commun.,* 129, 597, 1985.
147. **Burgen, A. S. V., Roberts, G. C. K., and Feeney, J.,** Binding of flexible ligands to macromolecules, *Nature (London),* 253, 753, 1975.
148. **Sachdev, G. P. and Fruton, J. S.,** Kinetics of action of pepsin on fluorescent peptide substrates, *Proc. Natl. Acad. Sci. U.S.A.,* 72, 3424, 1975.
149. **Kitagashi, K., Nakatani, H., and Hiromi, K.,** Static and kinetic studies on the binding between pepsin and *Streptomyces* pepsin inhibitor with a fluorescent probe, *J. Biochem. (Tokyo),* 87, 573, 1980.
150. **Hofmann, T., Dunn, B. M., and Fink, A. L.,** Appendix: mechanism of action of aspartyl proteinases, *Biochemistry,* 23, 5247, 1984.
151. **Kluger, R. and Chin, J.,** Carboxylic acid participation in amide hydrolysis. Evidence that separation of a nonbonded complex can be rate determining, *J. Am. Chem. Soc.,* 104, 2891, 1982.
152. **James, M. N. G. and Sielecki, A. R.,** Stereochemical analysis of peptide bond hydrolysis catalyzed by the aspartic proteinase penicillopepsin, *Biochemistry,* 24, 3701, 1985.
153. **James, M. N. G., Hsu, I.-N., and Delbaere, L. T. J.,** Mechanism of acid protease catalysis based on the crystal structure of penicillopepsin, *Nature (London),* 267, 808, 1977.
154. **Polgár, L.,** The mechanism of action of aspartic proteases involves "push-pull" catalysis, *FEBS Lett.,* 219, 1, 1987.
155. **Suguna, K., Padlan, E. A., Smith, C. W., Carlson, W. D., and Davies, D. R.,** Binding of a reduced peptide inhibitor to the aspartic proteinase from *Rhizopus chinensis:* implications for a mechanism of action, *Proc. Natl. Acad. Sci. U.S.A.,* 84, 7009, 1987.

Chapter 6

METALLOPROTEASES

I. INTRODUCTION

Metalloproteases, which have been studied in some detail, are all zinc-containing enzymes. The role of zinc in the zinc-containing enzymes has recently been reviewed.[1] In this chapter, we concentrate on two zinc proteases, carboxypeptidase A and thermolysin, whose tertiary structures have been determined at atomic resolution, thus providing a solid basis for mechanistic investigations. These two enzymes, although they possess similar active sites, have unrelated tertiary structures, i.e., they belong to different families of the metalloproteases (Chapter 2, Section I, Table 3). There are numerous other zinc-containing proteases which have been extensively studied. Notably, angiotensin-converting enzyme (ACE) has been investigated in detail because of its possible role in the regulation of blood pressure. Neutral endopeptidase 24.11 or "enkephalinase" has been carefully studied because it was thought to participate in the degradation of enkephalins. Aminopeptidases have been examined as possible modifiers of biologically active polypeptides. Collagenases have been studied because they may be important in the processes of human disease.

II. GENERAL FEATURES OF ZINC PROTEASES

A. Carboxypeptidase A

Carboxypeptidase A is the most extensively studied zinc-containing protease. The earlier mechanistic and crystallographic results have been reviewed in detail.[2,3] Most studies on carboxypeptidase A have been carried out with the bovine pancreatic enzyme which was first crystallized 50 years ago.[4] The porcine enzyme has also been isolated.[5,6]

Depending on the isolation procedure, bovine carboxypeptidase A exhibits four different forms, primarily as a result of different cleavages at the N-terminal portion when formed from procarboxypeptidase.[7] The relationship among the forms is illustrated in Figure 1. The γ- and δ-forms, which possess identical N-termini, differ significantly in two respects: (1) the δ-form is eight times more soluble in 1 M NaCl than is the γ-form;[8] (2) the enzymic activity changes reversibly upon removal and addition of zinc in the case of the δ-form, whereas the activity of the γ-form is only partially regenerated.[8,9]

It is interesting that two allelomorphs of carboxypeptidase A have been isolated.[10] These forms differ in three amino acids at positions 179, 228, and 305. The distribution of these forms appears to be Mendelian, since only animals having one of the two forms or an approximately equal mixture of the two forms have been found.

The amino acid sequence of carboxypeptidase A has been determined.[11] The α-form contains 307 amino acids and has a molecular mass of 34,500. The enzyme contains a relatively large number of aromatic residues and two cysteine residues. The thiol group of one cysteine was originally thought to be a zinc ligand, but X-ray diffraction studies have now shown that the two cysteine residues form a disulfide bond.[12]

As for the zymogen of carboxypeptidase A, the amino acid sequence of the rat proenzyme has recently been deduced from the corresponding nucleotide sequence.[13] Furthermore, the activation segment of the porcine zymogen has been isolated and characterized.[14] Its molecular mass is 11,500 to 12,000 daltons. It has a high content of hydrophobic and acidic amino acids, and lacks cysteine. A remarkable feature of the peptide is its strong inhibitory effect on carboxypeptidase A with a K_i in the nanomolar range. Physicochemical studies indicate that the activation segment constitutes a folded structural domain containing a high percentage of α-helix.[15]

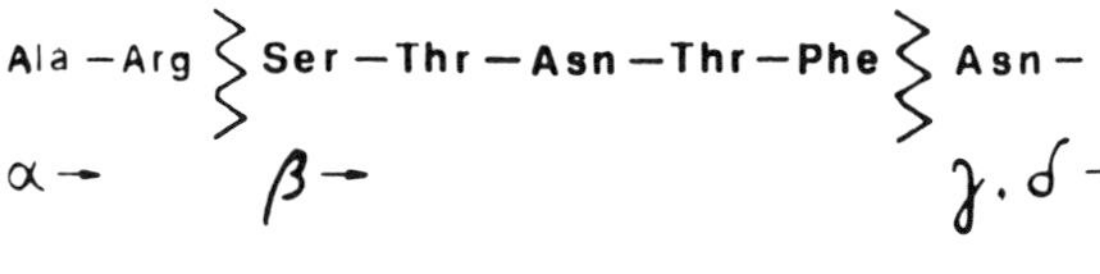

Ala —Arg $\lessgtr$ Ser —Thr —Asn —Thr—Phe $\lessgtr$ Asn —

$\alpha\rightarrow$ $\beta\rightarrow$ $\gamma.\delta\rightarrow$

FIGURE 1.   The N-termini of different forms of carboxypeptidase A.

The early experiments, in particular the inhibition by cysteine, indicated that carboxypeptidase A was a metalloprotease. The metal was later identified as zinc.[16] Additionally, it was deduced from several studies that there were two protein ligands to the zinc ion, one was a thiol group,[9,17,18] and the other was the N-terminal α-amino group.[18] However, these groups were eliminated as possible ligands when the crystal structure of the enzyme became known, and three other ligands to the zinc ion were identified, namely His 69, Glu 72, and His 196.[19]

The Zn^{2+} in carboxypeptidase A could be replaced by Co^{2+} or Ni^{2+}, resulting in small or negligible changes in the catalytic activity.[20] On the other hand, the substitution of Zn^{2+} by Hg^{2+} or Cd^{2+} abolished peptidase activity, but did not affect or even increase esterase activity of the enzyme as measured with O-hippuryl-L-β-phenyllactate.[20]

Experiments involving tyrosine modification in carboxypeptidase A have led to the conclusion that one or two such residues are involved in the catalytic activity of the enzyme.[21-25] Diazotation[24] and nitration[25] inactivated the enzyme toward peptide substrates in most cases, but the presence of the inhibitor β-phenylpropionate protected the enzyme. Most interestingly, the esterase activity was not abolished by tyrosine modification, but rather increased, for example, six times in the case of the acetylated enzyme.[23] The most recent site-directed mutagenesis studies have shown that the hydroxyl groups of the two tyrosine residues (198 and 248) at the active site region do not play a critical role in the catalysis.[25a]

Finally, two other zinc-containing carboxypeptidases are worthy of note. Thus, the pancreas also produces carboxypeptidase B, which is homologous to, and in many respects similar to, carboxypeptidase A.[26] However, carboxypeptidase B hydrolyzes basic amino acid residues (arginine and lysine) at the C-terminal of peptide substrates, whereas carboxypeptidase A prefers large hydrophobic residues such as phenylalanine.

Carboxypeptidase N (arginine carboxypeptidase, kininase I) resembles carboxypeptidase B in its specificity. However, this enzyme, which can be purified from human plasma,[27] is a tetramer of M_r 280,000. It is composed of two types of immunologically distinct subunits, a M_r 40,000 subunit that is responsible for the enzymic activity and a M_r 83,000 subunit that may stabilize the enzyme in blood.[28] Indeed, because carboxypeptidase N is the major blood-borne inactivator of potent peptides, such as kinins and anaphylatoxins, the preservation of its activity in blood is of great importance.

B. ACE and Neutral Endopeptidase 24.11

Because of its implication in blood pressure regulation, ACE (kininase II) has been the subject of extensive studies[29] (see also Chapter 2, Section III). The enzyme was isolated from a variety of animal tissues.[30-35] Membrane-bound converting enzyme can be purified by detergent extraction, retaining the membrane-binding sequence. On the other hand, isolation involving trypsin treatment leads to an enzyme derivative which is devoid of its membrane-binding sequence. Soluble-converting enzyme has been obtained from blood plasma and seminal plasma. High molecular mass 140 kdalton-forms have been isolated from a number of human tissues[35] by using a highly specific affinity chromatography ligand, Sepharose-bound lisinopril.[33,34] It is interesting that testis contained a 90 kdalton-form in addition to the 140 kdalton-form at a ratio of approximately 4:1.[35] The enzyme contains one gram-atom zinc ion per mole and carbohydrate up to about 30% by weight.[29]

ACE is a peptidyl dipeptidase that cleaves C-terminal dipeptides of diverse substrates, such as angiotensin I, bradykinin, enkephalin, and a series of N-terminal blocked synthetic tripeptides.[29] One such tripeptide, benzoyl-Gly-His-Leu (hippuryl-His-Leu), has been used most frequently for measuring enzymic activity by spectrophotometric determination of the extracted hippuric acid (benzoyl-Gly) product.[36] Alternatively, the released His-Leu dipeptide can be assayed fluorimetrically.[37] Continuous spectrophotometric assays have also been described using, for example, benzyloxycarbonyl-Phe(NO$_2$)-Gly-Gly[38] or furylacryloyl-Phe-Gly-Gly[39] as substrates. Internally quenched substrates, such as 2-aminobenzoyl-Gly-Phe(NO$_2$)-Pro[40] and others,[41,42] are useful for continuous spectrofluorimetric measurements. It is interesting that a chloride ion stimulates ACE in its reactions with many substrates so that rate assays are usually carried out in the presence of 0.1 to 0.3 M NaCl.[29] The chloride ion probably binds to a particular lysine residue.[43] It has been postulated that a converting enzyme also contains a second anion binding site, possibly a second lysine.[44]

Although ACE is a much larger single-chain protein (140 to 150 kdaltons) than carboxypeptidase A (34.5 kdaltons), the active sites of the two enzymes appear to be similar. This is indicated by the amino acid sequence of an active site peptide from the converting enzyme covalently labeled on the essential glutamic acid residue.[45] This sequence proved to be homologous with the corresponding sequences of carboxypeptidases A and B, but not with thermolysin which also contains an essential active site glutamic acid residue. Similarity is also apparent in the location of a tyrosine residue at or near the active site of both the converting enzyme[46] and carboxypeptidase A (see above). Furthermore, ACE can catalyze proton abstraction from an activated methylene group in a specific ketone substrate,[47] a reaction also observed with carboxypeptidase A.[48,49]

We have already mentioned that ACE can hydrolyze enkephalins (Tyr-Gly-Gly-Phe-Met and Tyr-Gly-Gly-Phe-Leu) very rapidly, but with a high K_m value. Another enzyme in striatum of mouse brain was found to hydrolyze enkephalins with much lower K_m, and it was thought that this second enzyme, a peptidyl dipeptidase, is the true "enkephalinase".[50,51] However, it has turned out that enkephalinase is identical with the membrane-bound neutral endopeptidase (EC 3.4.24.11),[52-54] originally purified from kidney.[55,56] Later, this enzyme was shown to be present in many tissues, and it was renamed endopeptidase 24.11.[57] The enzyme had been successfully released from the kidney membrane by toluene-trypsin treatment,[55] and the purified product was a monomer of M_r 93,000.[56] In contrast to most other microvillar hydrolases, this endopeptidase could not be released by treatment of the membrane with papain or trypsin, which suggested a different type of association with the membrane from that observed with "stalk proteins".[58] Studies with the detergent-solubilized form of the enzyme have shown that endopeptidase 24.11 possesses an unusually short stalk (2 nm),[59] which can explain why treatments with proteases did not release the enzyme. Recent analysis of the cDNA of rabbit kidney endopeptidase 24.11 has shown that the protein consists of 750 amino acids. It contains a short N-terminal cytoplasmic domain (27 amino acids), a single membrane-spanning segment (23 amino acids), and an extracellular domain that comprises most of the protein mass. The comparison of the primary structure with that of thermolysis indicates that most of the amino acid residues involved in zinc coordination and catalytic activity are found in highly homologous sequences in the two enzymes.[59a]

The activity of endopeptidase 24.11 can be determined by a coupled enzyme reaction in the presence of excess aminopeptidase M.[52] With glutaryl-Ala-Ala-Phe-2-naphthylamide, the reaction proceeds as shown by Equations 1a and 1b. The excess of aminopeptidase M assures the quantitative release of 2-naphthylamine, which can be monitored spectrofluorimetrically.

$$\text{Glt--Ala--Ala--Phe--2NA} \xrightarrow{\text{endopeptidase}} \text{Glt--Ala--Ala} + \text{Phe--2NA} \qquad (1a)$$

$$\text{Phe-2NA} \xrightarrow{\text{aminopeptidase M}} \text{Phe} + 2\text{NA} \tag{1b}$$

The internally quenched substrate, dansyl-D-Ala-Gly-Phe(NO$_2$)-Gly, may also be employed for continuous rate assay.[60] The increase in fluorescence can be followed by using an excitation wavelength of 342 nm and an emission wavelength of 562 nm.

Endopeptidase 24.11 cleaves peptides on the amino side of hydrophobic amino acids[52] and, in this respect, resembles thermolysin. Comparison of subsite specificities of endopeptidase 24.11 and thermolysin has shown that although the two enzymes have similar primary specificities apparently imposed by the hydrophobic pocket at the S$_1'$ subsite, they differ significantly at other subsites of the active site.[61,62] Notably, the mammalian enzyme showed the highest rate constants (k_{cat}/K_m) with substrates having the sequence -Phe-Gly-Phe- or -Phe-Ala-Phe- in positions -P$_2$-P$_1$-P$_1'$, while the sequence -Ala-Phe-Phe-, was the most favored by the bacterial enzyme.[61] Furthermore, a comparison of substrates in the free acid form with their corresponding amides showed that binding to the mammalian enzyme is dependent in part on an ionic interaction between the substrate carboxyl group and the enzyme, whereas such an ionic interaction was not observed with the bacterial enzyme.[62] Studies with the arginine-specific reagents, phenylglyoxal and butanedione,[63,64] suggested that an arginine residue on the enzyme is involved in substrate binding.

C. Other Zinc Proteases

One important group of metalloproteases includes aminopeptidases that catalyze the hydrolysis of the N-terminal peptide bonds in polypeptides. Aminopeptidases with similar or identical properties have been found in many tissues.[65] The vast literature on aminopeptidases is mainly concerned with the purification and characterization of enzymes from various sources. The relationship among enzymes isolated in different laboratories is often unclear. Leucine aminopeptidase (LAP) and alanine aminopeptidase (AAP) constitute the two most extensively studied types of this group of enzymes.

LAP was discovered and characterized earlier[65,66] than AAP. It is a hexameric enzyme with a subunit of M_r 54,000. In contrast to the enzymes of the carboxypeptidase A or the thermolysin family, it is a two-metal protease, inasmuch as it requires Mn^{2+} or Mg^{2+} for activation. Sulfhydryl groups may be involved in the binding of metal ions.[67] Its primary amino acid sequence has been reported, but no significant homology with other proteins could be detected.[68] Although efforts to study the mechanism of action are being made,[69-74] the results are not conclusive and permit only speculation about the basic features of the catalysis.

Distinctive differences between LAP and AAP have been reported from several laboratories. These included different substrate specificities,[75,76] different responses to inhibition by puromycin,[77] and different pH optima.[78] AAP has been designated by different names, such as leucine-β-naphthylamide splitting enzyme,[76] amino acid naphthylamidase,[79] aminopeptidase M,[80] and arylamidase,[81,82] just to mention a few. Some of these names are still in use.

The physiological role of the aminopeptidases is not clear. They may participate in various post-translational processing or modifying pathways for peptides with biological activity. In some cases, this processing involves the final activating step for a peptide and, in other cases, involves the inactivation of a peptide. For example, kinin-converting enzyme, an aminopeptidase, converts immediate precursors of bradykinin, i.e., methionyl-lysyl-bradykinin and lysyl-bradykinin (kallidin), to bradykinin by stepwise removal of methionine and lysine.[83] This enzyme is blocked when the penultimate N-terminal residue is proline, as in bradykinin, thus allowing for protection of the product from inactivation by the same kinin-converting enzyme. By contrast, another AAP may act as a kininase by removing the

N-terminal arginine residue from the nonapeptide bradykinin.[84] Furthermore, enkephalins may also be inactivated by aminopeptidases when their N-terminal Tyr residue is removed.[85,86]

Collagenases constitute another important group of zinc proteases which cleave peptide bonds located in the characteristic triple-helical region of collagen. Two types of collagenases, the bacterial and the tissue collagenases, may be distinguished. The basic results obtained with this group of enzymes have been reviewed in detail.[87]

Clostridium histolyticum secretes collagenases that contribute to the invasiveness of this highly pathogenic bacterium by destroying the connective tissue barrier of the host. The bacterium remains unaffected by the enzyme because it contains no collagen. Collagenases cause a large number of cleavages in native triple-helical collagen at the Xaa-Gly bond in the sequence Xaa$\perp$Gly-Pro-Xaa, where Xaa is frequently alanine or hydroxyproline, but can be any amino acid. Six individual collagenases have recently been isolated from the culture filtrate of *Clostridium histolyticum*.[88,89] Their M_r varies between 68,000 and 125,000.[89] It appears that the larger molecules evolved from the smaller ones by gene duplication followed by point mutations that altered their specificities.[90] On the basis of specificity, the enzymes fall into two classes. Collagenases of class I have high activity toward collagen and low activity toward several synthetic peptides, such as furylacryloyl-Leu$\perp$Gly-Pro-Ala, whereas collagenases of class II exhibit the opposite trend of activity. More recent specificity studies have shown that the class II enzymes have a broader specificity than the class I enzymes.[91] It is interesting to note that all six enzymes prefer aromatic amino acids in the P_1 position, even though such residues do not occur at this position in the native substrate. N-terminal elongation of synthetic substrates causes the k_{cat}/K_m values first to rise markedly and then to level off after occupancy of subsite S_6 for the class I enzymes and subsite S_3 for the class II enzymes. C-terminal elongation produces the best substrate for both classes of enzymes when subsites S_3' or S_4' are occupied by amino acids with free carboxyl groups.[92]

Tissue collagenases are found in amphibian and mammalian tissues undergoing growth or remodeling.[87] Notably, a large amount of collagen is resorbed in the tail fin of tadpoles during metamorphosis. High enzyme activity is also found in the uterus following pregnancy. In contrast to the bacterial collagenase, tadpole collagenase cleaves tropocollagen across its three chains at a unique site, giving rise to two triple-helical fragments of a molecular mass of 24,000 and 71,000.[93] The two fragments spontaneously unfold at body temperature and are then cleaved by other proteolytic enzymes.

There are a large number of metalloproteases of microbial origin.[94,95] Most of these enzymes qualify as metalloproteases by their sensitivities to metal-chelating agents such as ethylenediaminetetraacetate (EDTA) or *o*-phenanthroline and by their insensitivity to the usual serine and cysteine protease inhibitors. The direct involvement of zinc in the catalysis has been demonstrated in a few cases only. Among the best characterized enzymes, the *Bacillus subtilis* neutral protease and the *B. thermoproteolyticus* thermolysin are worthy of mention. Pure preparations of neutral proteases have been isolated from different strains of *B. subtilis*.[96-99] (The same microorganisms also produce the alkaline serine proteases, called subtilisins, already discussed in Chapter 3.) Thermolysin has also been obtained in pure form by using conventional methods[100,101] and also by affinity chromatography on Sepharose-Gly-D-Phe.[102] Its amino acid sequence[103] and X-ray crystallographic structure[104,105] are known. The N-terminal portions of the amino acid sequences of neutral proteases from *B. subtilis* have also been determined and were found to be homologous to that of thermolysin.[102] The molecular mass of thermolysin (34,600 daltons), as well as that of other neutral proteases, is greater than that of chymotrypsin or subtilisin. Thermolysin is an extremely stable protease, retaining about 50% of its activity after 1 hr incubation at 80°C.[100,106] Calcium ions play an important role in this stability since their removal by EDTA causes loss of the enzyme activity under similar conditions.[106] Calcium ions also considerably stabilize the *B. subtilis* neutral proteases.[97]

The specificity of neutral zinc proteases toward synthetic substrates has been systematically studied.[107-109] It was shown that these enzymes are specific for large hydrophobic residues in the P_1' position. The Leu-containing substrates are the best ones, whereas Tyr-containing substrates are hydrolyzed very poorly. The presence of a free α-carboxyl or α-amino group in a dipeptide substrate greatly diminishes the hydrolysis rate. Only slight differences were observed in the specificities of thermolysin and the *B. subtilis* neutral protease. A remarkable similarity in the specificities between these enzymes and that isolated from the venom of the snake, *Crotalus atrox,* was found.[110] This has recently been confirmed with two hemorrhagic zinc proteases purified from *C. atrox* venom.[111]

III. THE ACTIVE SITE STRUCTURE

A. Carboxypeptidase A

The structure of carboxypeptidase A at 0.2 nm resolution has been described in detail[1,2,12,19] and slightly modified in the light of later data obtained at 0.175[112] and 0.154 nm[113] resolution. The molecule has an ellipsoid shape with approximate dimensions $5.0 \times 4.2 \times 3.8$ nm. Nine helices, a twisted pleated sheet containing both parallel and antiparallel β-structures, and a random coil possessing a few hydrogen bonds and a disulfide bridge constitute the structure of the enzyme. The zinc ion is adjacent to the pleated sheet running through the center of the molecule. His 69, Glu 72, and His 196 are the ligands from the protein to the zinc. The refinement of the structure at 0.175 nm resolution has shown that the coordination number of the zinc is five: two imidazole ND1 nitrogens, the two carboxylate oxygens of Glu 72, and a water molecule.[113] Associated with the zinc is a cleft and a pocket which accommodate the substrate.

Information about the mode of substrate binding was obtained from difference electron density maps. With this technique, a poor substrate of carboxypeptidase A, Gly-Tyr, was first analyzed in detail.[12,113] There are several important interactions between the enzyme and the dipeptide. (1) The aromatic ring of Tyr is situated in a pocket in the enzyme. This pocket contains no specific binding groups and is large enough to accommodate a tryptophan side chain. (2) The C-terminal carboxylate group of the substrate, which is essential for catalysis, forms a salt bridge with the positively charged guanidinium group of Arg 145. (3) The carbonyl oxygen of the scissile peptide bond becomes a ligand to the zinc ion after replacing the bound water molecule. (4) The N-terminal amino group of Gly-Tyr statistically occupies two positions. Specifically, it binds to the zinc ion or alternatively to the carboxylate group of Glu 270. These two conformations are related by a rotation of approximately 180° about the glycine α-carbon-carbonyl carbon bond.[113] Apparently, the nonprotonated amino group interacts with the zinc ion and the protonated form with Glu 270. A previous conclusion that the coordination number of zinc is six when the amino group is a zinc ligand has been revised in the light of more refined X-ray crystallographic data.[114] Specifically, the coordination number has been found to be five rather than six as a consequence of the displacement of the zinc-bound water by the substrate amino terminus. The interactions between the enzymic groups and the amino group of Gly-Tyr, which are only possible with nonacylated dipeptide substrates, could account for the unusual stability of the enzyme-ligand complex. The other three interactions appear to be characteristic of a productive enzyme-substrate complex.

The difference Fourier electron density map for the enzyme-dipeptide complex also shows that a hydrogen bond is formed between the nitrogen atom of the scissile peptide bond and the phenolic hydroxyl group of Tyr 248.[1] This is the consequence of a large movement of the tyrosine side chain. In the native enzyme, this side chain is located on the surface of the molecule with the hydroxyl group 1.2 nm away from the position which it occupies in the complex. The relocation involves rotation of the side chain about its C_α–C_β bond by

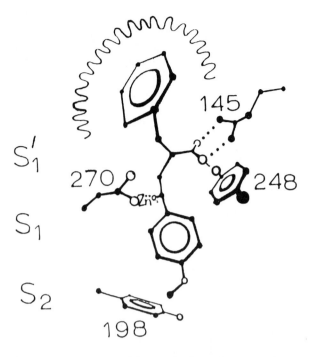

FIGURE 2. A scheme of the complex formed between carboxypeptidase A and (2-benzyl-3-*p*-methoxybenzoyl)proponic acid. (From Lipscomb, W. N., *Proc. Natl. Acad. Sci. U.S.A.*, 77, 3875, 1980. With permission.)

about 120°. Smaller changes (about 0.2 nm) were also observed in the positions of Arg 145 and Glu 270.[1] Conformational changes upon substrate binding were also indicated in several physicochemical studies prior to the crystallographic investigations. For example, the development of the Cotton effect at 295 nm in the optical rotatory dispersion (ORD) spectrum of the complex between carboxypeptidase A and β-phenylpropionate was attributed to a change in the environment of certain aromatic side chains in the enzyme as a result of a strong binding with the aromatic ring of the inhibitor.[115]

A binding mode similar to that found with Gly-Tyr was also found with a ketonic inhibitor of carboxypeptidase A (Figure 2).[116,117] This compound, 2-benzyl-3-*p*-methoxybenzoyl) propionic acid (form [1]), is an analog of the *N-p*-methoxybenzoyl-L-phenylalanine or *O-p*-methoxybenzoyl-L-phenyllactate; the latter is a good ester substrate of carboxypeptidase A. The ketonic inhibitor [1] contains a CH_2 group in place of the peptide NH or the ester oxygen which prevents cleavage of the inhibitor molecule. Most interestingly, one hydrogen atom of the 3-CH_2 group has been shown to undergo stereospecific exchange when labeled with ^2H at the hydrogen of R configuration.[48,49] The proton exchange to Glu 270 or a water molecule adjacent to Glu 270 could be facilitated by enolization promoted by the interaction between the ketonic carbonyl group and the zinc ion.

$$CH_3O-\bigcirc-\underset{3}{CO}-\underset{2}{CH_2}-\underset{1}{\overset{\overset{\displaystyle \bigcirc}{\overset{|}{CH_2}}}{CH}}-COO^-$$

[1]

```
                                  < Glu-Gln-His-Ala-Asp-Pro-Ile-


        8                12                        18
  -Cys-Asn-Lys-Pro-Cys-Lys-Thr-His-Asp-Asp-Cys-Ser-Gly-Ala-Trp-Phe-
    |                /                              | |
    |              /                                | |
    24              27                        34                39
  -Cys-Gln-Ala-Cys-Trp-Asn-Ser-Ala-Arg-Thr-Cys-Gly-Pro-Tyr-Val-Gly
```

FIGURE 3.   The amino acid sequence of potato carboxypeptidase inhibitor. Disulfide pairings are marked. The N- terminal residue is pyroglutamic acid.

Comparison of the complexes of carboxypeptidase A with Compound [1] and with Gly-Tyr revealed very similar binding modes for the two molecules, in particular for the location of the carboxyl, carbonyl, and phenyl groups. The large positional change of Tyr 248 and the smaller shifts of the zinc and the Arg 145 side chain were observed in both complexes.[116] A recent reinvestigation at 0.154 nm resolution of the structure shown in Figure 2 has shown that the carbonyl oxygen, which corresponds to that of an actual substrate, is not coordinated to the zinc ion, as shown in the figure. Instead, it is hydrogen bonded to the guanidinium group of Arg 127.[118]

Kinetic investigations have demonstrated that about five residues of the substrate (P_4-P_1') may influence both the binding and the catalytic parameters of the hydrolysis.[119] The structural details of binding at the S_2 and S_3 subsites have been established by crystallographic studies on the complex between carboxypeptidase A and the potato carboxypeptidase inhibitor (PCI).[120,121] This inhibitor is a single-chain polypeptide containing 39 amino acid residues, 3 disulfide bonds, a pyrrolidone carboxylic acid as the amino terminus, and a glycine as the carboxyl terminus (Figure 3).[122-124] The tertiary structure of the inhibitor is determined by the disulfide bridges and no α-helices or β-sheets are present. The four C-terminal residues of the inhibitor bind in the active site groove, defining subsites S_1', S_1, S_2, and S_3 on the enzyme. The C-terminal glycine is cleaved from the inhibitor in the complex, but remains trapped in the active site pocket (S_1' subsite).[120,121] Upon dissociation of the complex, free glycine and des-Gly 39 inhibitor are released.[124] With the isolated des-Gly 39 inhibitor, it has been shown that Gly 39 is not essential for inhibitor binding to carboxypeptidase A. Interestingly, and in contrast to the native inhibitor, the modified inhibitor is capable of binding to the enzyme reacted at Glu 270 with the affinity label bromo-N-acetyl-L-phenylalanine.[125] Perusal of the structure of the enzyme-inhibitor complex indicates that the active site pocket is of sufficient size to accommodate either the glycine or the modifying group, but not both species simultaneously.[121]

The carboxylate group of Val 38, which is generated by cleavage of the reactive site peptide bond, is coordinated to the zinc ion through one oxygen atom. The other oxygen atom seems to form a hydrogen bond with the phenol oxygen of Tyr 248 which undergoes a large conformational change as in the complexes of the enzyme with Gly-Tyr and the ketonic inhibitor. The amide nitrogen of Val 38 is also situated at a hydrogen bond distance from the same phenol oxygen atom. The side chain carbon atoms are in van der Waals contact with several residues, including Ile 247, Tyr 248, Ser 197, Tyr 198, and Phe 279.[121]

Besides Tyr 248, Arg 71 is the other enzymic residue that participates in hydrogen bonding contacts with the inhibitor main chain. Thus, at the S_2 subsite, the carbonyl oxygen of Tyr 37 of the inhibitor forms a hydrogen bond with the guanidinium group of Arg 71 of the enzyme. A few van der Waals contacts are also found between the Tyr 37 side chain and the S_2 binding site. Only van der Waals interactions are seen at the S_3 subsite with Pro 36 of the inhibitor. Interestingly, this side chain forms an intramolecular hydrogen bond with Trp 28 of the inhibitor. After Pro 36, the inhibitor peptide chain bends away from the active

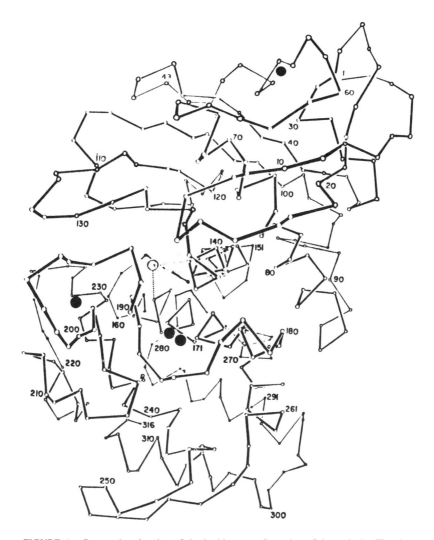

FIGURE 4. Perspective drawing of the backbone conformation of thermolysin. The zinc
ion is drawn stippled, and the calcium ions as solid circles. (From Colman, P. M., Jansonius,
J. N., and Matthews, B. W., *J. Mol. Biol.*, 70, 701, 1972. With permission.)

site groove. As a consequence, a fifth binding site, S_4, proposed on the basis of kinetic
studies,[119] is not observed in the potato inhibitor-carboxypeptidase A complex.[121] Besides
the interactions taking place in the active site groove, there are further interactions between
the enzyme and its inhibitor which contribute to the strong binding. The K_i for this complex
is about 2 nM.[124]

B. Thermolysin

The structure of thermolysin was initially determined at 0.23 nm resolution.[104,105,126,127]
Later refinement at 0.16 nm resolution caused only small adjustments in the atomic coordinates.[128] The overall structure of thermolysin consists of two roughly spherical domains
with a deep cleft across the middle of the molecule between the two domains (Figure 4).
The β-structures predominate in the N-terminal lobe and helices in the C-terminal lobe. The
catalytically essential zinc lies in the cleft. Two of the four calcium ions, Ca(1) and Ca(2)
in the C-terminal domain, bind very close together. They share the coordination of three
carboxyl groups (Glu 177, Asp 185, and Glu 190) which act as bridging ligands between

the two ions. Ca(1) also interacts with Asp 138 from the N-terminal domain, thereby linking together the two halves of the molecule. The coordination of both calcium ions is a distorted octahedral. The structural and functional roles of the metal ions in thermolysin have been reviewed.[129]

The structure of thermolysin does not show any special feature that would account for its enhanced thermostability. The bound calcium ions, however, clearly contribute to the overall stability of the molecule and may also help protect surface loops of the enzyme against autolysis.[127-129] Indeed, in the presence of EDTA and/or low calcium ion concentration, autolysis produces three fragments (1 to 196, 197 to 204, and 205 to 316), showing that limited proteolysis occurs at the loop involved in the binding of Ca(4).[130] Limited proteolysis of thermolysin by subtilisin was also found at exposed loops.[131,132] These loops were also characterized as possessing the highest crystallographic temperature factors determined with the native molecule.[127]

The overall dimensions of thermolysin (6.4 × 3.8 × 3.7 nm) are significantly different from those of carboxypeptidase A (5.0 × 4.2 × 3.8 nm). A remarkable difference between the two metalloproteases is that in carboxypeptidase A, the zinc ion lies in a shallow groove which terminates in a large pocket, whereas in thermolysin, it lies in a deep cleft open at each end. This is, of course, consistent with the fact that carboxypeptidase A is an exopeptidase, whereas thermolysin is an endopeptidase. Furthermore, a characteristic feature of the carboxypeptidase secondary structure is a twisted pleated sheet extending into the molecular interior and made up of eight polypeptide chains including residues widely separated in the amino acid sequence. No counterpart to this extended secondary structure is encountered in thermolysin. Finally, the three enzymic ligands to zinc in carboxypeptidase A are found at positions 69 (His), 72 (Glu), and 196 (His) in the amino acid sequence, whereas in thermolysin, the corresponding zinc ligands are grouped together in the sequence at positions 142 (His), 146 (His), and 166 (Glu).

The above dissimilarities clearly indicate that the two metalloproteases belong to different families. However, the active sites of thermolysin and carboxypeptidase A are remarkably similar, indicating convergent evolution for the two metalloproteases. Thus, the protein ligands (His, His, Glu) shown in Figure 5 are identical, although only one oxygen atom of the carboxylate group of glutamic acid is bound to the zinc in thermolysin,[127] whereas in carboxypeptidase A, both oxygen atoms of the corresponding carboxylate group were found as ligands to zinc.[112] Figure 5 also shows Glu 143 which can be involved in the attack of the substrate carbonyl carbon, either directly or through a water molecule; in this respect, it is similar to Glu 270 in carboxypeptidase A. Further similarities between the two proteases embrace a system of salt links from one of the histidine zinc ligands through an aspartic acid to an arginine residue. These are His 142-Asp 170-Arg 203 in thermolysin and His 69-Asp 142-Arg 145 in carboxypeptidase A. It is interesting that thermolysin, an endopeptidase, has Arg 203, an analog of Arg 145 of carboxypeptidase A, which binds the C-terminal carboxyl group of the substrate. Of course, the guanidinium group of Arg 203 in thermolysin could also be involved in substrate binding by forming a hydrogen bond with the P_1' carbonyl group.

There are two conspicuous differences between the active site regions of thermolysin and carboxypeptidase A. First, the NE2 atom of the imidazole ring of His 231 of thermolysin (Figure 5) is located 0.4 nm from the zinc, while the ND1 atom interacts with the carboxyl group of Asp 226. This amino acid couple has no counterpart in carboxypeptidase A. Second, the position of Tyr 157 in thermolysin is substantially different from that of Tyr 248 in native carboxypeptidase A.

The binding of several dipeptide analog inhibitors of thermolysin was examined in order to obtain information about the probable substrate binding modes.[133] Notably, difference Fourier maps have indicated that β-phenylpropionyl-Phe [2] binds to thermolysin with the

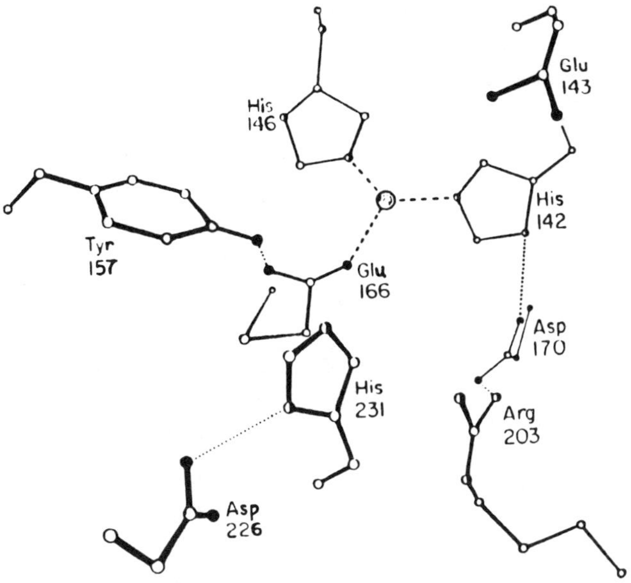

FIGURE 5. A scheme illustrating the position of some of the residues in the active site of thermolysin. The zinc ion is drawn stippled. (From Colman, P. M., Jansonius, J. N., and Matthews, B. W., *J. Mol. Biol.*, 70, 701, 1972. With permission.)

carbonyl oxygen of its peptide bond 0.21 nm from the zinc, displacing a water molecule, the fourth ligand to the metal ion. The carbonyl carbon is 0.39 nm from the side chain of Glu 143 and 0.38 nm from NE2 of His 231. The nitrogen of the inhibitor peptide bond forms a hydrogen bond with the carbonyl oxygen of Ala 113. These results could readily be incorporated into the model representing the binding mode of a specific substrate (Figure 6). The phenylalanyl phenyl ring of the inhibitor, which corresponds to R'_1 in Figure 6, was found to bind between Leu 202 and Val 139 in a cavity (S'_1 specificity pocket) lined entirely by nonpolar side chains. The other phenyl ring of the inhibitor lies with its plane almost parallel to that of Phe 114, at a distance of 0.37 nm, which corresponds to a favorable stacking interaction. The binding of the inhibitor causes little change in the conformation of the protein. Only localized alterations in the orientations of the side chains of Leu 202 and Asp 112 are worthy of mention. The side chain of Leu 202 rotates approximately 120° about its C_β–C_γ bond, away from the phenyl ring of the inhibitor. The side chain of Asp 112 also moves away from the inhibitor, but only slightly

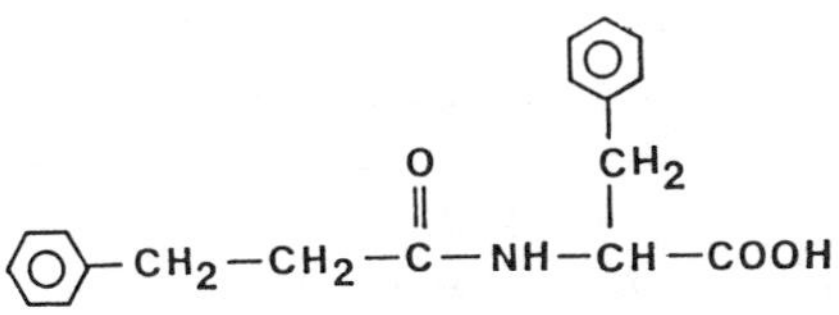

[2]

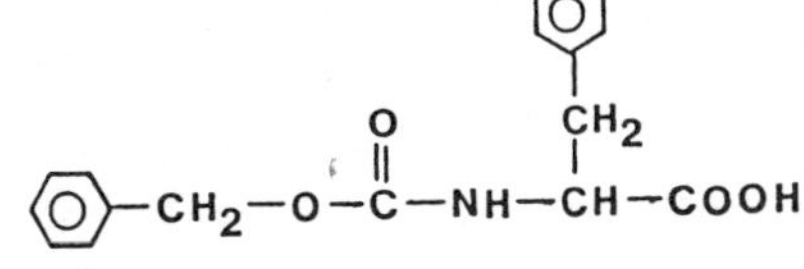

[3]

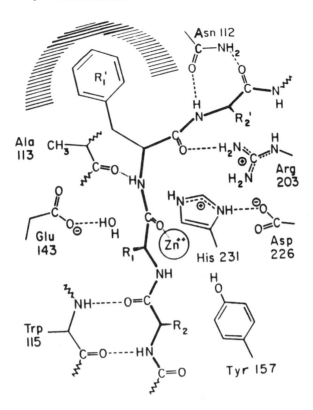

FIGURE 6.   Schematic drawing illustrating the binding of an extended substrate to thermolysin, inferred from the binding of inhibitors. (From Kester, W. R. and Matthews, B. W., *Biochemistry*, 16, 2506, 1977. With permission.)

It is of considerable interest that an almost isostructural inhibitor, benzyloxycarbonyl-Phe [3], which differs from β-phenylpropionyl-Phe [2] only in the substitution of an oxygen for the propionyl α-carbon, binds in a radically different manner with the two phenyl rings interchanged and the carboxyl group bond to the zinc.[133] It is important to note, however, that this binding mode cannot serve as a basis for constructing a model for substrate binding.[133] Indeed, the substrate binding depicted in Figure 6 was inferred from the binding mode of β-phenyl-propionyl-Phe, at least for the binding of residues P_1 and P_1'. The placement of residues P_2 and P_3 was determined from model building alone. It appears from the model building that the substrate backbone at the P_2 residue forms a pair of antiparallel β-sheet hydrogen bonds with the protein backbone at Trp 115. This is reminiscent of the binding of polypeptide substrates to serine and cysteine proteases, although in the serine enzymes, it is the P_3 rather than the P_2 residue that forms a pair of backbone hydrogen bonds to the protein (Chapter 3, Section III.B). No such binding mode was observed in the case of carboxypeptidase A, although a number of similarities exists in the immediate vicinity of the scissile bond.

Direct evidence for the geometry of binding of residues P_1' and P_2' was provided by the analysis of phosphoramidon-inhibited thermolysin.[134,135] Phosphoramidon (N-(α-L-rhamnopyranosyl-oxyhydroxyphosphinyl)-Leu-Trp), shown by Structure [4], is a potent inhibitor of thermolysin (K_i = 28 nM).[136,137] Its binding to the enzyme is analogous to the binding of β-phenylpropionyl-Phe, rhamnose , Leu, and Trp occupying the S_1, S_1', and S_2' subsites, respectively.[134] A pair of S_1'-P_2' hydrogen bonds exists between the inhibitor and the side chain of Asn 112. In addition, the tryptophan ring nitrogen forms a hydrogen bond to the backbone carbonyl oxygen of Asn 111. Thus, the direction of the extended substrate backbone appears to be antiparallel to that of the protein backbone between Asn 111 and Asn 116. The side chain of Arg 203 is also involved in hydrogen bonding with the substrate backbone (Figure 6).

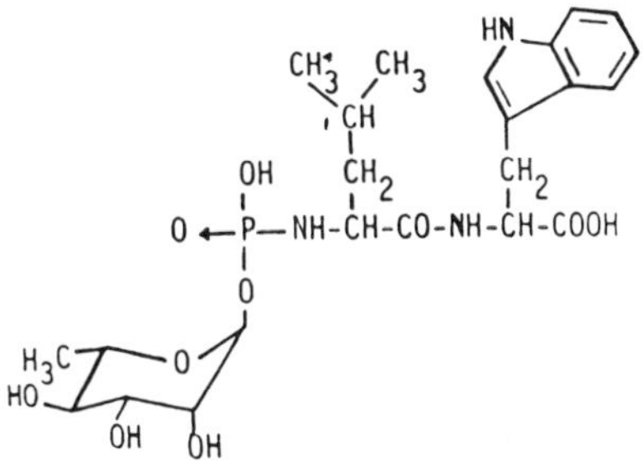

[4]

It is of interest that the contribution to binding of the sugar is negligible. Notably, the phosphoramidon analog, N-phosphoryl-Leu-Trp, which lacks the rhamnose moiety, binds to the thermolysin even somewhat more tightly than does phosphoramidon.[137] However, the phosphoryl group appears to be crucial in the binding of the inhibitor. Indeed, the simple phosphoramidate N-phosphoryl-Leu-NH$_2$ is an excellent inhibitor of thermolysin (K$_i$ = 1.3 μM).[138] In general, the addition of substituents to the phosphoramidate group tends to decrease the effectiveness of the inhibitor. For instance, the binding of O-methylphosphoryl-Leu-NH$_2$ is weaker by about 100-fold relative to phosphoryl-Leu-NH$_2$, and O-methyl-O-phenylphosphoryl-Leu-NH$_2$ does not inhibit thermolysin.[138] Crystallographic analysis of the geometry of the phosphoryl group in the phosphoramidon-thermolysin complex has suggested that the geometry of the phosphoryl group resembles that of the presumed tetrahedral intermediate formed during the hydrolysis of peptides.[134] Figure 7 shows the interactions between the active site of thermolysin and two inhibitors, phosphoryl-Leu-NH$_2$ (Figure 7A) and phosphoramidon (Figure 7B). The high resolution (0.16 nm) data revealed that phosphoramidon binds to the zinc ion with a single oxygen of the phosphoramidate moiety. Together with the three ligands to the metal from the protein, the resultant complex has an approximately tetrahedral geometry. On the other hand, phosphoryl-Leu-NH$_2$ binds to the zinc with two of the phosphoramidate oxygens, leading to a pentacoordinated complex.[135] The zinc ion with five ligands in thermolysin was also observed with hydroxamates[139] and N-(1-carboxy-3-phenyl)propyl-Leu-Trp [5].[140] The latter inhibitor binds to thermolysin with both oxygens of the carboxymethyl group liganded to the zinc, as shown in Figure 8. The importance of pentacoordination of the zinc ion in the transition-state complex will be discussed in Section IV.

[5]

There is a third protease whose structure has recently been determined at medium resolution.[141] This enzyme is a zinc-containing D-alanyl-D-alanine-cleaving carboxypeptidase (Zn^{2+} G peptidase) which participates in bacterial cell wall metabolism. It is isolated from *Streptomyces albus*. As compared with carboxypeptidase A or thermolysin, Zn^{2+} G peptidase is a significantly smaller protein containing only 212 amino acid residues. The molecule consists of two distinct globular domains, a small N-terminal part (76 residues), and the

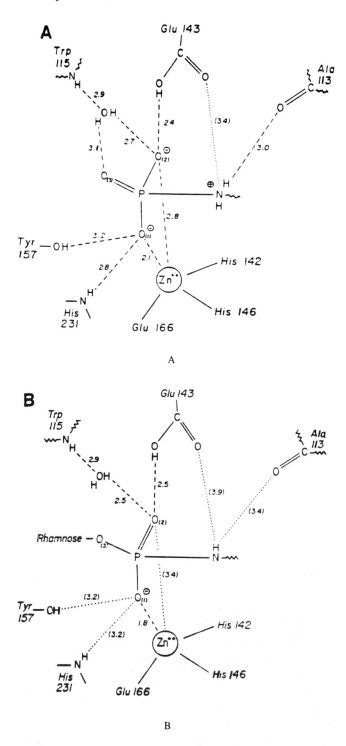

FIGURE 7.   Interactions between thermolysin and two phosphoramidate inhibitors. Scheme A shows the interactions for phosphoryl-Leu-NH$_2$; scheme B illustrates the interactions for phosphoramidon. Presumed hydrogen bonds and interactions with the zinc are drawn as broken lines, other close approaches are drawn as dotted lines with the distances indicated in parentheses. Distances are given in Å (1 Å = 0.1 nm). (From Tronrud, D. E., Monzingo, A. F., and Matthew, B. W., *Eur. J. Biochem.*, 157, 261, 1986. With permission.)

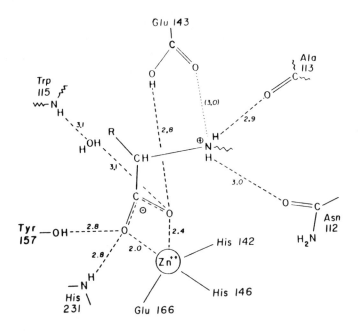

FIGURE 8. A scheme for the interactions and interaction distances between *N*-(1-carboxy-3-phenylpropyl)-Leu-Trp and thermolysin in the vicinity of the zinc ion. Presumed hydrogen bonds are drawn as broken lines, and the close contact between the nitrogen atom of the inhibitor and Glu 143 is indicated by a dotted line with the distance in parentheses. (From Monzingo, A. F. and Matthews, B. W., *Biochemistry*, 23, 5724, 1984. With permission.)

larger C-terminal domain (136 residues) which holds the active site. The active site cleft cuts the domain into two parts with the zinc ion bound inside and coordinated with three protein ligands, His 152, His 193, and His 196. Instead of the catalytically competent Glu of carboxypeptidase A (residue 270) and thermolysin (residue 143), an Asp (159 or 192) fulfills the corresponding role in the action of Zn^{2+} G peptidase. The folding of the molecule indicates that Zn^{2+} G peptidase belongs to a family different from from that of carboxypeptidase A or thermolysin.

IV. INHIBITORS

A. Low Molecular Mass Inhibitors

Besides their use in mechanistic studies, inhibitors to metalloproteases have also attracted considerable interest as potential therapeutic agents. In particular, the inhibition of ACE, neutral endopeptidase 24.11, and vertebrate collagenase have been studied. The kinetic and structural results obtained with carboxypeptidase A have served primarily as the basis for inhibitor design when the structure of the respective enzyme was unknown.

Several simple carboxypeptidase inhibitors, such as 3-phenylpropionic acid and D- and L-phenylalanine, have been examined by kinetic and X-ray crystallographic methods.[2] 3-Phenylpropionic acid is one of the best of the simple inhibitors ($K_i \sim 0.1$ mM). Under some conditions, it shows mixed inhibition, and crystallographic studies have shown that it occupies two distinct loci in the active site region. In contrast, D- and L-phenylalanine have been shown to be bound at a single, but not entirely identical locus.[2]

Among the simple compounds, the dicarboxylic acid, L-benzylsuccinic acid, proved to be an extremely effective reversible inhibitor ($K_i = 0.45$ μM).[142,143] The presence of the two carboxyl groups in the given positions appears to be critical; the corresponding malonic and glutaric acid derivatives have been found to be considerably less effective. The structural relationship between the carboxyl groups of L-benzylsuccinic acid [6] resembles that of the collected products of peptide hydrolysis [7] at the moment of their formation, or as they combine with the enzyme in the reverse, peptide-forming reaction.

$$COO^-$$
$$|$$
$$Ph-CH_2-CH-CH_2-COO^-$$

[6]

$$COO^- \qquad\qquad R$$
$$| \qquad\qquad\qquad |$$
$$Ph-CH_2-CH-NH_2 \quad {}^-OOC-CH-NH-peptide$$

[7]

Efforts to determine the structure of the complex of benzylsuccinate with carboxypeptidase A have been frustrated by disordering and solution of enzyme crystals soaked in solutions containing the inhibitor.[143] However, from studies on the presumably analogous complex formed between benzylsuccinate and thermolysin, a possible binding mode for benzylsuccinate and carboxypeptidase A could be inferred.[144] According to this model, the benzyl group occupies the S_1' active site pocket, the P_1' carboxyl group forms a salt bridge with Arg 145, and the other carboxyl group is liganded to the zinc ion.

pH-Dependence studies have indicated that benzylsuccinic acid exhibits maximum inhibitory effect toward carboxypeptidase A at pH values where the inhibitor bears a single negative charge in solution.[143] With the use of inhibitors enriched with ^{13}C at each of the carboxyl groups, alternatively, it could be demonstrated by NMR measurements that both carboxyl groups are ionized in the enzyme-bound inhibitor.[145] This is consistent with the above structural model [7] showing that each carboxyl group interacts with a positive ion, i.e., with a zinc and a guanidinium ion, respectively. The apparent inconsistency with the pH-dependence studies showing only one carboxyl group in dissociated form may be reconciled by postulating that the hydrogen ion of the nondissociated carboxyl group is taken up by the protein as the inhibitor is bound.[145]

The enzymic group accepting the proton from the inhibitor is most probably the carboxylate group of Glu 270.[145] This follows from the model derived from X-ray crystallography[144] which indicated that the carboxyl group of benzylsuccinate liganded with one oxygen to the zinc ion, points with the other oxygen directly toward the side chain carboxylate of Glu 270. Juxtaposition of two carboxylate groups would be expected to create a repulsive electrostatic potential. However, addition of a proton to the carboxylate of Glu 270 can eliminate the repulsive potential. The resulting carboxylate-carboxylic acid hydrogen bond would explain the observed coupling between proton uptake and binding of the inhibitor with two dissociated carboxyl groups.

Benzylsuccinic acid has served as a point of departure for the design of drugs antagonistic to ACE[29,146-147a] and as an affinity ligand for the isolation of carboxypeptidases.[148] Replacement of the zinc-binding carboxyl group, as found with benzylsuccinic acid, by a more effective thiol ligand led to captopril (SQ 14,225), the first orally active antihypertensive drug.[146,149] This potent competitive inhibitor ($K_i = 1.7$ nM) may be considered as a biologically stable analog of the optimal C-terminal dipeptide, Ala-Pro, of substrates or inhibitors which also incorporates a strong zinc ligand [8]. Several analogs of captopril have been reported.[29] For instance, D-cysteinyl-L-proline [9], which contains an amino group in place of the methyl substituent of captopril, is only two to three times less effective in vitro as an ACE inhibitor than captopril.[150]

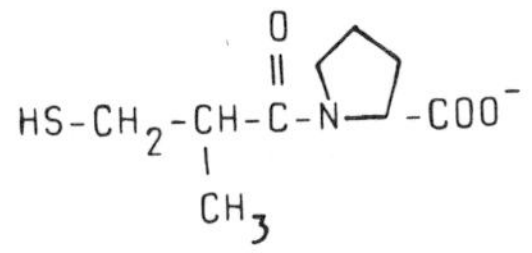

$$HS-CH_2-\underset{\underset{CH_3}{|}}{CH}-\overset{\overset{O}{\|}}{C}-N \overset{\frown}{\diagdown}-COO^-$$

[8]

$$HS-CH_2-\underset{\underset{^+NH_3}{|}}{CH}-\overset{\overset{O}{\|}}{C}-N \overset{\frown}{\diagdown}-COO^-$$

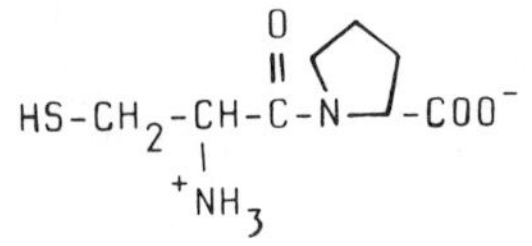

[9]

N-Phosphoryl-Ala-Pro [10] is another type of inhibitor of ACE[151-153] The binding modes of the N-phosphoryl peptides to thermolysin have been discussed in the preceding section. This compound, however, does not appear to be very potent in vivo.[29]

Patchett and co-workers[147] have synthesized substituted N-carboxymethyl dipeptides, representing another class of inhibitors of ACE. Some of these inhibitors, such as enalapril or enalaprilat [11] and lisinopril [12], are more potent than captopril [8] and do not contain a free thiol group that causes side effects. The carboxyl function linked to the phenylpropyl group is assumed to be liganded to the zinc ion upon binding. Nevertheless, the ethyl ester of [11] (MK-421) has been shown to be an orally effective inhibitor of the converting enzyme, presumably because it is hydrolyzed in vivo.[154]

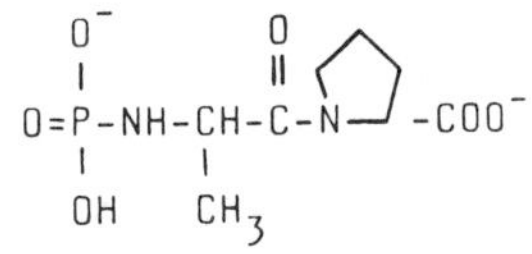

$$\underset{\underset{OH}{|}}{\overset{\overset{O^-}{|}}{O}}=P-NH-\underset{\underset{CH_3}{|}}{CH}-\overset{\overset{O}{\|}}{C}-N \overset{\frown}{\diagdown}-COO^-$$

[10]

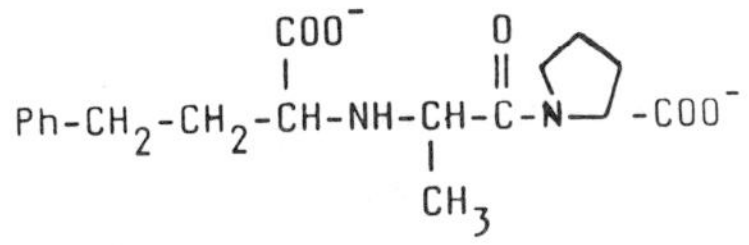

$$Ph-CH_2-CH_2-\underset{\underset{CH_3}{|}}{\overset{\overset{COO^-}{|}}{CH}}-NH-\underset{}{CH}-\overset{\overset{O}{\|}}{C}-N \overset{\frown}{\diagdown}-COO^-$$

[11]

$$Ph-CH_2-CH_2-\underset{\underset{\underset{\underset{^+NH_3}{|}}{(CH_2)_4}}{|}}{\overset{\overset{COO^-}{|}}{CH}}-NH-\overset{}{CH}-\overset{\overset{O}{\|}}{C}-N \overset{\frown}{\diagdown}-COO^-$$

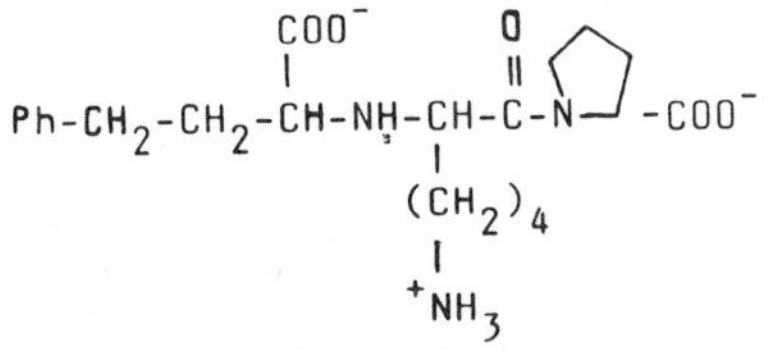

[12]

It is worth noting that enalapril, lisinopril, as well as captopril are slow binding inhibitors.[155,156] With this kind of inhibitor, the equilibrium between enzyme, inhibitor, and enzyme-inhibitor complex is not reached before substantial depletion of the substrate has taken place.[157] In other words, at inhibitor concentrations that cause moderate inhibition, the steady state is reached much more slowly than in the absence of the inhibitor. Slow inhibition associated with tight binding has been observed in several other cases, for example, pepsin with pepstatin[158] and cathepsin with leupeptin[159] interact through a slow binding mechanism.

Natural peptide inhibitors of ACE have been isolated from the venom of *Bothrops jararaca*, a South American pit viper.[160,161] The most potent of these inhibitors is the nonapeptide bradykinin-potentiating peptide 9a (BPP$_{9a}$) or SQ 20,881 [13]. The other extensively studied inhibitor is the pentapeptide BPP$_{5a}$ or SQ 20,475 [14]. The N-terminal residue is pyroglutamic acid (L-2-pyrrolidone-5-carboxylic acid) for both venom peptides. The C-terminal tripeptide parts of the inhibitors compete with the substrates, angiotensin I, and bradykinin for the S$_1$, S$_1'$, and S$_2'$ subsites. Replacement of any of the four proline residues of BPP$_{9a}$ by dehydroproline produced a 20- to 100-fold increase in inhibitory activity.[162]

<Glu–Trp–Pro–Arg–Pro–Gln–Ile–Pro–Pro

[13]

<Glu–Lys–Trp–Ala–Pro

[14]

Strong inhibition of ACE by aldehyde and ketone substrate analog[163-166] and α-fluoro ketones[167] has been reported. These compounds in their hydrated forms could act as transition-state inhibitors.

Endopeptidase 24.11 or "enkephalinase" is, in many respects, similar to ACE, but its hydrophobic pocket is at the S$_1'$ subsite rather than at the S$_1$ subsite. This specificity difference has been exploited in the design of specific inhibitors. Thus, thiorphan [15], a structural analog of captopril, is about 30 times more active toward "enkephalinase" than toward ACE,[168] whereas captopril displays an inverse specificity pattern. Other 3-thiolpropanoyl peptide derivatives were synthesized to study further details of the active sites of "enkephalinase" and the converting enzyme.[169,170] Substituted *N*-carboxymethyl dipeptides[171,172] and other peptide derivatives,[173] including hydroxamic acids and phosphoramidon, were also found to be potent "enkephalinase" inhibitors.

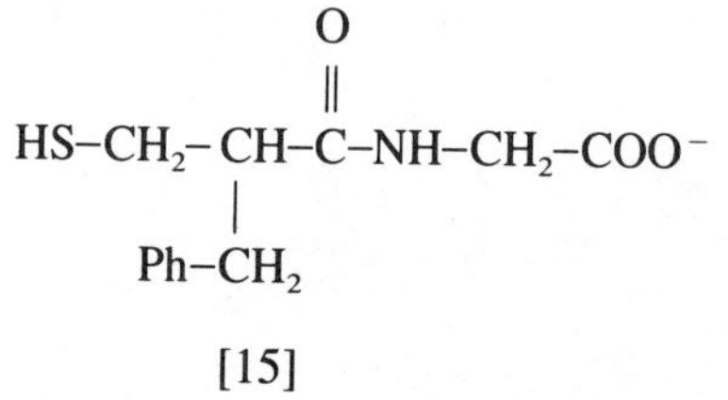

$$\begin{array}{c} O \\ \parallel \\ HS{-}CH_2{-}\underset{\underset{Ph{-}CH_2}{|}}{CH}{-}C{-}NH{-}CH_2{-}COO^- \end{array}$$

[15]

Inhibitors to other metalloproteases have also been designed. In particular, effective inhibitors to aminopeptidases have been reported to include amino aldehydes,[73,174] amino ketones, and derivatives,[175,176] boronic acid derivatives,[177,178] amino acid thiols,[179] and amino acid hydroxamates.[70,180,181] Screening of *Streptomyces* culture filtrates for inhibitors to aminopeptidases has led to the discovery of bestatin [16][182,183] and amastatin [17].[184]

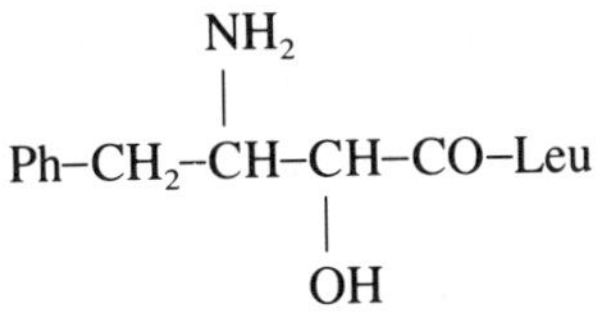

$$NH_2$$
$$|$$
$$Ph-CH_2-CH-CH-CO-Leu$$
$$|$$
$$OH$$

[16]

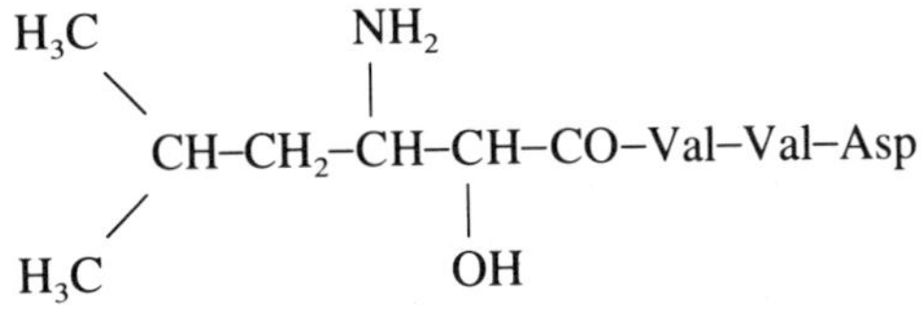

$$H_3C \qquad\qquad NH_2$$
$$\diagdown \qquad\qquad |$$
$$CH-CH_2-CH-CH-CO-Val-Val-Asp$$
$$\diagup \qquad\qquad |$$
$$H_3C \qquad\qquad OH$$

[17]

B. Protein-Protease Inhibitors

In Section II.A, we discussed the small PCI. There is another type of metalloprotease inhibitor, the vertebrate collagenase inhibitor, which has extensively been investigated because of its physiological importance. A collagenase inhibitor separate from α_2-macroglobulin was first detected in human serum.[185] It inhibited collagenases purified from several human tissues, but clostridial collagenase and thermolysin were not inhibited. The purified inhibitor, a glycoprotein, exhibited an apparent molecular mass of 30 kdaltons by sodium dodecyl sulfate/polyacrylamide-gel electrophoresis.[186]

Collagenase inhibitors were also purified from other sources, such as amniotic fluid,[187,188] human skin fibroblast culture medium,[189] human sinovial fluid,[190] and bovine cartilage and body fluids.[191] It is likely that all these proteins are identical.[186]

The collagenase inhibitor isolated from leukocytes may be different from the inhibitors just discussed.[192-194] This inhibitor has a free thiol group which is necessary for the inhibition of collagenase, which possesses a reactive disulfide bond. Upon formation of the enzyme-inhibitor complex, a thiol-disulfide interchange takes place. From the resulting covalent complex, the enzyme can be reactivated by the addition of disulfide compounds, such as oxidized glutathione. The thiol-disulfide interchange mechanism was also demonstrated for the inhibition by the plasma collagenase inhibitor,[193] but later studies with a plasma inhibitor preparation did not confirm the existence of the essential thiol group.[186]

The amino acid sequence of the human amniotic fluid inhibitor, also called tissue inhibitor of metalloproteinases (TIMP), has been determined from the corresponding cDNA sequence.[195] The inhibitor contains 207 amino acids, 23 of which constitute the signal sequence. The predicted molecular mass of the unglycosylated mature TIMP is 20,685. Two potential N-glycosylation sites are seen in the amino acid sequence. The amino acid sequence of TIMP was unexpectedly found to be identical to that of a protein which specifically stimulates the growth of peripheral blood- and bone marrow-derived erythroid precursors.[196]

V. MECHANISM OF ACTION

In the case of chymotrypsin, the basic features of the mechanism of action, including the catalytically competent Ser and His residues, were established without the knowledge of the tertiary structure (Chapter 3, Section I). By contrast, the important mechanistic features of the carboxypeptidase A mechanism have not been unambiguously ascertained even in the light of the three-dimensional structure.[2,3] Moreover, one of the catalytic groups, the general acid catalyst, was assigned erroneously to Tyr 248. Mechanistic suggestions based on kinetic and modification studies carried out before the crystal structure determination of carboxypeptidase A have been reviewed.[2] In the following, we shall consider the mechanistic proposals which rely on the three-dimensional structures of carboxypeptidase A and thermolysin.

Valuable structural information for the carboxypeptidase A mechanism was obtained from X-ray analyses of enzyme-substrate complexes.[2,3,117,197,198] These studies, discussed partly in Section II.A, have shown that the enzymic groups, which are near enough to the scissile peptide bond to be directly involved in the catalysis, are the zinc ion coordinated with a water molecule, Glu 270, and Tyr 248. The mechanistic suggestions deduced from the crystallographic data are depicted in Figure 9. Although consideration of the active site structure severely limits the possible mechanisms, it does not yield a unique interpretation. Thus, the carboxylate group of Glu 270 could function either as a nucleophile by forming an acyl-enzyme (anhydride) intermediate with the carbonyl group of the substrate (Figure 9B), or as a general base by promoting the attack of a water molecule on the same carbonyl carbon atom (Figure 9C). A zinc-bound water or hydroxide ion was also considered as a possible nucleophile.[117] The zinc ion interacts with the carbonyl oxygen and serves to polarize the C=O bond. The phenolic hydroxyl of Tyr 248 was originally thought to be a general acid catalyst donating a proton to the leaving group NH of the scissile bond.[2,3,117,197,198] However, site-directed mutagenesis has recently shown that this is not the case.[199,199a] Tyr 248 was replaced by Phe, and the new variant possessed an essentially unchanged catalytic constant (k_{cat}) with various peptide and ester substrates. However, the Michaelis constant K_m of peptide substrates and the inhibition constant (K_i) of the PCI increased 6- and 70-fold, respectively. These data indicate that Tyr 248 does not serve as a general acid catalyst, but could participate in ligand binding. As for the nature of the general acid catalyst, we shall return to this point after the discussion of the relevant mechanistic features of thermolysin catalysis.

The active site of thermolysin is, in many respects, similar to that of carboxypeptidase A (Section II.B). However, a difference between the active site clefts of the two enzymes renders it possible to distinguish between the acyl-enzyme and the general base mechanisms in the case of thermolysin. Specifically, Glu 143 of thermolysin, the counterpart of Glu 270 of carboxypeptidase A, is located within a rather narrow cleft which does not permit the approach of the scissile bond sufficiently close enough to form a covalent adduct without large conformational changes in the protein.[133] In the case of carboxypeptidase A, the active site is more open so that the same structural restrictions do not apply. Furthermore, upon binding to thermolysin, the transition-state analog inhibitor, phosphoramidon, provides a structural analog of the tetrahedral intermediate that would be formed following the attack of a water molecule on the carbonyl carbon.[134]

Another mechanistically significant difference between carboxypeptidase A and thermolysin is that the latter does not possess a Tyr corresponding to Tyr 248 of carboxypeptidase A. Instead, thermolysin has His 231 which was originally thought to contribute a proton to the leaving NH group of the scissile peptide bond,[133,134,200-202] just as Tyr 248 was proposed to do it in carboxypeptidase A. However, later structure analyses at high resolution have revealed novel features of the enzyme-inhibitor interaction that were not anticipated in previous studies. In particular, transition-state analog inhibitors, such as a substituted N-carboxymethyl peptide (Figure 8)[140] and a phosphoramidate (Figure 7),[135] have indicated that the inhibitor nitrogen atom corresponding to the leaving nitrogen atom of the tetrahedral intermediate is close enough to Glu 143 so that this residue may contribute a proton to the leaving group during catalysis, whereas His 231 is too far away to do so.[135,140,203] Consequently, it has been proposed that Glu 143 is protonated first by accepting the proton from the water molecule attacking the substrate carbonyl carbon atom and subsequently donates the same proton to the leaving nitrogen.[135,140,203]

A further important result of the structural studies of enzyme-inhibitor complexes concerns

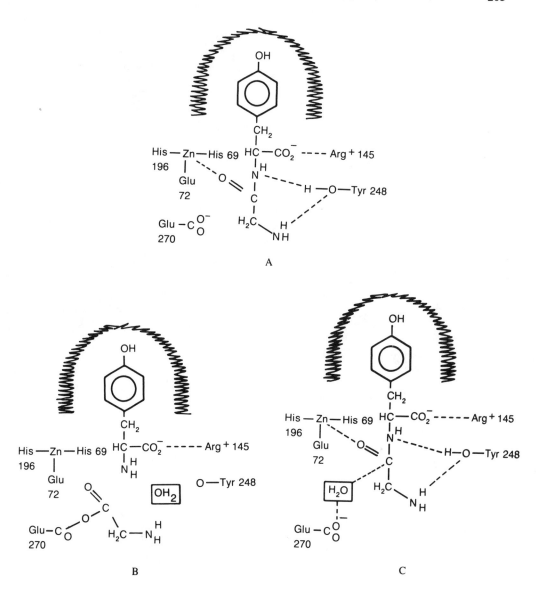

FIGURE 9. Possible intermediates in the hydrolysis of Gly-Tyr by carboxypeptidase A. Michaelis complex (A), covalent anhydride intermediate (B), and noncovalent intermediate (C). (From Quiocho, F. A. and Lipscomb, W. N., *Adv. Protein Chem.*, 25, 1, 1971. With permission.)

the number of the zinc ligands which are present upon formation of the tetrahedral intermediate. Several such structures have been shown to involve a pentacoordinate zinc complex (Figures 7 and 8)[135,139,140] supporting the participation of such intermediates in the mechanism of action of thermolysin. On the basis of these results, involving (1) general acid-base catalysis by Glu 143, (2) binding of the tetrahedral intermediate as a bidentate structure, and (3) binding interactions at different subsites, an interactive computer graphics study of thermolysin catalysis was undertaken by Hangauer et al.[203] They examined two possible structures for the Michaelis complex and three for the tetrahedral intermediate. The most probable mechanistic features which they deduced are depicted in Figure 10A and 10B.

It is seen from Figure 10A that the Michaelis complex does not involve coordination of the scissile peptide to the zinc ion. This is in contrast to many earlier suggestions, but model building of the Michaelis complex indicates that such a coordination would result in unfa-

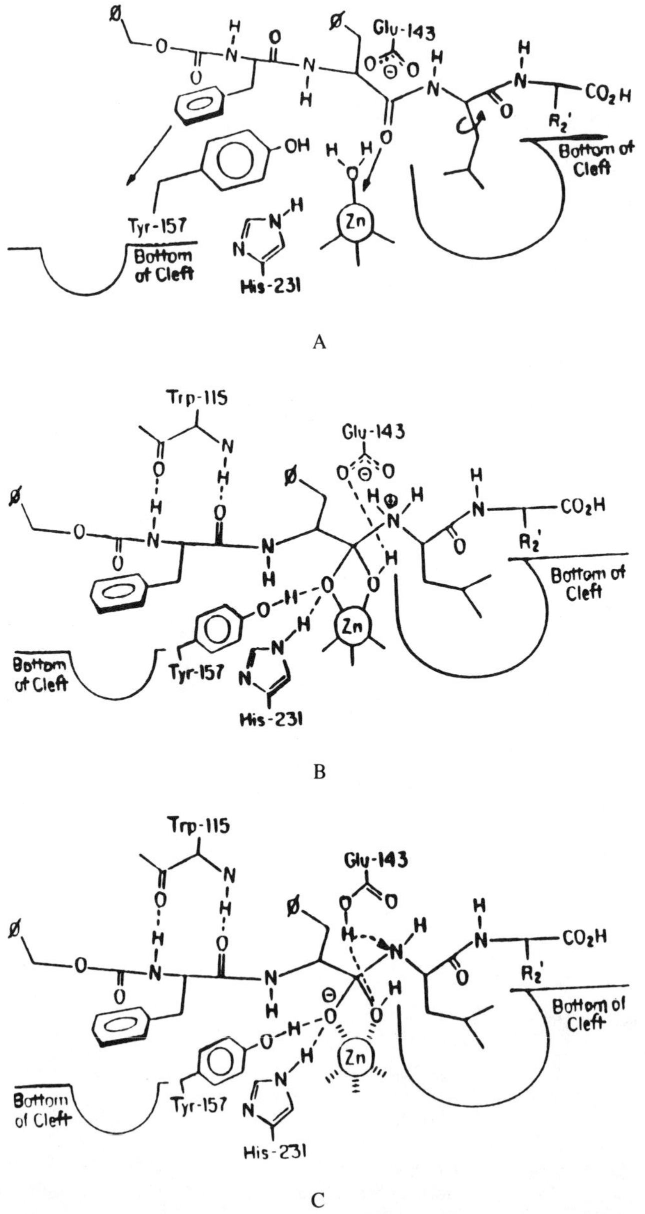

FIGURE 10.   Schematic representation of proposed intermediates in the catalysis by thermolysin. Schemes A and B indicate the Michaelis complex and the tetrahedral intermediate, respectively. (From Hangauer, D. G., Monzingo, A. F., and Matthews, B. W., *Biochemistry,* 23, 5730, 1984. With permission.) Scheme C is a modified version of the tetrahedral intermediate.[203a] The route of proton transfer from the zinc-bound water to the leaving nitrogen atom is shown by a dashed arrow.

vorable contacts at several subsites. As the scissile peptide moves toward the metal ion, the zinc-bond water molecule attacks the carbonyl carbon, and this process is promoted by general base catalysis by Glu 143. At the same time, the carbonyl oxygen becomes liganded to the zinc, which altogether results in the formation of the tetrahedral intermediate bound to the zinc ion as a bidentate complex (Figure 10B). In this structure, Tyr 157 and His 231 are within hydrogen bond distance of the oxyanion and may help stabilize the tetrahedral intermediate. The negative charge is not shown in Figure 10B, but indicated in Figure 10C, illustrating an alternative structure of the tetrahedral intermediate. The principal differences between the mechanisms depicted in Figure 10B and 10C is the decomposition of the tetrahedral intermediate, i.e., the C–N bond cleavage. The mechanism of Figure 10B involves the transfer of both protons of the attacking water to the leaving nitrogen. The transfer of one proton leads to the structure of Figure 10B. Hangauer et al.[203] proposed that Glu 143 once again functions as a general base-acid catalyst by abstracting the remaining proton on the oxygen derived from the nucleophilic water molecule and delivering it to the positively charged nitrogen, with the subsequent release of the protonated amine. However, protonation of the protonated amine is apparently not feasible, not only because of repulsion between the positive charges, but also because the protonated nitrogen does not have a free electron pair which could accept the second proton. Therefore, the one-proton transfer mechanism (Figure 10C) appears to be more reliable. This involves a general base-catalyzed formation of the tetrahedral intermediate by Glu 143 as shown in Figure 10C, and this is followed by the general acid-catalyzed bond cleavage. The generated amine product then accepts a proton from solution at neutral pH.[203a]

Single proton transfer to the leaving nitrogen has also been found in the catalysis by serine, cysteine, and aspartic proteases, indicating that the second proton transfer is unnecessary. Moreover, removal of the second proton from the zinc-bound carboxyl group would make the release of the acyl moiety difficult. Specifically, after the C–N bond cleavage, the negatively charged carboxylate group would bind strongly to the zinc ion as a bidentate ligand. On the other hand, the protonated carboxyl group may bind as a monodentate ligand permitting the coordination of a new water molecule, which in turn displaces the carboxyl group, probably facilitated by the hydrogen bonds from Tyr 157 and His 231 to the carboxyl group. The same residues can also stabilize the tetrahedral intermediate in a manner similar to the oxyanion binding site residues of serine proteases.

By analogy with thermolysin, it has been proposed that a related mechanism should be considered for peptide cleavage by carboxypeptidase A.[140] In such a mechanism, Glu 270 of carboxypeptidase A would promote the attack of the zinc-bound water molecule on the carbonyl carbon of the substrate. Glu 270 would then transfer the proton to the leaving nitrogen of the pentacoordinate complex. Thus, Tyr 248 of carboxypeptidase A would not act as a proton donor.[140] This proposal has recently been confirmed by site-directed mutagenesis.[199] The following X-ray crystallographic studies of carboxypeptidase-inhibitor complexes, using aldehyde,[204] ketone,[205] and phosphonamidate[206] inhibitors, have also clearly indicated that Tyr 248 is involved in substrate binding, rather than in proton donation to the leaving group, and that proton donation could be mediated by Glu 270. However, it was emphasized by the authors[204,205] that the other pathway, formation of the covalent anhydride intermediate, cannot be ruled out simply on the basis of the structure of the enzyme-inhibitor complexes. These investigations have also shown that Arg 127 can assist in the stabilization of the oxyanion of the tetrahedral intermediate by hydrogen bonding and/or electrostatic stabilization. Thus, Arg 127 would functionally correspond to His 231 of thermolysin.

The most recent X-ray crystallographic studies[206a] support a mechanism for carboxypeptidase A action similar to that illustrated in Figure 10C for the thermolysin action, except that the protonation of the amine product occurs inside the enzyme-product complex from the product carboxyl mediated by Glu 270. As pointed out above, this would inhibit the release of the negatively charged product from the zinc ion.

A major issue of the carboxypeptidase A mechanism concerns ester hydrolysis. A number of reports have suggested that this process occurs through the formation of a covalent intermediate involving Glu 270. [207-211] The experiments supporting this conjecture are based mainly on studies providing indirect kinetic evidence. None of these studies, including gel filtration at subzero temperature in the presence of cryosolvent,[210] demonstrated that the intermediate found was of a true covalent species. Even resonance Raman measurements at subzero temperature could not confirm the existence of an anhydride intermediate.[212] Nevertheless, on the basis of similar stereochemistries and pH-rate profiles for the ester and peptide reactions, it was claimed that not only esters but also peptides are hydrolyzed by the anhydride mechanism.[210,213] The difference between the two mechanisms would be that formation of the mixed anhydride is rate determining in peptide hydrolysis, while deacylation of the mixed anhydride is rate limiting in ester hydrolysis.[213] Other studies, however, have led to the opposite conclusion, stating that if an acyl-enzyme intermediate is involved in the ester catalysis, then the rate of acylation must be slower than the rate of deacylation.[214] Indeed, the experimental data are insufficient to prove a covalent anhydride intermediate even in the case of ester hydrolysis by carboxypeptidase A, although this possibility cannot be ruled out.

Intermediates, but not of the covalent types, have been demonstrated in Co^{2+}-carboxypeptidase A catalysis. [215-219] Two intermediates were observed during the hydrolysis of both peptides and depsipeptides (the corresponding esters). The second intermediate on the reaction path has been characterized as an enzyme-substrate rather than an enzyme-product complex.[218] The visible absorption spectra of the complexes formed with the Co^{2+}-enzyme were significantly different for the peptide and the corresponding ester substrates.[216]

The cobalt-substituted carboxypeptidase A proved to be very useful in the studies of interactions of substrates and inhibitors with the active site. As demonstrated by X-ray crystallography, the coordination structure of the metal ion in Co^{2+}-carboxypeptidase A is essentially identical with that of the native Zn^{2+}-enzyme.[220] The Co^{2+}-enzyme is twice as active as the native enzyme toward peptides, but its activity toward esters is nearly the same.[221] The pH dependence of peptidase activity of the Co^{2+}-enzyme and the spectral changes in absorptivity at 625 nm are superimposable, indicating that the ionization of a single group ($pK_a = 8.8$) is responsible for these two effects.[222]

Besides Co^{2+}, other metal ions (Mn^{2+}, Cd^{2+}) have also been used to probe the active site of carboxypeptidase A.[221] The peptide and ester hydrolyses were affected differently by the various metal ions. Changes in the k_{cat} and K_m values indicated that in the case of peptides, the metal is required for catalysis, but not for the binding, and in the case of esters, it is critical for both binding and catalysis.[221] Indeed, the peptides bind tightly to the apoenzyme lacking the metal ion, whereas esters do not bind.[221]

The involvement of zinc ion in the hydrolysis of peptides is supported by the studies using an oligopeptide substrate containing a thiopeptide linkage as the scissile bond.[223] This substrate was not hydrolyzed by carboxypeptidase A, an analogous finding to that observed in the reactions of serine proteases with thionesters (Chapter 3, Section VI.B). It is interesting that substitution of the active site Zn^{2+} by Cd^{2+} yields an enzyme inactive toward ordinary peptide substrates, but active toward the thiopeptide substrate. The mechanisms, as indicated by the similar pH dependencies of the catalyses, appear to be identical for the zinc and the cadmium enzymes, but the rate constants are considerably smaller with the cadmium enzyme and the thiopeptide than with the native carboxypeptidase A and the ordinary peptide substrates.[223] Further studies are needed to explain this phenomenon.

An index of the Lewis acidity of the active site zinc ion has been provided by a spectrophotometric study of the binding of a potent competitive inhibitor to carboxypeptidase A. Upon ligation to the active site, the pK_a of the azophenol moiety of the inhibitor was changed from 8.76 to 4.9.[224] Hence, the active site zinc ion stabilizes a coordinated phenolate anion by a factor of $\geq 10^4$, thereby suppressing its proton affinity by ≥ 4 pK_a units.[224] The proton affinity of the oxyanion of the catalytic intermediate could also decrease to a similar extent.

Extensive kinetic studies have been performed with carboxypeptidase A.[2,207] Nonetheless, it is difficult to correlate the pK_a values obtained from kinetics with the dissociation of the active site groups known from structural determination. The pK_a values reported in the literature for dipeptide hydrolysis are variable, presumably because of the complex kinetic behavior (substrate activation) of these compounds, which generally has been incompletely analyzed. More reliable results are obtained with longer peptides showing bell-shaped pH dependence for k_{cat}/K_m which reflects two catalytically important groups, an acid and a base, on the free enzyme (Chapter 2, Section VIII.C). The pK_a values are about 6 and 9 with carboxypeptidase A [223] and somewhat lower, about 5 and 8, with thermolysin.[201] The glutamic acid (residue 270 of carboxypeptidase and 143 of thermolysin) and the zinc-bound water both have been considered as candidates for the ionizing group of lower pK_a. The zinc-bound water has been also suggested as the functional group of higher pK_a. Tyr 248 of carboxypeptidase was originally considered to be the acid group when it was thought to be catalytically important as proton donor to the leaving group. The most probable assignment of the lower pK_a now appears to be the glutamate hydrogen bonded to the zinc-bound water both in carboxypeptidase A[224] and in thermolysin.[201] However, in these cases, no assignment was made for the higher pK_a value. A simple explanation of the two pK_a values for the bell-shaped pH-rate profile is offered by Equation 2. This equation implies three different protonation forms of the active site. The double protonated form at low pH (both the glutamic acid and the zinc-bound water are nondissociated species) and the nonprotonated form at high pH (both groups are dissociated) are inactive catalytically. Only the monoprotonated form is catalytically competent. This form probably represents an equilibrium as shown by Equation 2. Kinetic isotope investigations have led to controversial results whether or not an isotope effect occurs.[2,207] Therefore, it is difficult to decide about the predominant species in the equilibrium. It should be pointed out in this respect that because the water molecule is considerably activated by the zinc ion, general base catalysis with metalloproteases is of less importance than is with serine or aspartic proteases.

The above proposal for the active monoprotonated species is analogous to the thiolate-imidazolium ion-pair formation in cysteine proteases (Chapter 4, Section IV). In both cases, decomposition of the active species by proton uptake and proton release controls the catalytic activity characterized by the two pK_a values associated with the bell-shaped pH-rate profiles.

Summarizing, we conclude that several important features of the catalysis by zinc-containing proteases have recently been clarified. It is most probable that the zinc-bound water is the nucleophile which attacks peptide substrates, and that this process is facilitated by general base catalysis by a glutamate residue. One of the most important pieces of evidence in favor of the general base mechanism, as opposed to the covalent anhydride intermediate, arises from transpeptidation reactions carried out in $H_2^{18}O$, as we have discussed in Chapter 5, Section VI.B, concerning pepsin catalysis. Determination of ^{18}O incorporation into trans-peptidation products has also indicated a general base mechanism for the thermolysin catalysis.[225] These conclusions must be valid even though the water molecule is coordinated to the zinc ion. In this conjunction, it is worth noting that this water molecule should be readily exchangeable because it is a weak ligand of zinc. Its position is much less definite, indeed, than those of the buried or certain surface-bound water molecules.[113]

The activation of a water molecule by zinc ion is not unique for metalloproteases. Its existence has been demonstrated in model reactions (Chapter 1, Section XI) as well as in other enzymic reactions, in particular, in the catalysis by carbonic anhydrase.[226,227]

Another important feature of the catalysis by zinc-containing proteases is the formation of a pentacoordinated zinc ion, which is generated upon the formation of the tetrahedral adduct and involves two oxygen atoms, as ligands, originating from the carbonyl oxygen and the attacking water molecule. The decomposition of the tetrahedral intermediate could proceed by general acid catalysis, i.e., by donation of proton from the carboxyl group to the leaving nitrogen atom.

There are still considerable uncertainties associated with further details of the mechanism.

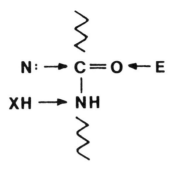

FIGURE 11. Possible sites and reactants implicated in peptide-bond cleavage. N:, E, and XH stand for the nucleophile, electrophile, and proton donor, respectively.

Thus, it is not known to what extent, if at all, the ligation of the carbonyl oxygen and the nucleophilic attack by water are parallel processes. Also, the product release mechanism following the breakdown of the tetrahedral adduct is unclear. Furthermore, possible differences between the mechanisms of carboxypeptidase A and thermolysin, as well as the apparent alterations in the ester and peptide hydrolyses, remain to be established. However, it is possible to make a comparison between the mechanisms of zinc proteases and the other three types of proteases, and this follows in the next section.

VI. COMMON FEATURES OF THE FOUR BASIC PROTEASE MECHANISMS

We have seen in Chapter 1 that the hydrolysis of the peptide bond is an addition-elimination reaction. A nucleophile (water or hydroxide ion) adds to the carbonyl carbon atom (Figure 11) which is followed by expulsion of the leaving group (the amine) from the resultant tetrahedral intermediate. The first step could be facilitated by an electrophile, such as the proton in the acid-catalyzed hydrolysis. This assistance is important when water, a weak nucleophile, attacks the carbon atom. The contribution of the electrophile is less significant when the hydroxide ion, a strong nucleophile, is the reacting species. The second step requires the protonation of the leaving group because the amine anion is an extremely poor leaving group.

As for the protease catalysis, two basic types of peptide hydrolysis may be distinguished: double addition-elimination and single addition-elimination mechanisms. The former involves the formation of an acyl-enzyme intermediate, as in the catalysis by serine and cysteine proteases, the latter is a direct hydrolysis by water, as in aspartic and zinc protease catalyses. The formation and breakdown of the acyl-enzymes proceed by similar mechanisms using the same catalytic machinery in both steps.

A comparison of the addition-elimination reactions associated with the four types of protease catalysis exhibits several important features which they have in common, as well as some interesting differences. Water, as a nucleophile, participates in all four enzymic reactions. Since it is a weak nucleophile, its reaction requires facilitation. Indeed, in all cases, the nucleophilic attack is promoted by general base catalysis, carried out by an imidazole group in the double displacement mechanisms and a carboxylate group in the simple addition-elimination mechanisms. It should be pointed out, however, that the water molecule in the zinc protease catalysis is a considerably better nucleophile than it is in the catalyses by the other three types of proteases. This is a consequence of its binding to the metal ion. An even better nucleophile is the thiolate ion in the acyl-enzyme formation with cysteine proteases. This reaction is a direct nucleophilic attack not facilitated by general base catalysis. The acyl-enzyme formation with serine proteases, on the other hand, requires general base assistance, just as do the water reactions.

The most significant differences encountered among the four types of proteases concern electrophilic catalysis. In the catalysis by serine proteases, electrophilic catalysis is provided by hydrogen bonds from the oxyanion binding site to the oxyanion of the tetrahedral intermediate. A similar contribution appears to be less significant in the case of cysteine proteases as discussed in detail in Chapter 4, Section VI. However, a positively charged imidazolium ion appears to be very important in stabilizing the negatively charged tetrahedral intermediate both in the serine and the cysteine protease catalyses. Electrophilic catalysis by positive charge is also found in the action of zinc proteases where the metal ion stabilizes the intermediate. In addition, other side chains, such as His 231 and Tyr 157 of thermolysin and Arg 127 of carboxypeptidase A, might also promote the catalysis as electrophiles. However, their contributions as compared to that of the zinc ion, should be estimated from future studies in which they are replaced by site-directed mutagenesis.

In contrast to the other three types of proteases, aspartic proteases do not contain a positively charged electrophile in the neighborhood of the tetrahedral intermediate. Instead, they utilize the simplest and most effective form of electrophilic catalysis, namely, a complete proton transfer to the incipient tetrahedral adduct. This process represents a typical general acid catalysis, which leads to a neutral intermediate, unique among the four protease mechanisms.

Proton donation to the leaving group is mediated by an imidazole group in serine and cysteine protease catalysis or by carboxyl group(s) in the aspartic and zinc protease reactions. Examination of these proton transfers permits us to postulate the following general rules for protease catalysis.[203a]

(1) The catalytic groups always constitute a functional unit. These are the Ser . . . His . . . Asp triad, the thiolate-imidazolium ion-pair, the Asp . . . Asp diad, and the Zn^{2+} . . . H_2O . . . Glu complex, in the serine, cysteine, aspartic, and zinc protease catalyses, respectively. This cooperation makes possible the precatalytic (cysteine and zinc proteases) and catalytic (serine proteases) activation of the nucleophile, or provides the appropriate conditions for the "push-pull" type catalysis (aspartic proteases).

(2) The same proton that originates from the nucleophile is transferred to the leaving group. This is also valid for cysteine proteases, where the nucleophile is in a dissociated form, but the dissociated proton is preserved in the thiolate-imidazolium functional unit until it is donated to the leaving group. This rule poses restrictions on the possible mechanisms. Specifically, it is inconsistent with proton donated to the leaving group by the bulk water, as has been proposed for the aspartic protease catalysis, or by His 231 and Tyr 248, as suggested earlier for the thermolysin and the carboxypeptidase A catalysis, respectively.

(3) The same group conveys the proton from the nucleophile to the leaving group in two subsequent steps, with the formation of a tetrahedral adduct between them. The proton carrier is an imidazole, a carboxyl group, or a carboxyl diad. The formation of the tetrahedral intermediate abolishes the strong resonance stabilization (Chapter 1, Section II) associated with the peptide bond. The partial positive charge located on the peptide nitrogen is also removed upon formation of the tetrahedral intermediate. This makes the donation of the proton to the nitrogen atom easier.

General base-acid catalysis by one and the same enzymic group ensures regeneration of the enzyme at the end of each catalytic turn. If two different general catalysts were involved in the reaction, regeneration of the catalytic groups would require a more complicated system. The simplicity of the catalytic strategy thus explains why all four protease groups utilize the same basic concept. We may conclude that the protease can be regarded as an enzyme that conveys the proton from the nucleophile to the peptide nitrogen, which in turn results in the CO–NH bond cleavage. Nature has elaborated four different ways to accomplish this proton transfer, resulting in the four different mechanisms of action of proteases.

REFERENCES

1. **Vallee, B. L. and Galdes A.,** The metallobiochemistry of zinc enzymes, *Adv. Enzymol. Relat. Areas Mol. Biol.,* 56, 283, 1984.
2. **Hartsuck, J. A. and Lipscomb, W. N.,** Carboxypeptidase A, in *The Enzymes,* Vol. 3., 3rd ed., Boyer, P. D., Ed., Academic Press, New York, 1971, 1.
3. **Quiocho, F. A. and Lipscomb, W. N.,** Carboxypeptidase A: a protein and an enzyme, *Adv. Protein Chem.,* 25, 1, 1971.
4. **Anson, M. L.,** Carboxypeptidase. I. The preparation of crystalline carboxypeptidase, *J. Gen. Physiol.,* 20, 663, 1937.
5. **Folk, J. E. and Schirmer, E. W.,** The porcine pancreatic carboxypeptidase A system. I. Three forms of the active enzyme, *J. Biol. Chem.,* 238, 3884, 1963.
6. **Folk, J. E.,** The porcine pancreatic carboxypeptidase A system. II. Mechanism of the conversion of carboxypeptidase A_1 to carboxypeptidase A_2, *J. Biol. Chem.,* 238, 3895, 1963.
7. **Sampath Kumar, K. S. V., Clegg, J. B., and Walsh, K. A.,** The N-terminal sequence of bovine carboxypeptidase A and its relation to zymogen activation, *Biochemistry,* 3, 1728, 1964.
8. **Allan, B. J., Keller, P. J., and Neurath, H.,** Procedures for the isolation of crystalline bovine pancreatic carboxypeptidase A. I. Isolation from acetone powders of pancreas glands, *Biochemistry,* 3, 40, 1964.
9. **Vallee, B. L., Coombs, T. L., and Hoch, F. L.,** The "active site" of bovine carboxypeptidase A, *J. Biol. Chem.,* 235, PC45, 1960.
10. **Petra, P. H., Bradshaw, R. A., Walsh, K. A., and Neurath, H.,** Identification of the amino acid replacements characterizing the allotypic forms of bovine carboxypeptidase A, *Biochemistry,* 8, 2762, 1969.
11. **Bradshaw, R. A., Ericsson, L. H., Walsh, K. A., and Neurath, H.,** The amino acid sequence of bovine carboxypeptidase A, *Proc. Natl. Acad. Sci. U.S.A.,* 63, 1389, 1969.
12. **Reeke, G. N., Hartsuck, J. A., Ludwig, M. L., Quiocho, F. A., Steitz, T. A., and Lipscomb, W. N.,** The structure of carboxypeptidase A. VI. Some results at 2.0-Å resolution, and the complex with glycyl-tyrosine at 2.8-Å resolution, *Proc. Natl. Acad. Sci. U.S.A.,* 58, 2220, 1967.
13. **Quinto, C., Quiroga, M., Swain, W. F., Nikovits, W. C., Jr., Standring, D. N., Pictet, R. L., Valenzuela, P., and Rutter, W. J.,** Rat preprocarboxypeptidase A: cDNA sequence and preliminary characterization of the gene, *Proc. Natl. Acad. Sci. U.S.A.,* 79, 31, 1982.
14. **San Segundo, B., Martinez, M. C., Vilanova, M., Cuchillo, C. M., and Aviles, F. X.,** The severed activation segment of porcine pancreatic procarboxypeptidase A is a powerful inhibitor of the active enzyme. Isolation and characterization of the activation peptide, *Biochim. Biophys. Acta,* 707, 74, 1982.
15. **Aviles, F. X., San Segundo, B., Vilanova, M., Cuchillo, C. M., and Turner, C.,** The activation segment of procarboxypeptidase A from porcine pancreas constitutes a folded structural domain, *FEBS Lett.,* 149, 257, 1982.
16. **Vallee, B. L. and Neurath, H.,** Carboxypeptidase, a zinc metalloprotein, *J. Am. Chem. Soc.,* 76, 5006, 1954.
17. **Williams, R. J. P.,** Binding of zinc in carboxypeptidase, *Nature (London),* 188, 322, 1960.
18. **Coombs, T. L., Omote, Y., and Vallee, B. L.,** The zinc-binding groups of carboxypeptidase A, *Biochemistry,* 3, 653, 1964.
19. **Lipscomb, W. N., Hartsuck, J. A., Quiocho, F. A., and Reeke, G. N., Jr.,** The structure of carboxypeptidase A. IX. The X-ray diffraction results in the light of the chemical sequence, *Proc. Natl. Acad. Sci. U.S.A.,* 64, 28, 1969.
20. **Coleman, J. E. and Vallee, B. L.,** Metallocarboxypeptidases: stability constants and enzymatic characteristics, *J. Biol. Chem.,* 236, 2244, 1961.
21. **Simpson, R. T. and Vallee, B. L.,** Iodocarboxypeptidase, *Biochemistry,* 5, 1760, 1966.
22. **Roholt, O. A. and Pressman, D.,** The sequence around the active-center tyrosyl residue of bovine pancreatic carboxypeptidase A, *Proc. Natl. Acad. Sci. U.S.A.,* 58, 280, 1967.
23. **Riordan, J. F. and Vallee, B. L.,** Acetylcarboxypeptidase, *Biochemistry,* 2, 1460, 1963.
24. **Sokolovsky, M. and Vallee, B. L.,** Azocarboxypeptidase: functional consequences of tyrosyl and histidyl modification, *Biochemistry,* 6, 700, 1967.

25. **Riordan, J. F., Sokolovsky, M., and Vallee, B. L.,** The functional tyrosyl residues of carboxypeptidase A. Nitration with tetranitromethane, *Biochemistry,* 6, 3609, 1967.

25a. **Gardell, S. J., Hilvert, D., Barnett, J., Kaiser, E. T., and Rutter, W. J.,** Use of directed mutagenesis to probe the role of tyrosine 198 in the catalytic mechanism of carboxypeptidase A, *J. Biol. Chem.,* 262, 576, 1987.

26. **Folk, J. E.,** Carboxypeptidase B, in *The Enzymes,* Vol. 3, 3rd ed., Boyer, P. D., Ed., Academic Press, New York, 1971, 57.

27. **Plummer, T. H., Jr. and Erdös, E. G.,** Human plasma carboxypeptidase N, *Methods Enzymol.,* 80, 442, 1981.

28. **Levin, Y., Skidgel, R. A., and Erdös, E. G.,** Isolation and characterization of the subunits of human plasma carboxypeptidase N (kininase I), *Proc. Natl. Acad. Sci. U.S.A.,* 79, 4618, 1982.

29. **Ondetti, M. A. and Cushman, D. W.,** Enzymes of the renin-angiotensin system and their inhibitors, *Annu. Rev. Biochem.,* 51, 283, 1982.

30. **Dorer, F. E., Kahn, J. R., Lentz, K. E., Levine, M., and Skeggs, L. T.,** Purification and properties of angiotensin converting enzyme from hog lung, *Circ. Res.,* 31, 356, 1972.

31. **Stewart, T. A., Weare, J. A., and Erdös, E. G.,** Human peptidyl dipeptidase (converting enzyme, kininase II), *Methods Enzymol.,* 80, 450, 1981.

32. **Weare, J. A., Gafford, J. T., Lu, H. S., and Erdös, E. G.,** Purification of human kidney angiotensin I converting enzyme using reverse-immunoadsorption chromatography, *Anal. Biochem.,* 123, 310, 1982.

33. **El-Dorry, H. A., Bull, H. G., Iwata, K., Thornberry, N. A., Cordes, E. H., and Soffer, R. L.,** Molecular and catalytic properties of rabbit testicular dipeptidyl carboxypeptidase, *J. Biol. Chem.,* 257, 14128, 1982.

34. **Bull, H. G., Thornberry, N. A., and Cordes, E. H.,** Purification of angiotensin-converting enzyme from rabbit lung and human plasma by affinity chromatography, *J. Biol. Chem.,* 260, 2963, 1985.

35. **Lanzillo, J. J., Stevens, J., Dasarathy, Y., Yotsumota, H., and Fanburg, B. L.,** Angiotensin-converting enzyme from human tissues. Physicochemical, catalytic and immunological properties, *J. Biol. Chem.,* 260, 14938, 1985.

36. **Cushman, D. W. and Cheung, H. S.,** Spectrophotometric assay and properties of the angiotensin converting enzyme of rabbit lung, *Biochem. Pharmacol.,* 20, 1637, 1971.

37. **Piquilloud, Y., Reinharz, A., Roth, M.,** Studies on the angiotensin converting enzyme with different substrates, *Biochim. Biophys. Acta,* 206, 136, 1970.

38. **Stevens, R. L., Micalizzi, E. R., Fessler, D. C., and Pals, D. T.,** Angiotensin I converting enzyme of calf lung. Method of assay and partial purification, *Biochemistry,* 11, 2999, 1972.

39. **Holmquist, B., Bünning, P., and Riordan, J. F.,** A continuous spectrophotometric assay for angiotensin converting enzyme, *Anal. Biochem.,* 95, 540, 1979.

40. **Carmel, A. and Yaron, A.,** An intramolecularly quenched flourescent tripeptide as a fluorogenic substrate of angiotesin-I-converting enzyme and of bacterial dipeptidyl carboxypeptidase, *Eur. J. Biochem.,* 87, 265, 1978.

41. **Persson, A. and Wilson, I. B.,** A fluorogenic substrate for angiotensin-converting enzyme, *Anal. Biochem.,* 83, 296, 1977.

42. **Hersh, L. B., Gafford, J. T., Powers, J. C., Tanaka, T., and Erdös, E. G.,** Novel substrates for angiotensin I converting enzyme. *Biochem. Biophys. Res. Commun.,* 110, 654, 1983.

43. **Bünning, P. and Riordan, J. F.,** Activation of angiotensin converting enzyme by monovalent anions, *Biochemistry,* 22, 110, 1983.

44. **Weare, J. A.,** Activation/inactivation of human angiotensin I converting enzyme following chemical modifications of amino groups near the active site, *Biochem. Biophys. Res. Commun.,* 104, 1319, 1982.

45. **Harris R. B. and Wilson, I. B.,** Sequencing of an active-site peptide of angiotensin I-converting enzyme containing an essential glutamic acid residue, *J. Biol. Chem.,* 260, 2208, 1985.

46. **Bünning, P., Holmquist, B., and Riordan, J. F.,** Functional residues at the active site of angiotensin converting enzyme, *Biochem. Biophys. Res. Commun.,* 83, 1442, 1978.

47. **Spratt, T. E. and Kaiser, E. T.,** Catalytic versatility of angiotensin converting enzyme: catalysis of an α,β-elimination reaction, *J. Am. Chem. Soc.,* 106, 6440, 1984.

48. **Sugimoto, T. and Kaiser, E. T.,** Carboxypeptidase A catalyzed enolization of a ketonic substrate. A new stereochemical probe for an enzyme-bound nucleophile, *J. Am. Chem. Soc.,* 100, 7750, 1978.

49. **Sugimoto, T. and Kaiser, E. T.,** Elucidation of the stereochemistry of the carboxypeptidase A catalyzed enolization of 2-benzyl-3-p-methoxybenzoylpropionate, a ketone substrate, *J. Am. Chem. Soc.,* 101, 3946, 1979.

50. **Malfroy, B., Swerts, J. P., Guyon, A., Roques, B. P., and Schwartz, J. C.,** High-affinity enkephalin-degrading peptidase in brain is increased after morphine, *Nature (London),* 276, 523, 1978.

51. **Schwartz, J. C., Malfroy, B., and De La Baume, S.,** Minireview. Biological inactivation of enkephalins and the role of enkephalin-dipeptidyl-carboxypeptidase (''enkephalinase'') as neuropeptidase, *Life Sci.,* 29, 1715, 1981.

52. **Orlowski, M. and Wilk, S.,** Purification and specificity of a membrane-bound metalloendopeptidase from bovine pituitaries, *Biochemistry,* 20, 4942, 1981.

53. **Fulcher, I. S., Matsas, R., Turner, A. J., and Kenny, A. J.,** Kidney neutral endopeptidase and the hydrolysis of enkephalin by synaptic membranes show similar sensitivity to inhibitors, *Biochem. J.,* 203, 519, 1982.

54. **Almenoff, J. and Orlowski, M.,** Membrane-bound kidney neutral metalloendopeptidase: interaction with synthetic substrates, natural peptides and inhibitors, *Biochemistry,* 22, 590, 1983.

55. **Kerr, M. A. and Kenny, A. J.,** The purification and specificity of a neutral endopeptidase from rabbit kidney brush border, *Biochem. J.,* 137, 477, 1974.

56. **Kerr, M. A. and Kenny, A. J.,** The molecular weight and properties of a neutral metallo-endopeptidase from rabbit kidney brush border, *Biochem. J.,* 137, 489, 1974.

57. **Matsas, R., Fulcher, I. S., Kenny, A. J., and Turner, A. J.,** Substance P and (Leu)enkephalin are hydrolysed by an enzyme in pig caudate synaptic membranes that is identical with the endopeptidase of kidney microvilli, *Proc Natl. Acad. Sci. U.S.A.,* 80, 3111, 1983.

58. **Kenny, A. J. and Maroux, S.,** Topology of microvillar membrane hydrolases of kidney and intestine, *Physiol. Rev.,* 62, 91, 1982.

59. **Kenny, A. J., Fulcher, I. S., McGill K. A., and Kershaw, D.,** Proteins of the kidney microvillar membrane. Reconstitution of endopeptidase in liposomes shows that it is a short-stalked protein, *Biochem. J.,* 211, 755, 1983.

59a. **Devault, A., Lazure, C., Nault, C., Le Moual, H., Seidah, N. G., Chrétien, M., Kahn, P., Powell, J., Mallet, J., Beaumont, A., Roques, B. P., Crine, P., and Boileau, G.,** Amino acid sequence of rabbit kidney neutral endopeptidase 24.11 (enkephalinase) deduced from a complementary DNA, *EMBO J.,* 6, 1317, 1987.

60. **Florentin, D., Sassi, A., and Roques, B. P.,** A highly sensitive fluorometric assay for "enkephalinase", a neutral metalloendopeptidase that releases tyrosine-glycine-glycine from enkephalins, *Anal. Biochem.,* 141, 62, 1984.

61. **Pozsgay, M., Michaud, C., Liebman, M., and Orlowski, M.,** Substrate and inhibitor studies of thermolysin-like neutral metalloendopeptidase from kidney membrane fractions. Comparison with bacterial thermolysin, *Biochemistry,* 25, 1292, 1986.

62. **Hersh, L. B. and Morihara, K.,** Comparison of the subsite specificity of the mammalian neutral endopeptidase 24.11 (enkephalinase) to the bacterial neutral endopeptidase thermolysin, *J. Biol. Chem.,* 261, 6433, 1986.

63. **Malfroy, and Schwartz, J. C.,** Properties of "enkephalinase" from rat kidney: comparison of dipeptidylcarboxypeptidase and endopeptidase activities, *Biochem. Biophys. Res. Commun.,* 106, 276, 1982.

64. **Jackson, D. G. and Hersh, L. B.,** Reaction of neutral endopeptidase 24.11 (enkephalinase) with arginine reagents, *J. Biol. Chem.,* 261, 8649, 1986.

65. **DeLange, R. J. and Smith, E. L.,** Leucine aminopeptidase and other N-terminal exopeptidases, in *The Enzymes,* Vol. 3, 3rd ed., Boyer, P. D., Ed., Academic Press, New York, 1971, 81.

66. **Hanson, H. and Frohne, M.,** Crystalline leucine aminopeptidase from lens (α-aminoacyl-peptide hydrolase; EC 3.4.11.1.), *Methods Enzymol,* 45, 504, 1977.

67. **Cuypers, H. T., van Loon-Klaassen, L. A. H., Vree Egberts, W. T. M., de Jong, W. W., and Bloemendal, H.,** Sulfhydryl content of bovine eye lens leucine aminopeptidase. Determination of the reactivity of the sulfhydryl group of the zinc metalloenzyme, of the enzyme activated by Mg^{2+}, Mn^{2+}, and Co^{2+}, and of the metal-free apoenzyme, *J. Biol. Chem.,* 257, 7086, 1982.

68. **Cuypers, H. T., van Loon-Klaassen, L. A. H., Vree Egberts, W. T. M., de Jong, W. W., and Bloemendal, H.,** The primary structure of leucine aminopeptidase from from bovine eye lens, *J. Biol. Chem.,* 257, 7077, 1982.

69. **Lin, S. H. and Van Wart, H. E.,** Effect of cryosolvents and subzero temperatures on the hydrolysis of L-leucine-p-nitroanilide by porcine kidney leucine aminopeptidase, *Biochemistry,* 21, 5528, 1982.

70. **Chan, W. W.-C., Dennis, P., Demmer, W., and Brand, K.,** Inhibition of leucine aminopeptidase by amino acid hydroxamates, *J. Biol. Chem.,* 257, 7955, 1982.

71. **Taylor, A., Sawan, S., and James, T. L.,** Structural aspects of the inhibitor complex formed by N-(leucyl)-o-aminobenzenesulfonate and manganese with Zn^{2+}-Mn^{2+} leucine aminopeptidase (EC 3.4.11.1), *J. Biol. Chem.,* 257, 11571, 1982.

72. **Allen, M. P., Yamada, A. H., and Carpenter, F. H.,** Kinetic parameters of metal-substituted leucine aminopeptidase from bovine lens, *Biochemistry,* 22, 3778, 1983.

73. **Andersson L., MacNeela, J., and Wolfenden R.,** Use of secondary isotope effects and varying pH to investigate the mode of binding inhibitory amino aldehydes by leucine aminopeptidase, *Biochemistry,* 24, 330, 1985.

74. **Shenvi, A. B.,** α-Aminoboronic acid derivatives: effective inhibitors of aminopeptidase, *Biochemistry,* 25, 1286, 1986.

75. **Patterson, E. K., Hsaio, S. H., and Keppel, A.,** Studies on dipeptidases and aminopeptidases. I. Distinction between LAP and enzymes that hydrolyze L-leucyl-β-naphthylamide, *J. Biol. Chem.* 238, 3611, 1963.

76. **Hanson, H., Hütter, H.-J., Mannsfeldt, H.- G., Kretschmer, K. and Sohr, Ch.,** Zur Darstellung und Substratspezifität einer von der Leucin-aminopeptidase unterscheidbaren Aminopeptidase aus Nierenpartikeln, *Hoppe-Seyler's Z. Physiol. Chem.,* 348, 680, 1967.

77. **Ellis, S. and Perry, M.,** Inhibition of thio-activated aminopeptidase by puromycin, *Biochem. Biophys. Res. Commun.,* 15, 502, 1964.

78. **Fleischer F. A., Pankow, M., and Warmka, C.,** Leucine aminopeptidase in human serum: comparison of hydrolysis of L-leucylglycine and L-leucyl-β-naphthylamide, *Clin. Chim. Acta,* 9, 259, 1964.

79. **Smith, E. E., Kaufman, J. T., and Rutenburg, A. M.,** The partial purification of an amino acid naphthylamidase from human liver, *J. Biol. Chem.,* 240, 1718, 1965.

80. **Femfert, U., Cichocki, P., and Pfleiderer, G.,** On the mechanism of amino bond cleavage catalyzed by aminopeptidase M. Enzymatic properties of nitroaminopeptidase M, *FEBS Lett.,* 26, 39, 1972.

81. **Mahadevan, S. and Tappel, A. L.,** Arylamidase of rat liver and kidney, *J. Biol. Chem.,* 242, 2369, 1967.

82. **Behal, F. J. and Story, M. N.,** Arylamidase of human kidney, *Arch. Biochem. Biophys.,* 131, 74, 1969.

83. **Sirodowicz, W., Zownir, O., and Behal, F. J.,** Action of human pancreas alanine aminopeptidase on biologically active peptides: kinin converting activity, *Clin. Chim. Acta,* 111, 69, 1981.

84. **Ryan, J. W., Roblero, J., and Stewart, J. M.,** Inactivation of bradykinin in rat lung, *Adv. Exp. Med. Biol.,* 8, 263, 1970.

85. **Hersh, L. B.,** Solubilization and characterization of two rat brain membrane-bound aminopeptidases active on Met-enkephalin, *Biochemistry,* 20, 2345, 1981.

86. **Hersh, L. B.,** Degradation of enkephalins: the search for enkephalinase, *Mol. Cell. Biochem.,* 47, 35, 1982.

87. **Seifter, S. and Harper, E.,** The collagenases, in *The Enzymes,* Vol. 3, 3rd ed., Boyer, P. D., Ed., Academic Press, New York, 1971, 649.

88. **Bond, M. D. and Van Wart, H. E.,** Purification and separation of individual collagenases from *Clostridium histolyticum* using red dye ligand chromatography, *Biochemistry,* 23, 3077, 1984.

89. **Bond, M. D. and Van Wart, H. E.,** Characterization of individual collagenases from *Clostridium histolyticum, Biochemistry,* 23, 3085, 1984.

90. **Bond, M. D. and Wart, H. E.,** Relationship between the individual collagenases of *Clostridium histolyticum:* evidence for evolution by gene duplication, *Biochemistry,* 23, 3092, 1984.

91. **Van Wart, H. E. and Steinbrink, D. R.,** Complementary substrate specificities of class I and class II collagenases from *Clostridium histolyticum, Biochemistry,* 24, 6520, 1985.

92. **Mookhtiar, K. A., Steinbrink, D. R., and Van Wart, H. E.,** Mode of hydrolysis of collagen-like peptides by class I and class II *Clostridium histolyticum* collagenases: evidence for both endopeptidase and tripeptidylcarboxypeptidase activities, *Biochemistry,* 24, 6527, 1985.

93. **Gross, J. and Nagai, Y.,** Specific degradation of the collagen molecule by tadpole collagenolytic enzyme, *Proc. Natl. Acad. Sci. U.S.A.,* 54, 1197, 1965.

94. **Matsubara, H. and Feder, J.,** Other bacterial, mold, and yeast proteases, in *The Enzymes,* Vol. 3., 3rd ed., Boyer, P. D., Ed., Academic Press, New York, 1971, 765.

95. **Morihara, K.,** Comparative specificity of microbial proteinases, *Adv. Enzymol. Relat. Areas Mol. Biol.,* 41, 200, 1974.

96. **Tsuru, D., McConn, J. D., and Yasunobu, K. T.,** *B. Subtilis* neutral protease a zinc enzyme of high activity, *Biochem. Biophys. Res. Commun.,* 15, 367, 1964.

97. **McConn, J. D., Tsuru, D., and Yasunobu, K. T.,** *Bacillus subtilis* neutral proteinase. I. A zinc enzyme of high specific activity, *J. Biol. Chem.,* 239, 3706, 1964.

98. **Tsuru, D., Yamamoto, T., and Fukumoto, J.,** Studies on bacterial protease. XIII. Purficiation, crystallization and some enzymatic properties of neutral protease of *Bacillus subtilis* var. *amylosacchariticus, Agr. Biol. Chem.,* 30, 651, 1966.

99. **Tsuru, D. Kira, H., Yamamoto, T., and Fukumoto, J.,** Studies on bacterial protease. XV. Some physicochemical properties and substrate specificity of neutral protease of *Bacillus subtilis* var. *amylosacchariticus, Agr. Biol. Chem.,* 30, 1164, 1966.

100. **Endo, S.,** Studies on protease produced by thermophilic bacteria, *J. Ferment. Technol. (Tokyo),* 40, 346, 1962.

101. **Matsubara, H.,** Purification and assay of thermolysin, *Methods Enzymol.,* 19, 642, 1970.

102. **Pangburn, M. K., Burstein, Y., Morgan P. M., Walsh, K. A., and Neurath, H.,** Affinity chromatography of thermolysin and of neutral proteases from *B. subtilis, Biochem. Biophys. Res. Commun.,* 54, 371, 1979.

103. **Titani, K., Hermodson, M. A., Ericsson, L. H., Walsh, K. A., and Neurath, H.,** Amino acid sequence of thermolysin, *Nature (London) New Biol.,* 238, 35, 1972.

104. **Matthews, B. W., Jansonius, J. N., Colman, P. M., Schoenborn, B. P., and Dupourque, D.,** Three-dimensional structure of thermolysin, *Nature (London) New Biol.,* 238, 37, 1972.
105. **Colman, P. M., Jansonius, J. N., and Matthews B. W.,** The structure of thermolysin: an electron density map at 2.3 Å resolution, *J. Mol. Biol.,* 70, 701, 1972.
106. **Ohta, Y.,** Thermostable protease from thermophilic bacteria. II. Studies on the stability of the protease, *J. Biol. Chem.,* 242, 509, 1967.
107. **Feder, J.,** Studies on the specificity of *Bacillus subtilis* neutral protease with synthetic substrates, *Biochemistry,* 6, 2088, 1967.
108. **Morihara, K., Tsuzuki, H., and Oka, T.,** Comparison of the specificities of various neutral proteinases from microorganisms, *Arch. Biochem. Biophys.,* 123, 572, 1968.
109. **Morihara, K. and Tsuzuki, H.,** Comparative study of various neutral proteinases from microorganisms: specificity with oligopeptides, *Arch. Biochem. Biophys.,* 146, 291, 1971.
110. **Pfleiderer, G. and Krauss, A.,** Die Wirkungsspezifität von Schlangengift-Proteasen *Crotalus atrox, Biochem. Z.,* 342, 85, 1965.
111. **Fox, J. W., Campbell, R., Beggerly, L., and Bjarnson, B.,** Substrate specificities and inhibition of two hemorrhagic zinc proteases Ht-c and Ht-d from *Crotalus atrox* venom, *Eur. J. Biochem.,* 156, 65, 1986.
112. **Rees, D. C., Lewis, M., Honzatko, R. B., Lipscomb, W. N., and Hardman, K. D.,** Zinc environment and *cis* peptide bonds in carboxypeptidase A at 1.75Å resolution, *Proc. Natl. Acad. Sci. U.S.A.,* 78, 3408, 1981.
113. **Rees, D. C., Lewis, M., and Lipscomb, W. N.,** Refined crystal structure of carboxypeptidase A at 1.54 Å resolution, *J. Mol. Biol.,* 168, 367, 1983.
114. **Christianson, D. W. and Lipscomb, W. N.,** X-ray crystallographic investigation of substrate binding to carboxypeptidase A at subzero temperature, *Proc. Natl. Acad. Sci. U.S.A.,* 83, 7568, 1986.
115. **Fujioka, H. and Imahori, K.,** Studies on the binding of carboxypeptidase A and several competitive inhibitors, *J. Biol. Chem.,* 237, 2804, 1962.
116. **Rees, D. C., Honzatko, R. B., and Lipscomb, W. N.,** Structure of an actively exchanging complex between carboxypeptidase A and a substrate analog, *Proc. Natl. Acad. Sci. U.S.A.,* 77, 3288, 1980.
117. **Lipscomb, W. N.,** Carboxypeptidase A mechanisms, *Proc. Natl. Acad. Sci. U.S.A.,* 77, 3875, 1980.
118. **Christianson, D. W., Kuo, L. C., and Lipscomb, W. N.,** Novel structure of the complex between carboxypeptidase A and a ketonic substrate analogue, *J. Am. Chem. Soc.,* 107, 8281, 1985.
119. **Abramowitz, N., Shechter, I., and Berger, A.,** On the size of the active site in proteases. II. Carboxypeptidase A, *Biochem. Biophys. Res. Commun.,* 29, 862, 1967.
120. **Rees, D. C. and Lipscomb, W. N.,** Structure of the potato inhibitor complex of carboxypeptidase A at 2.5 Å resolution, *Proc. Natl. Acad. Sci. U.S.A.,* 77, 4633, 1980.
121. **Rees, D. C. and Lipscomb, W. N.,** Refined crystal structure of the potato inhibitor complex of carboxypeptidase A at 2.5 Å resolution, *J. Mol. Biol.,* 160, 475, 1982.
122. **Hass, G. M., Nau, H., Biemann, K., Grahn, D. T., Ericsson, L. H., and Neurath, H.,** The amino acid sequence of a carboxypeptidase inhibitor from potatoes, *Biochemistry,* 14, 1334, 1975.
123. **Leary, T. R., Grahn, D. T., Neurath, H., and Hass, G. M.,** Structure of potato carboxypeptidase inhibitor: disulfide pairing and exposure of aromatic residues, *Biochemistry,* 18, 2252, 1979.
124. **Hass, G. M. and Ryan, C. A.,** Carboxypeptidase inhibitor from potatoes, *Methods Enzymol.,* 80, 778, 1981.
125. **Hass, G. M. and Ryan, C. A.,** Cleavage of the carboxypeptidase inhibitor from potatoes by carboxypeptidase A, *Biochem. Biophys. Res. Commun.,* 97, 1481, 1980.
126. **Matthews, B. W., Colman, P. M., Jansonius, J. N., Titani, K., Walsh, K. A., and Neurath, H.,** Structure of thermolysin, *Nature (London) New Biol.,* 238, 41, 1972.
127. **Matthews, B. W., Weaver, L. H., and Kester, W. R.,** The conformation of thermolysin, *J. Biol. Chem.,* 249, 8030, 1974.
128. **Holmes, M. A. and Matthews, B. W.,** Structure of thermolysin refined at 1.6 Å resolution, *J. Mol. Biol.,* 160, 623, 1982.
129. **Roche, R. S. and Voordouw, G.,** The structural and functional roles of metal ions in thermolysin, *CRC Crit. Rev. Biochem.,* 5, 1, 1978.
130. **Fassina, G., Vita, C., Dalzoppo, D., Zamai, M., Zambonin M., and Fontana, A.,** Autolysis of thermolysin. Isolation and characterization of a folded three-fragment complex, *Eur. J. Biochem.,* 156, 221, 1986.
131. **Vita, C., Dalzoppo, D., and Fontana, A.,** Limited proteolysis of thermolysin by subtilisin: isolation and characterization of a partially active enzyme derivative, *Biochemistry,* 24, 1798, 1985.
132. **Fontana, A., Fassina, G., Vita, C., Dalzoppo, D., Zamai, M., and Zambonin, M.,** Correlation between sites of limited proteolysis and segmental mobility in thermolysin, *Biochemistry,* 25, 1847, 1986.
133. **Kester, W. R. and Matthews, B. W.,** Crystallographic study of the binding of dipeptide inhibitors to thermolysin: implications for the mechanism of catalysis, *Biochemistry,* 16, 2506, 1977.

134. **Weaver, L. H., Kester, W. R., and Matthews, B. W.,** A crystallographic study of the complex of phosphoramidon with thermolysin. A model for the presumed catalytic transition state and for the binding of extended substrates, *J. Mol. Biol.,* 114, 119, 1977.

135. **Tronrud, D. E., Monzingo, A. F., and Matthews, B. W.,** Crystallographic structural analysis of phosphoramidates as inhibitors and transition state analogs of thermolysin, *Eur. J. Biochem.,* 157, 261, 1986.

136. **Suda, H., Aoyagi, T., Takeuchi, T., and Umezawa, H.,** A thermolysin inhibitor produced by Actinomycetes: phosphoramidon, *J. Antibiot.,* 26, 621, 1973.

137. **Komiyama, T., Suda, H., Aoyagi, T., Takeuchi, T., Umezawa, H., Fujimoto, K., and Umezawa, S.,** Studies on inhibitory effect of phosphoramidon and its analogs on thermolysin, *Arch. Biochem. Biophys.,* 171, 727, 1975.

138. **Kam, C.-M., Nishino, N., and Powers, J. C.,** Inhibition of thermolysin and carboxypeptidase A by phosphoramidates, *Biochemistry,* 18, 3032, 1979.

139. **Holmes, M. A. and Matthews, B. W.,** Binding of hydroxamic acid inhibitors to crystalline thermolysin suggests a pentacoordinate zinc intermediate in catalysis, *Biochemistry,* 20, 6912, 1981.

140. **Monzingo, A. F. and Matthews, B. W.,** Binding of N-carboxymethyl dipeptide inhibitors to thermolysins determined by X-ray crystallography: a novel class of transition-state analogues for zinc peptidases, *Biochemistry,* 23, 5724, 1984.

141. **Dideberg, O., Charlier, P., Dive, G., Joris, B., Frere, J. M., and Ghuysen, J. M.,** Structure of a Zn^{2+}-containing D-alanyl-D-alanine-cleaving carboxypeptidase at 2.5 Å resolution, *Nature (London),* 299, 469, 1982.

142. **Byers, L. D. and Wolfenden, R.,** A potent reversible inhibitor of carboxypeptidase A, *J. Biol. Chem.,* 247, 606, 1972.

143. **Byers, L. D. and Wolfenden, R.,** Binding of the by-product analog benzylsuccinic acid by carboxypeptidase A, *Biochemistry,* 12, 2070, 1973.

144. **Bolognesi, M.C. and Matthews, B.W.,** Binding of the biproduct analog L-benzylsuccinic acid to thermolysin determined by X-ray crystallography, *J. Biol. Chem.,* 254, 634, 1979.

145. **Palmer, A. R., Ellis, P. D., and Wolfenden, R.,** Extreme state of ionization of benzylsuccinate bound by carboxypeptidase A, *Biochemistry.* 21, 5056, 1982.

146. **Cushman, D. W. Cheung, H. S., Sabo, E. F., and Ondetti, M. A.,** Design of potent competitive inhibitors of angiotensin-converting enzyme. Carboxyalkanoyl and mercaptoalkanoyl amino acids, *Biochemistry,* 16, 5484, 1977.

147. **Patchett, A. A. et al.,** A new class of angiotensin-converting enzyme inhibitors, *Nature (London),* 288, 280, 1980.

147a. **Patchett, A. A. and Cordes, E. H.,** The design and properties of N-carboxymethyl inhibitors of angiotensin-converting enzyme, *Adv. Enzymol. Relat. Areas Mol. Biol.,* 57, 1, 1985.

148. **Peterson, L. M., Sokolovsky, M., and Vallee, B. L.,** Purification and cystallization of human carboxypeptidase A, *Biochemistry,* 15, 2501, 1976.

149. **Ondetti, M. A., Rubin, B., and Cushman, D. M.,** Design of specific inhibitors of angiotensin-converting enzyme: new class of orally active antihypertensive agents, *Science,* 196, 441, 1977.

150. **Harris, R. B., Ohlsson, J. T., and Wilson, I. B.,** Inhibition and affinity chromatography of human serum angiotensin converting enzyme with cysteinyl-proline derivatives, *Arch. Biochem. Biophys.,* 206, 105, 1981.

151. **Galardy, R. E.,** Inhibition of angiotensin converting enzyme with N^{α}-phosphoryl-L-alanyl-L-proline and N^{α}-phosphoryl-L-valyl-L-tryptophan, *Biochem. Biophys. Res. Commun.,* 97, 94, 1980.

152. **Galardy, R. E.,** Inhibition of angiotensin converting enzyme by phosphoramidates and polyphosphates, *Biochemistry,* 21, 5777, 1982.

153. **Thorsett, E. D., Harris, E. E., Peterson, E. R., Greenlee, W. J., Patchett, A. A. Ulm, E. H., and Vassil, T. C.,** Phosphorus-containing inhibitors of angiotensin-converting enzyme, *Proc. Natl. Acad. Sci. U.S.A.,* 79, 2176, 1982.

154. **Sweet, C. S., Gross, D. M., Arbegast, P. T., Gaul, S. L., Britt, P. M., Ludden, C. T., Weitz, D., and Stone, C. A.,** Anthihypertensive activity of N-[(S)-1-ethoxycarbonyl-3-phenylpropyl]-L-Ala-L-Pro (MK-421), on orally active converting enzyme inhibitor, *J. Pharmacol. Exp. Ther.,* 216, 558, 1981.

155. **Shapiro, R. and Riordan, J. F.,** Inhibition of angiotensin converting enzyme: dependence on chloride, *Biochemistry,* 23, 5234, 1984.

156. **Bull, H. G., Thornberry, N. A., Cordes, M. H. J., Patchett, A. A., and Cordes, E. H.,** Inhibition of rabbit lung angiotensin-converting enzyme by N^{α}-[(S)-l-carboxy-3-phenylpropyl]-L-lysyl-L-proline and N^{α}-[(S)-l-carboxy-3-phenylpropyl]-L-lysl-L-proline, *J. Biol. Chem.,* 260, 2952, 1985.

157. **Morrison, J. F.,** The slow-binding and slow, tight-binding inhibition of enzyme-catalysed reactions, *Trends Biochem. Sci.,* 7, 102, 1982.

158. **Rich, D. H. and Sun, E. T. O.,** Mechanism of inhibition of pepsin by pepstatin. Effect of inhibitor structure on dissociation constant and time-dependent inhibition, *Biochem. Pharmacol.,* 29, 2205, 1980.

159. **Baici, A. and Gyger-Marazzi, M.,** The slow, tight-binding inhibition of cathepsin B by leupeptin. A hysteretic effect. *Eur. J. Biochem.,* 129, 33, 1982.

160. **Ferreira, S. H., Bartelt, D. C., and Green, L. J.,** Isolation of bradykinin-potentiating peptides from *Bothrops jararaca* venom, *Biochemistry.* 9, 2583, 1970.

161. **Ondetti, M. A., Williams, N. J., Sabo, E. F., Pluščec, C. J., Weaver, E. R., and Kocy, O.,** Angiotensin-converting enzyme inhibitors, form the venom of *Bothrops jararaca.* Isolation, elucidation of structure, and synthesis, *Biochemistry,* 10, 4033, 1971.

162. **Fischer, G. H. and Ryan, J. W.,** Superactive inhibitors of angiotensin converting enzyme. Analogs of BPP_{9a} containing dehydroproline, *FEBS Lett.,* 107, 273, 1979.

163. **Almquist, R. G., Chao, W. R., Ellis, M. E., and Johnson, H. L.,** Synthesis and biological activity of a ketomethylene analogue of a tripeptide inhibitor of angiotensin converting enzyme, *J. Med. Chem.,* 23, 1392, 1980.

164. **Gordon, E. M., Natarajan, S., Pluščec, J., Weller, H. N., Godfrey, J. D., Rom, M. B., Sabo, E. F., Engebrecht, J., and Cushman, D. W.,** Ketomethyldipeptides II. Effects of modifications of the α-aminoketone portion on inhibition of angiotensin converting enzyme, *Biochem. Biophys. Res. Commun.,* 124, 148, 1984.

165. **Natarajan, S., Gordon, E. M., Sabo, E. F., Godfrey, J. D., Weller, H. N., Pluščec, J., Rom, M. B., and Cushman, D. W.,** Ketomethyldipeptides I. A new class of angiotensin converting enzyme inhibitor, *Biochem. Biophys. Res. Commun.,* 124, 141, 1984.

166. **Grobelny, D. and Galardy, R. E.,** Inhibition of angiotensin converting enzyme by aldehyde and ketone substrate analogues, *Biochemistry,* 25, 1072, 1986.

167. **Gleb, M. H., Svaren, J. P., and Abeles, R. H.,** Fluoro ketone inhibitors of hydrolytic enzymes, *Biochemistry.* 24, 1813, 1985.

168. **Roques, B. P., Fournié-Zaluski, M. C., Soroca, E., Lecomte, J. M., Malfroy, B., Llorens, C., and Schwartz, J.-C.,** The enkephalinase inhibitor thiorphan shows antinociceptive activity in mice, *Nature (London),* 288, 286, 1980.

169. **Roques, B. P., Fournié-Zaluski, M. C., Florentin, D., Waksman, G., Sassi, A., Chaillet, P., Collado, H., and Costentin, J.,** New enkephalinase inhibitors as probes to differentiate "enkekphalinase" and angiotensin-converting enzyme active sites, *Life Sci.,* 31, 1749, 1982.

170. **Gordon, E. M., Cushman, D. W., Tung, R., Cheung, H. S., Wang, F. L., and Delaney, N. G.,** Rat brain enkephalinase; characterization of the active site using mercaptopropanoyl amino acid inhibitors, and comparison with angiotensin-converting enzyme, *Life Sci.,* 33, 113, 1983.

171. **Mumford, R. A., Zimmerman, M., ten Broeke, J., Taub, D., Joshua, H., Rothrock, J. W., Hirshfield. J. M., Springer, J. P., and Patchett, A. A.,** Inhibition of porcine kidney "enkephalinase" by substituted-N-carboxymethyl dipeptides, *Biochem. Biophys. Res. Commun.,* 109, 1302, 1982.

172. **Almenoff, J. and Orlowski, M.,** Membrane-bound kidney neutral metalloendopeptidase: interaction with synthetic substrates, natural peptides, and inhibitors, *Biochemistry,* 22, 590, 1983.

173. **Hudgin, R. L., Charleson, S. E., Zimmerman, M., Mumford, R., and Wood, P. L.,** Enkephalinase: selective peptide inhibitors, *Life Sci.,* 29, 2593, 1981.

174. **Andersson, L., Isley, T. C., and Wolfenden, R.,** α-Amino-aldehydes: transition state analogue inhibitors of leucine aminopeptidase, *Biochemistry,* 21, 4177, 1982.

175. **Birch, P. L., EL-Obeid, H. A., and Akhtar, M.,** The preparation of chloromethylketone analogues of amino acids: inhibition of leucine aminopeptidase, *Arch. Biochem. Biophys.,* 148, 447, 1972.

176. **Kettner, C., Glover, G. I., and Prescott, J. M.,** Kinetics of inhibition of *Aeromonas* aminopeptidase by leucine methyl ketone derivatives, *Arch. Biochem. Biophys.,* 165, 739, 1974.

177. **Baker, J. O., Wilkes, S. H., Bayliss, M. E., and Prescott, J. M.,** Hydroxamates and aliphatic boronic acid: marker inhibitors for aminopeptidase, *Biochemistry,* 22, 2098, 1983.

178. **Shenvi, A. B.,** α-Aminoboronic acid derivatives: effective inhibitors of aminopeptidases, *Biochemistry,* 25, 1286, 1986.

179. **Chan, W. W.-C.,** L -Leucinthiol — a potent inhibitor of leucine aminopeptidase, *Biochem. Biophys. Res. Commun.,* 116, 297, 1983.

180. **Wilkes, S. H. and Prescott, J. M.,** Stereospecificity of amino acid hydroxamate inhibition of aminopeptidases, *J. Biol. Chem.,* 258, 13517, 1983.

181. **Chan, W. W.-C., Dennis, P., Demmer, W., and Brand, K.,** Inhibition of leucine aminopeptidase by amino acid hydroxamates, *J. Biol. Chem.,* 257, 7955, 1986.

182. **Umezawa, H., Aoyagi, T., Suda, H., Hamada, M., and Takeuchi, T.,** Bestatin, an inhibitor of aminopeptidase B, produced by actinomycetes, *J. Antibiot.,* 29, 97, 1976.

183. **Suda, H., Takita, T., Aoyagi, T., and Umezawa, H.,** The structure of bestatin, *J. Antibiot.,* 29, 100, 1976.

184. **Aoygai, T., Tobe, H., Kojima, F., Hamada, M., Takeuchi, T., and Umezawa, H.,** Amastatin, an inhibitor of aminopeptidase A, produced by actinomycetes, *J. Antibiot.,* 31, 636, 1978.

185. **Woolley, D. E., Roberts, D. R., and Evanson, J. M.,** Small molecular weight β_1 serum protein which specifically inhibits human collagenases, *Nature (London),* 261, 325, 1976.
186. **Cawston, T. E., Noble, D. N., Murphy, G., Smith, A. J., Woodley, C., and Hazelman, B.,** Rapid purification of tissue inhibitor of metalloproteinases from human plasma and identification as a γ-serum protein, *Biochem. J.,* 238, 677, 1986.
187. **Murphy, G., Cawston, T. E., and Reynolds, J. J.,** An inhibitor of collagenase from human aminotic fluid. Purification, characterization and action on metalloproteinase, *Biochem. J.,* 195, 167, 1981.
188. **Welgus, H. G. and Stricklin, G. P.,** Human skin fibroblast collagenase inhibitor. Comparative studies in human connective tissues, serum, and aminotic fluid, *J. Biol. Chem.,* 258, 12259, 1983.
189. **Stricklin, G. P. and Welgus, H. G.,** Human skin fibroblast collagenase inhibitor. Purification and biochemical characterization, *J. Biol. Chem.,* 258, 12252, 1983.
190. **Mercer, E., Cawston, T. E., De Silva, M., and Hazleman, B. L.,** Purification of a metalloproteinase inhibitor from human rheumatoid synovial fluid, *Biochem. J.,* 231, 505, 1985.
191. **Bunning, R. A. D., Murphy, G., Kumar, S., Phillips, P., and Reynolds, J. J.,** Metalloproteinase inhibitors from bovine cartilage and body fluids, *Eur. J. Biochem.,* 139, 75, 1984.
192. **Macartney, H. W. and Tschesche, H.,** The collagenase inhibitor from human polymorphonuclear leucocytes. Isolation, purification and characterisation, *Eur. J. Biochem.,* 130, 79, 1983.
193. **Macartney, H. W. and Tschesche, H.,** Characterisation of β_1-anticollagenases from human plasma and its reaction with polymorphonuclear leukocyte collagenase by disulfide/thiol interchange, *Eur. J. Biochem.,* 130, 85, 1983.
194. **Macartney, H. W. and Tschesche, H.,** Interaction of β_1-anticollagenase from human plasma with collagenases from various tissues and competition with α_2-macroglobulin, *Eur. J. Biochem.,* 130, 93, 1983.
195. **Docherty, A. J. P., Lyons, A., Smith, B. J., Wright, E. M., Stephens, P. E., Harris, T. J. R., Murphy, G., and Reynolds, J. J.,** Sequence of human tissue inhibitor of metalloproteinases and its identity to erythroid-potentiating activity, *Nature (London),* 318, 66, 1985.
196. **Gasson, J. C., Golde, D. W., Kaufman, S. E., Westbrook, C. A., Hewick, R. M., Kaufman, R. J., Wond, G. G., Temple, P. A., Leary, A. C., Brown, E. L., Orr, E. C., and Clark, S. C.,** Molecular characterization and expression of the gene encoding human erythroid-potentiating activity, *Nature (London),* 315, 768, 1985.
197. **Lipscomb, W. N.,** Acceleration of reactions by enzymes, *Acc. Chem. Res.,* 15, 232, 1982.
198. **Lipscomb, W. N.,** Structure and catalysis of enzymes, *Annu. Rev. Biochem.,* 52, 17, 1983.
199. **Gardell, S. J., Craik, C. S., Hilvert, D., Urdea, M. S., and Rutter, W. J.,** Site-directed mutagenesis shows that tryosine 248 of carboxypeptidase A does not play a crucial role in catalysis, *Nature (London),* 317, 551, 1985.
199a. **Hilvert, D., Gardell, S. J., Rutter, W. J., and Kaiser, E. T.,** Evidence against a crucial role for the phenolic hydroxyl of Tyr-248 in peptide and ester hydrolyses catalyzed by carboxypeptidase A: comparative studies of the pH dependencies of the native and Phe-248-mutant forms, *J. Am. Chem. Soc.,* 108, 5298, 1986.
200. **Pangburn, M. K. and Walsh, K. A.,** Thermolysin and neutral protease: mechanistic considerations, *Biochemistry,* 14, 4050, 1975.
201. **Kunugi, S., Hirohara, H., and Ise, N.,** pH and temperature dependences of thermolysin catalysis. Catalytic role of zinc-coordinated water, *Eur. J. Biochem.,* 124, 157, 1982.
202. **Holmes, M. A., Tronrud, D. E., and Matthews, B. W.,** Structural analysis of the inhibition of thermolysion by an active-site-directed irreversible inhibitor, *Biochemistry,* 22, 236, 1983.
203. **Hangauer, D. G., Monzingo, A. R., and Matthews, B. W.,** An interaction computer graphics study of thermolysin-catalyzed peptide cleavage and inhibition by N-carboxymethyl dipeptides, *Biochemistry,* 23, 5730, 1984.
203a. **Polgár, L.,** All four types of proteases are designed to achieve proton transfer from the attacking nucleophile to the substrate leaving group, *Acta Biochim. Biophys. Acad. Sci. Hung.,* in press, 1988.
204. **Christianson, D. W. and Lipscomb, W. N.,** Binding of possible transition state analogue to the active site of carboxypeptidase A, *Proc. Natl. Acad. Sci. U.S.A.,* 82, 6840, 1985.
205. **Christianson, D. W. and Lipscomb, W. N.,** The complex between carboxypeptidase A and a possible transition-state analogue: mechanistic inferences from high-resolution X-ray structures of enzyme-inhibitor complexes, *J. Am. Chem.* 108, 4998, 1986.
206. **Christianson, D. W. and Lipscomb, W. N.,** Structure of the complex between an unexpectedly hydrolyzed phosphoramidate inhibitor and carboxypeptidase A, *J. Am. Chem. Soc.,* 108, 545, 1986.
206a. **Christianson, D. W., David, P. R., and Lipscomb, W. N.,** Mechanism of carboxypeptidase A: hydration of a ketonic substrate analogue, *Proc. Natl. Acad. Sci. U.S.A.,* 84, 1512, 1987.
207. **Kaiser, E. T. and Kaiser, B. L.,** Carboxypeptidase A: a mechanistic analysis, *Acc. Chem. Res.,* 5, 219, 1972.
208. **Makinen, M. W., Yamamura, K., and Kaiser, E. T.,** Mechanism of action of carboxypeptidase A in ester hydrolysis, *Proc. Natl. Acad. Sci. U.S.A.,* 73, 3882, 1976.

209. **Makinen, M. W., Kuo, L. C., Dymowski, J. J., and Jaffer, S.,** Catalytic role of the metal ion of carboxypeptidase A in ester hydrolysis, *J. Biol. Chem.,* 254, 356, 1979.

210. **Makinen, M. W., Fukuyma, J. M., and Kuo, L. C.,** Evidence by gel filtration at subzero temperatures for the covalent reaction intermediate of carboxypeptidase A in ester hydrolysis, *J. Am. Chem. Soc.,* 104, 2667, 1982.

211. **Suh, J., Hong, S.-B., and Chung, S.,** Common acylcarboxypeptidase A intermediates for ester substrates containing different leaving alcohols, *J. Biol. Chem.,* 261, 7112, 1986.

212. **Hoffman, S. J., Chu, S. S.-T., Lee, H., Kaiser, E. T., and Carey, P. R.,** Multichannel resonance Raman experiments on carboxypeptidase A catalyzed ester hydrolysis under cryoenzymological conditions, *J. Am. Chem. Soc.,* 105, 6971, 1983.

213. **Kuo, L. C., Fukuyama, J. M., and Makinen, M. W.,** Catalytic conformation of carboxypeptidase A. The structure of a true reaction intermediate stabilized at subzero temperature, *J. Mol. Biol.,* 163, 63, 1983.

214. **Bunting, J. W. and Kabir, S. H.,** Nonspecific esterase activity of carboxypeptidase A. Specificity for the alcohol moiety of p-nitrobenzoate esters, *J. Am.. Chem. Soc.,* 99, 2775, 1977.

215. **Galdes, A., Auld, D. S., and Vallee, B. L.,** Cryokinectic studies of the intermediates in the mechanism of carboxypeptidase A, *Biochemistry,* 22, 1888, 1983.

216. **Geoghegan, K. F., Galdes, A., Martinelli, R. A., Holmquist, B., Auld, D. S., and Vallee, B. L.,** Cryospectroscopy of intermediates in the mechanism of carboxypeptidase A, *Biochemistry,* 22, 2255, 1983.

217. **Auld, D. S., Galdes, A., Geoghegan, K. F., Holmquist, B., Martinelli, R. A., and Vallee, B. L.,** Cryospectrokinetic characterization of intermediates in biochemical reactions: carboxypeptidase A, *Proc. Natl. Acad. Sci. U.S.A.,* 81, 5041, 1984.

218. **Galdes, A., Auld, D. S., and Vallee, B. L.,** Elucidation of the chemical nature of the steady-state intermediates in the mechanism of carboxypeptidase A, *Biochemistry,* 25, 646, 1986.

219. **Geoghedan, K. F., Galdes, A., Hanson, G., Holmquist, B., Auld, D. S., and Vallee, B. L.,** Hydrolysis of peptides by carboxypeptidase A: equilibrium trapping of the ES_2 intermediate, *Biochemistry,* 25, 4669, 1986.

220. **Hardman, K. D. and Lipscomb, W. N.,** Structures of nickel(II) and cobalt(II) carboxypeptidase A, *J. Am. Chem. Soc.,* 106, 463, 1984.

221. **Auld, D. S. and Holmquist, B.,** Carboxypeptidase A. Differences in the mechanisms of ester and peptide hydrolysis, *Biochemistry,* 13, 4355, 1974.

222. **Latt, S. A. and Vallee, B. L.,** Spectral properties of carboxypeptidase. The effects of substrates and inhibitors, *Biochemistry,* 10, 4263, 1971.

223. **Mock, W. L., Chen, J.-T., and Tsang, J. W.,** Hydrolysis of a thiopeptide by cadmium carboxypeptidase A, *Biochem. Biophys. Res. Commun.,* 102, 389, 1981.

224. **Mock, W. L. and Tsay, J.-T.,** A probe of the active site acidity of carboxypeptidase A, *Biochemistry,* 25, 2920, 1986.

225. **Antonov, V. K., Ginodman, L. M., Rumsh, L. D., Kapitannikov, Yu. V., Barshevskaya, T. N., Yavashev, L. P., Gurova, A. G., and Volkova, L. I.,** Studies on the mechanism of action of proteolytic enzymes using heavy oxygen exchange, *Eur. J. Biochem.,* 117, 195, 1981.

226. **Davis, R. P.,** Carbonic anhydrase, in *The Enzymes,* Vol. 5., 2nd ed., Boyer, P. D., Lardy, H., and Myrbäck, K., Eds., Academic Press, New York, 1961, 545.

227. **Argos, P., Garavito, R. M., Eventoff, W., Rossmann, M. G., and Brändén, C. I.,.** Similarities in active center geometries of zinc-containing enzymes, proteases and dehydrogenases, *J. Mol. Biol.,* 126, 141, 1978.

INDEX